JUNGLE

JUNGLE

*How Tropical Forests
Shaped the World—and Us*

PATRICK ROBERTS

BASIC BOOKS

New York

Basic Books
Hachette Book Group
1290 Avenue of the Americas, New York, NY 10104
www.basicbooks.com

Printed in the United States of America

First US Edition: September 2021

Published by Basic Books, an imprint of Perseus Books, LLC, a subsidiary of Hachette Book Group, Inc. The Basic Books name and logo is a trademark of the Hachette Book Group.

The Hachette Speakers Bureau provides a wide range of authors for speaking events. To find out more, go to www.hachettespeakersbureau.com or call (866) 376-6591.

The publisher is not responsible for websites (or their content) that are not owned by the publisher.

Print book interior design by Jeff Williams

Library of Congress Cataloging-in-Publication Data
Names: Roberts, Patrick, 1991– author.
Title: Jungle : how tropical forests shaped the world-and us / Patrick Roberts.
Description: First US edition. | New York, NY : Basic Books, 2021. | Includes
    bibliographical references and index.
Identifiers: LCCN 2021000308 | ISBN 9781541600096 (hardcover) | ISBN
    9781541600102 (ebook)
Subjects: LCSH: Rain forests—History.
Classification: LCC QH86 .R63 2021 | DDC 578.734—dc23
LC record available at https://lccn.loc.gov/2021000308

ISBNs: 978-1-5416-0009-6 (hardcover), 978-1-5-416-0010-2 (ebook)

LSC-C

Printing 1, 2021

For Rhys, Ida, and Livia—
that they may yet walk through their own "Jungles"

CONTENTS

ILLUSTRATIONS

PREFACE

Battling through the heart of the Brazilian Neotropics in July 2019, my student Victor Caetano Andrade and I fended off poisonous snakes, gigantic mosquitoes that can bite through the tough skin of crocodiles, and heat that made sweat cascade like waterfalls into our eyes. Over the previous ten years, my fieldwork in the world's tropical forests had taken me from the leech-filled rainforests of Sri Lanka to the flammable dry forests of Australia and the volcano-skirted misty cloud forests of Mexico. But nothing had prepared me for the teeming life, stifling climate, and vast array of green that flanks the Amazon, the largest river in the world by volume. Just to reach our field site and host village required a 523-kilometer "slow ferry" (*recreio*) journey that took a staggering thirty-six hours, followed by another two-hour ride in a small, open boat, which we undertook while loaded down by several awkward boxes of scientific equipment, a rudimentary toolkit for on-site maintenance, and whatever clothes the remaining space and will-power would allow us to carry. Over the course of his summer research, Victor would suffer a bout of dengue fever, a pus-producing infected boil the size of a small plum on his hand, and the failure of an airplane engine during landing in a local city. This all likely sounds fairly familiar to those of you, particularly in Europe and North America, who have experienced "jungles" only in terms of the films and novels that are set there. From the trials of *Apocalypto* to Mowgli running with animals, most of us see tropical forests as a kind of terra incognita. They provide artistic metaphors and natural resources, but they are fundamentally different from our own ideas of "home." In fact, the overwhelming implication of most Euro-American books, series, and blockbusters set in the "jungle" is that tropical forests are fundamentally incapable of sustainably and safely supporting large human societies.

This notion informs not just popular but also academic thought. Tropical forests are often ignored in discussions of our human story and the history of life on Earth. The dominant narrative of human evolution, for example, is that our hominin ancestors left behind the perils and slim pickings of the forest as soon as they were able, striding out onto open savannahs with novel tools ready to exploit the rich game opportunities that these environments provided. The search for the origins of our own species, *Homo sapiens*, and its rapid "routes" of dispersal around the globe has similarly focused on open grasslands or, alternatively, coastal settings. Likewise, discussions of the origins of "agriculture" or cities almost never involve tropical forests. We have deemed them "unproductive" and assume that they have uniformly poor soils, deadly natural hazards, elusive animals, and extreme climates that would make it impossible to maintain the types of agriculture and cities we associate with our own, purportedly "complex" human societies. Given the seemingly inevitable devastation inflicted by industrial farming and urban populations on these environments today, we wonder how these habitats could ever have supported sweeping monoculture croplands, vast pastures, and bustling metropoles. Instead, accounts of communities living in tropical forests tend to describe small, often supposedly "isolated" groups that rely on hunting and gathering to survive. These assumptions shape not only how we understand the history of tropical forests but also how we go about trying to protect them. Traditional conservation approaches to tropical forests assume that humans are simply incapable of living in them sustainably and that the best way of protecting them is to treat them as intact ecological "wildernesses," with minimal human disturbance and presence. Even the word commonly used in English to refer to tropical forests—"jungle," from the Hindi word *jangal*—was originally meant to classify something outside the realm of human settlement and home comforts.[1]

Yet we should be skeptical of the stories we have so often been told about tropical forests. During my life-changing expedition to the Amazon Basin, two specific moments—far less expected, but far more intense, than any of the struggles Victor and I had under the canopy—stayed with me. These encounters highlight not only the longevity and intimacy of human interactions with these majestic environments but also their ongoing significance to all of us—whether we live in the tropics or not. First: One morning, as we awoke to the gentle shaking of the aging *recreio* and the sounds of parakeets, Victor, a seasoned traveler in the Amazon and a Brazilian national, pointed at the treetops and said, "We will see a village soon." I tried to follow his finger but could see no sign of people, houses, or even a clearing that might imply some kind of human presence; everything simply appeared wild and green. Then Victor pointed toward the *types* of vegetation that had suddenly begun to dominate the riverbanks. On closer inspection, in contrast to the previous mass of forest, a dense concentration of two particular types of plants became clear: *açai* berry–bearing palms and Brazil nut (or, more correctly, Amazon nut) trees. Long-term work as an ecologist and frequent visitor to settlements up and down the banks of the Amazon Basin meant that Victor knew these plants acted as a living signpost for human populations. Sure enough, a village—fittingly named Ponta da Castanha (or point of Amazon nut trees)—slowly emerged from the tangle of dense vegetation on the riverbank. As Victor and the local people, including our generous host Jucelino, well knew, today's river-shore settlements are almost always located on top of prehistoric occupation sites. Here, past human societies modified the fertility of the earth and the composition of vegetation over millennia—so much so that these spots still attract local Amazonian food producers today. If Amazonian tropical forests were indeed "untouched" and largely free of human history, how could I be standing in a location that had repeatedly seen settlement for thousands of years?

Second: At the end of our visit, we pushed off from the shore in a small motorboat. Riding just above the waterline and gripping our belongings and equipment in our laps for dear life, I looked up into the sky. I saw a fast-moving blanket of clouds, sailing above the Amazon rainforest and its human settlements, so close you could nearly touch it. In this moment, perhaps for the first time, I truly appreciated the regional, continental, and global significance of tropical forests. If these forests disappeared, the loss of water evaporating from billions upon billions of leaves would turn this blanket of cloud into a threadbare rag. Few of us realize that tropical forests are globally responsible for a significant portion of the planet's terrestrial rainfall. In the case of the Amazon Basin, if its tropical forests disappeared, rain would decrease not just locally but also across vast swathes of the South American continent. Thanks to the network of ocean and atmosphere circulation systems that cross the globe, climates as far away as Europe could be impacted. These same forests are also often major carbon sinks, performing over one-third of the planet's photosynthesis and trapping and storing around a quarter of the world's land-based carbon. If they disappeared, this carbon would be released into the atmosphere. Meanwhile, less atmospheric CO_2 would be recaptured into the biosphere in their absence. In the context of a growing urgency to address the role of emissions in human-induced climate change, I am sure you can imagine what that would mean for temperatures all around our planet.[2]

These two experiences provide an entry point into the vital but often remarkably neglected place of tropical forests in human history and the functioning of the Earth that I hope to reveal to you in this book. Tropical forests are not "green hells," hostile to human habitation and only useful for distant, relentless extraction and clearance. They can be productively lived with and lived in (a fact that I believe casts strong doubts on the assumptions we often make about how human societies, economies, and settlements should inevitably be organized). Indeed, given their importance to the entire planet, we are all going to have to find ways to do this sooner rather than later. As a multidisciplinary archeologist working in the tropics, I have seen how the discoveries of my predecessors and colleagues have increasingly highlighted the adaptive flexibility of human societies living in these environments on a global scale, from our earliest ancestors to the inhabitants of some of the largest urban areas ever to have existed prior to industrialization. I have seen how biologists and ecologists have long emphasized the importance of these, the oldest land-based ecosystems on the planet, to the evolution and sustenance of the world's greatest concentration of plant and animal diversity. And I have seen how Indigenous populations have long advocated for the critical role of tropical forests in their economic and cultural survival, as well as the importance of their active stewardship to the well-being of these ecosystems. In light of all this, it is remarkable that many of us still think of tropical forests as so inherently separate from human existence. We are happy to hear about Tarzan living among wild animals, to watch searches for "lost" cities by directionally challenged explorers, and to let the dulcet tones of David Attenborough wash over us as he describes a fabulous exotic bird attempting to dazzle an unimpressed mate. Yet the remoteness, biodiversity, and increasingly dramatic disappearance of tropical forests, while fascinating and shocking, also feel disconnected from many of our day-to-day lives. Even when we call for their protection, we tend to call for humans to be moved out of them rather than searching for ways in which humans might live *alongside* them. The plights facing these environments today are undoubtedly poignant, but their apparent distance, isolation, and exoticism all too often allow us to ultimately ignore them.[3]

This book is a history of the world according to the tropics and their "jungles." It begins with a journey through the colorful cast of tropical forests that have inhabited our planet for nearly 400 million years (Chapters 1 and 2), from the first treelike organisms emerging into a warm, wet world to forests that would be recognizable to those inhabiting (or visiting) the tropical portions of our planet today. We will see how, from their arrival on the Earth's surface, tropical forests have shaped—and been shaped by—barrages of climate change from above and crunching forces of tectonic plates from below. Tropical forests sculpt our planet's atmosphere, water cycle, and soils and have played a major role in the evolution of life on Earth. We will explore the role of these remarkable environments in hosting the first flowering plants and the earliest four-legged land-based creatures on Earth (Chapter 2), in impacting the evolution of the dinosaurs (Chapter 3), and in shaping the survival and evolution of many of the main ancestral lines of mammals that we flock to zoos to see today (Chapter 4). Tropical forests also played an essential role in human evolution. They were the leafy "cradles" where the first hominins appeared in Africa, splitting off from the last common ancestor between other great apes and ourselves (Chapter 5). They were also one of the diverse environments in which our own species, *Homo sapiens*, emerged in Africa, before we went on to occupy nearly all of the planet's continents between 300,000 and 12,000 years ago (Chapter 6).

Despite the remarkable position of tropical forests in these major evolutionary processes, many of us tend to see them as somewhat detached from our story: as impossible to farm in the absence of deforestation, as deserted, isolated islands, or as unattractive and fragile in the face of urban developments. I want to show you that these preconceptions are often the product of Euro-American ideas of what "agriculture" or "cities" should be. Many of the crops and animals we rely on today were, in fact, first managed and domesticated in the tropics (Chapter 7). Many different human societies grappled with sustainable living on tropical islands, managing local resources alongside new crops and animals brought from elsewhere (Chapter 8). And tropical forests were also home to some of the largest, and also comparatively most successful, preindustrial urban populations ever to have existed (Chapter 9). This, in turn, forces us to ask the question: given all of this evidence of previous innovative human habitation and management of tropical forest environments around the world, why do we commonly think of them as "empty" and "vulnerable" to human presence today? The answer is to be found in the historical processes of the last five hundred years, where Europe and the tropical world collided. Disease, warfare, and murder ravaged the cities and villages of Indigenous populations (Chapter 10). New profit-driven approaches to tropical landscapes saw mining and monoculture plantations strip away forests, erode soils, and act as the sites of forcibly abducted and transported labor as part of the transatlantic slave trade (Chapter 11). The spread of imperial and capitalist forces around the tropics resulted in degraded landscapes, in wealth imbalances between the western half of Europe and northern North America and the tropical world, in racial discrimination and violence, in a marginalization of Indigenous knowledge, and in the onset of significant human-induced climate change, all of which continue to represent perhaps the most major ecological, social, political, and economic challenges facing our societies in the twenty-first century (Chapters 11 and 12). They also led many of us to falsely conclude that these habitats cannot, and do not, support large numbers of people.

Although a number of readers may already care deeply about the environment and the decline of tropical forests, it is highly likely that, to some extent, you still feel a certain safe estrangement from some of the world's oldest land-based environments. Indeed, harsh working conditions, dense vegetation, and difficult navigation—as encountered by Victor and me on our Amazon expedition—have often hindered exploration of tropical forests and their histories. However, this book looks at how the latest scientific advancements, from laser scanning from the air to plant genetics in the lab, can take us through the canopy to show us how these habitats actually penetrate every inch of our lives, wherever we are. Our kitchen cupboards are filled with groceries that owe their beginnings to tropical forests. Our travel to work depends on latex originally tapped from tropical trees. Our consumer decisions about everything from furniture to beauty products continue to influence the nature and extent of tropical environments. Extinction, deforestation, and out-of-control fires in these environments can seem a world away from many of us as we hear about them in the news. But the loss of tropical forests and the colonial legacies that shape their plight today impact not just the lives and climates of people on the other side of the world, but also our own weather, politics, societies, and economies—from Manila to Munich, Colombo to Cardiff, and Nairobi to New York. This book is an attempt to convince you that the history of tropical forests is your history too. But before we get ahead of ourselves, it is time to look back beyond 500 million years ago. Back when plants had not yet graced the land surfaces of the Earth and things looked very, very different to how they do today.

Chapter 1

INTO THE LIGHT–
THE BEGINNING OF THE WORLD
AS WE KNOW IT

S chool tours of natural history museums around the globe frequently rush past the fossilized remains of early plants on the
way to the reconstructed dinosaur skeletons or stuffed blue
whales that often take the plaudits as the heavyweight "stars" of evolution.
Novels and films such as *Jurassic Park* have brought the work of paleontologists, picking away at long-extinct bones to determine the evolution
of fascinating creatures, into the public arena. But no such dramatization
has emerged for the explorers of the world's ancient plants. This speaks
to a more general apathy. As we go about our daily lives, we rarely spare
a thought for the mosses growing on our pavements, the grasses covering our fields, the flowers occupying our gardens, and the trees that, if
we're lucky, line our streets. We often take for granted that plants have
always been on our planet and always will be. There are also fewer well-
known documentaries that explore the emergence, evolution, and conservation status of plants. Plants are traditionally considered less "exciting"
than other life-forms. They are harder to empathize with than animals,

with their emotional eyes and semblance of what we might recognize as families. Plants also do not make noises audible to the human ear, apart from perhaps some accidental creaking in the wind, making it unlikely that they will feature in a viral YouTube video "talking." Yet, without plants, the world as we know it today would not exist.

Indeed, recent research makes it seem like we might have been too quick to dismiss our photosynthesizing friends. For example, we now know that plants actually share some features that make us so able to relate to other animals. Time-lapse cameras have revealed the amazing, dynamic way in which plants constantly grow and move in different directions to access light. While they do not feel pain in a traditional sense, plants being eaten by caterpillars will release defense chemicals that make them less appealing to their attacker. Similarly, trees being eaten by giraffes in eastern Africa will release compounds into the air that warn other trees to protect their own leaves. Such "communication" has also been demonstrated in forests where neighboring trees will try to route nutrients to their damaged or ailing companions, making use of mass fungal "networks" that are some of the largest organisms in the world. In an attempt to maximize growth, trees will also sometimes gamble on the shedding of their leaves in temperate winter environments, with daring individuals paying the potentially fatal penalty of frost damage if they keep their leaves for too long. However, by far the most impressive part of these organisms is the way in which they engineer our world to make it hospitable to almost all other life. The sheer scale of this accomplishment is only truly apparent if we explore what the Earth was like *without* plants, which, thanks to recent scientific advances, we can now do.[1]

Let's travel back to the early Cambrian period (between 538.8 million and 509.0 million years ago). While it might seem an odd place to start in a book about tropical forests—its name derives from *Cambria*, the Latin word for Wales—this famous geological period witnessed a diversification of complex life, often termed the "Cambrian explosion," that eventually produced the evolutionary lineages of all multicellular animals left on Earth today. Were we to dive into the significantly warmer oceans of the early Cambrian period, we would see many different species of trilobites (the characteristic wood-louse-like fossil arthropods many of you will be familiar with) and other invertebrates practicing predation, scavenging, and filter feeding as marine ecosystems began to take on a familiar form. However, if we left this marine cornucopia and stepped out onto the land, a seemingly deserted, and frankly apocalyptic, scene would greet us. The continents would be covered by dry, rocky landscapes, and the main type of life present would be in the form of patchy films of microbes smeared across the surface. Beyond that, the only visible signs of life on land would be the occasional slug-like mollusk precariously venturing out of the oceans to try and scratch a living on the slim pickings available. During the early Cambrian there was no variety of terrestrial life at all, let alone something we might associate with a tropical forest. However, the world was certainly "tropical" by common climate classification systems that define a tropical climate as one with monthly average temperatures of approximately 18°C or more. In the Cambrian, the *global* average temperature was around 19°C, a staggering 5°C higher than the present day, and it still remained at an average of approximately 18°C at the beginning of the Devonian (from 419.0 million years ago).[2]

It was into this "tropical" world that the first land plants arrived to perform their greatest trick, with an alchemical skill we can only dream of. All of you will have had a high school introduction to photosynthesis, the process by which most plants and certain bacteria can use the energy of the Sun to convert carbon dioxide (CO_2) in the air or oceans into sugars and oxygen (O_2). However, the fundamental nature of this process to life is easily forgotten. In the absence of land plants, even with some bacteria and aquatic algae performing photosynthesis, early Cambrian atmospheric CO_2 concentration was around 4,500 parts per million by volume. That's ten times higher than it is today. Meanwhile, oxygen made up only about 7 percent of the Earth's atmosphere, around three times less than today. As a result, on our tour of the early Cambrian land surface, we would require oxygen tanks to survive. Furthermore, in the absence of plants, there were no root systems to efficiently break down bedrock, and nutrients that are crucial to most food chains today remained locked away in the Earth's tough crust. We will now see how the broadly "tropical" state of our planet between the Cambrian and the Devonian represented the perfect "greenhouse" for the emergence of not only the first land-based plants but also the first trees and forests. While these pioneering organisms actually put an end to widespread warm conditions, causing significant drops in CO_2, global cooling, and even the formation of polar glaciers, they also formed the building blocks for the first complex, land-based ecosystems that were to emerge around the remaining warm, wet regions of the equatorial tropics in the Carboniferous period (359.3 to 298.9 million years ago). These forests fixed soils, released nutrients, and stabilized the climate and the composition of the atmosphere. They also provided ideal homes for a growing variety of interacting animals that would eventually lead to the amphibian, reptilian, and mammalian life that surrounds us today. It is no exaggeration to say that they changed the face of the world forever.[3]

THE PROCESS OF photosynthesis and the extension of roots into the very surface of the Earth meant that, from their first arrival, plants had the potential to be significant movers and shakers in the context of our planet's atmosphere, climate, and geology. Professor Timothy "Tim" Lenton, director of the Global Systems Institute at the University of Exeter, is interested in how the first plants on land might have influenced the various "systems" (e.g., atmosphere, geosphere, hydrosphere, and biosphere) that interact to shape the Earth's climate. As he puts it, "There is a complex interplay between our oceans, the proportions of different gases in the air around us, and, crucially, what covers the land that affects how the Sun's energy is stored and moved around the world." Earth systems scientists such as Tim rely on powerful computers to "model" how changes to the amounts of different gases in our atmosphere, to the speed and direction of ocean currents, and to the distribution of different types of vegetation will alter the overall condition of the planet and whether such changes will impact different parts of the Earth in different ways. The outputs from what might be best described as an earth scientist's version of *Minecraft* can then ultimately be tested against available records of CO_2 and O_2 concentration and temperature from ancient sedimentary or ice-core records to see whether they "work" and can effectively explain the state of the Earth at a given point in time.

One of the main uses of this climate "sandbox" has been to make predictions about how twenty-first-century greenhouse gas emissions will impact global temperatures in the decades to come. However, it has also been used to explore how the arrival of the first plant life would have altered the state of the planet. Yet, in order to do this properly, these computer whizzes must turn to the more "classic" approaches of exploring ancient plants, or "paleobotany," to answer the key question of when the first true land plants (or Embryophyta) appeared. It is generally accepted that the first land plants were nonvascular, or "lower plants," meaning that they lacked complex structures, such as the transport network of xylem cells that carry water and phloem cells that convey nutrients from a plant's roots to the rest of its body. These plants evolved from aquatic algae and have long been thought to have resembled the modern liverworts, hornworts, and mosses (or bryophytes) you will have likely seen covering large patches of ground, rocks, and trees in your backyard. These spore-producing plants are, today, well known for their ability to retain water and can also act as important sources of food for animals. However, suggestions that liverworts, in particular, were the earliest diverging "sister" lineage of land plants meant that climate models often neglected the impacts of the earliest land plants on the Earth's surface and atmosphere. This is because liverworts lack a number of physical features, such as stomata—the tiny holes in leaves that open and close to help plants control their uptake of CO_2 and loss of water during photosynthesis—that are found in more complex, "vascular" land plants. As a result, they seemed to represent a perfect halfway house between "primitive" aquatic algae and complex, terrestrial vascular plants. Nevertheless, as you can imagine, directly testing this assumption by actually finding and studying the earliest of these plants has posed a major challenge to even the most enthusiastic of paleobotanists.[4]

The most direct evidence for the appearance of land plants comes from fossilized impressions of their bodies or the spores they used to reproduce, which were left behind in ancient sediments. Following a series of high-pressure geological processes, these traces were eventually converted into rock to be preserved to this day. The earliest embryophyte-like spores come from sediments in Argentina dating to around 470 million years ago, during another geological period with a Welsh connection, the Ordovician (486.9 million to 443.1 million years ago), which was named after an ancient group (the Ordovicii) subjugated by the Romans in what is now North Wales. More recently, five bryophyte

Figure 1.1. An image of the liverwort *Marchantia polymorpha*. The earliest land plants are thought to have resembled living "Bryophytes" such as liverworts, hornworts, and mosses. Liverworts, such as *Marchantia polymorpha*, have often been considered the oldest lineage of land plants, acting as a model system for a transition between aquatic algae and complex vascular land plants. However, the exact nature and physiological features of the earliest plants to colonize the land have remained unknown until recently.   *Silvia Pressel*

clades (a clade is a group of organisms considered to all be the descendants of a single common ancestor) have been argued to be represented by fossil imprints from a sinkhole that was excavated during the creation of the Douglas Dam in Tennessee, which date to c. 460 million years ago. This would be significantly earlier than one of the most widely accepted earliest fossils with a land-plant body structure, from Ireland, with a minimum age of approximately 427 million years. Nevertheless, if you have ever touched a bryophyte while working in your garden or taking a walk in the woods, you will have noticed its fragility, and it will often simply crumble away in your hands. As a consequence, existing fossil discoveries are almost certainly underestimates of the true age of the first land plants. Comparisons of modern and fossil plants can be used to estimate how long it would have taken for different "morphologies," or body shapes, to have emerged in the past. However, this is imprecise, as evolution can occur at markedly different rates. Furthermore, the incredible patchiness of the fossil record of plants, particularly during this early time frame, means that there is frustratingly little to go on when attempting to understand when and how land-plant lineages emerged and proliferated, as well as what they looked like and what implications they might have had for the planet.[5]

Some scientists have therefore adopted different approaches to exploring the evolutionary history of land plants. For example, they have sampled the modern genetic data of different families of liverworts, mosses, and other land plants to try and produce a kind of "molecular clock." These "clocks" work on the principle that mutations within the genetic or protein sequences of plants will have occurred at a constant, calculable rate. With this in mind, the modern genetic sequences of different plant groups can then be compared to each other, with the amount of difference providing some kind of idea as to how long ago the different lineages formed and separated from each other. This work has allowed scientists to determine that vascular plants, or "tracheophytes," which include the trees, shrubs, flowers, ferns, and fern allies that characterize the green plant life most visible to us today, indeed emerged *after* the liverworts and mosses. More significantly, it has also been demonstrated that liverworts are not actually the earliest land-plant lineage, calling into question the relevance of their physical features for characterizing the earliest plants to colonize the land. Nevertheless, the earliest evolutionary history of plants remains hazy, and estimating a robust date for the first land plants has been incredibly difficult. Given the complexity of genetic mutations and differences in the rate of change between populations, species, and even plant families, dates based on modern genetic data alone are notoriously unreliable, providing a rough ballpark figure at best. If neither the date for the first plants nor the nature of their biological features can be determined, then working out when and if they had impacts on the Earth's climate obviously becomes a major challenge.[6]

Fortunately, in 2018, scientists decided to combine the strengths and weaknesses of the different fossil and genetic approaches to the origins of land plants to develop a more accurate and comprehensive model for their appearance. This team, headed by scholars at the University of

Bristol, combined the genetic and morphological estimates of rates of plant evolution with the available "fixed points" provided by actual remains in the fossil record. According to Dr. Silvia Pressel, head of the Algae, Fungi and Plants Division at the Natural History Museum of London and coauthor of the resulting study, this work allowed "initially rough evolutionary models to be tested and refined using the evidence that we can hold in our hands." Moreover, by repeating these models thousands and thousands of times, the researchers could work out just how likely it was that they represented reality. All of their analyses converged to indicate that the first land plants actually appeared about 500 million years ago, slap bang in the middle of the Cambrian period, just after the evolution of multicellular animal species had started to take off underwater. Even more significantly, biomolecular analysis of different plant lineages undertaken by a separate multidisciplinary team of paleobotanists indicated that the ancestor of land plants was probably more complex than assumed so far, already possessing rootlike filaments and stomata on its surfaces. These features would have allowed it to process more soil and more carbon dioxide than previously thought.[7]

Silvia's work has also added a further dimension to this emerging picture. You might have read about how plants and fungi combine to produce a close symbiotic, or "mycorrhizal," relationship. The fungi, interlocking with the roots of the plant, provide it with more water and soil nutrients, helping it to break down bedrock and soils and even protecting it from pathogens. Meanwhile, the plants provide the fungi with sugars for sustenance from photosynthesis. We now know that most land plants make use of this win-win partnership. About 90 percent of plant species today are supported by fungal helpers, and the mapping of species of fungi and bacteria between the roots of trees across the world has come to be known as the "wood wide web," enabling flows of nutrients across entire ecosystems. This relationship was already thought to be ancient because of the presence of fungal structures, including little "treelike" structures referred to as arbuscules, in the cells of 400-million-year-old fossil plants. However, some bryophytes, namely mosses, lack these interactions, leaving the question open as to whether the earliest plants on land had also already established fungal partnerships. By studying the diversity and function of fungi across the bryophyte groups, and by sequencing the modern genomes of mycorrhizal fungi, researchers like Silvia have confirmed that this "subterranean cooperation between plant and fungus extends back to the roots of plant evolution on land," as she puts it, and likely involved a greater diversity of fungi than previously thought. It is becoming increasingly clear then that the first land plants emerged earlier and, together with their fungal aides, had more significant impacts on the planet than often considered.[8]

Back in 2012, Tim's team had used their simulation-based approach to make two significant arguments about the first land plants. They performed modern experiments showing that modern mosses, even with limited root systems, significantly boost the weathering of rocks to nearly the same degree as more complex, "higher" vascular plants. They do so by releasing acids from their tissues onto the surrounding land surface. The weathering of rocks traps CO_2 from the atmosphere in acidic solutions with ions (bicarbonate, calcium), while at the same time CO_2 is absorbed from the atmosphere by photosynthesis. Applying these new weathering estimates—which would be significantly increased by the findings of the 2018 study and the discovery that, unlike mosses, the earliest land plants would have already established a joint venture with fungi—enables an estimation of the planet's CO_2 concentrations and temperatures during the Ordovician. Although these computer models are inevitably imprecise to some degree, assuming that nonvascular land plants had managed to colonize just 15 percent of the current vegetated parts of the Earth's surface between 475 million and 460 million years ago (plants make up around 32 percent of the Earth's surface today), they suggested a staggering *halving* of atmospheric CO_2 and a global cooling of at least 4°C to 5°C. The modern aim to limit global warming to 1.5°C in order to avoid a series of devastating tipping points for human societies around the world should put the scale of this planetary change into some perspective. The earlier dating of the first land plants to c. 500 million years ago could mean that these changes may eventually trace their origins to the late Cambrian period c. 490 million years ago. Nevertheless, it is in the Ordovician period that land plants become more widespread across the surface of the Earth. It is also in the Ordovician period that their predicted modeled impacts on earth systems become clearly visible in the available climate records.[9]

First, the reconstructed cooling was enough to trigger an "ice age" by the time of the Ordovician-Silurian geological transition 443.1 million years ago. The Silurian is yet another name derived from Wales. Although the chilly conditions at the start of this period may be more in keeping with popular ideas of a trip to the Brecon Beacons, the real reason so many of these periods have names deriving from this country is that a lot of formative work in the discipline of geology was done by British people in Britain, with the names of classic blocks of time first identified on these isles frequently being imposed upon the rest of the world. The drop in temperatures at the Ordovician-Silurian boundary was so extreme that it was one of the coldest periods in the last half billion years of the Earth's history (over 1,500 times longer than we have existed on the planet!). Second, enhanced weathering would have released significant nutrients, and specifically the element phosphorous, from the previously barren rocky land cover. Phosphorus is critical to the growth of organisms. However, too much phosphorus can be a bad thing, particularly in aquatic settings, and today phosphorous pollution can stimulate a rapid growth of toxin-producing algal blooms. Back in the Ordovician, the washing of newly weathered phosphorus into the oceans would have had a similar impact. Although these "blooms" would have supported more and more marine life, this would also have gradually sucked the limited oxygen supply from the water. Ultimately, suffocation and a mass extinction of marine life characterized the end of the Ordovician. The latest botanical and earth sciences research is therefore clear in showing that even the most "primitive" of land plants could, and did, dramatically alter the climate, atmosphere, and surface of the Earth. These plants survived their self-induced ice age, huddling in the warmer equatorial areas of the planet. There they waited, ready to grow and change the world to an even greater extent. While the oceans initially suffered, the land was bracing itself for an unprecedented expansion of life.[10]

ALTHOUGH THE FIRST plants certainly affected the Earth's systems, these changes only began to stick with the arrival of the most complex of vascular plants, the trees. The earliest "higher" vascular (or "tracheophyte") fossil body plant form, with xylem, phloem, and complex, invasive

rooting systems that fit with our more traditional school-based definitions of "plants," dates to around 420 million years ago in Canada, nearly
80 million years after the first plants colonized the land, and represents
the beginning of the plethora of complex green plant forms we are familiar with today. Shortly afterward, during the Devonian (419.0 million to
359.3 million years ago), the first trees appeared on the planet. Trees had
certain evolutionary benefits over other plants that enabled their rapid
success. Unsurprisingly, their sturdy structure enabled them to grow
taller and gain better access to light for photosynthesis. More photosynthesis meant more body mass, and their root systems could search
more deeply and more widely for nutrients, with weathering from roots
and their fungal collaborators increasing the rate of soil formation and
greatly expanding the area in which subsequent plants could take root.
More photosynthesis also increased the amount of CO_2 absorbed from
the atmosphere. As much of this CO_2 became stored in robust lignin-
based and, later, woody trunks resistant to decomposition after death,
it also became locked away from the atmosphere, increasing the relative
proportion of O_2 in the air that remained. The true giants of vegetation
cover, and thereby its impacts on the Earth's systems, had arrived.[11]

Until recently, the story of our planet's trees was thought to begin
at a stone works near Albany, New York. Here, at Riverside Quarry in
Gilboa, workers extracting stone needed for a nearby dam project in
the 1920s discovered hundreds of log-like shapes, one after the other,
sticking out upright from the rock they were chipping away. New York
State's first female paleontologist, Professor Winifred Goldring (1888–
1971), was called to the scene, and her research on these amazing finds
shaped discussions of ancient forests for a nearly a century. The trunks
were identified as *Eospermatopteris*, a plant very similar to modern tree
ferns. While it was considered to have a broadly "treelike" (what paleobotanists call "arborescent") structure, *Eospermatopteris* lacked proper
leaves. Instead, photosynthesis occurred on fronds (like those of palms)
inserted into almost vertical branches. The trunks of *Eospermatopteris*,
dating to around 380 million years ago, were also not composed of wood
but rather consisted of a tough, hollow lignin structure. Later work in
the 2000s identified a woody "aneurophytalean" vine at Gilboa that, with
mosses, apparently climbed the *Eospermatopteris* tree forms. While

Eospermatopteris has subsequently often been heralded as the first anatomical "tree," the aneurophytalean vines were actually considered a
better biological example of a "progymnosperm"—an early ancestor of
the woody, seed-bearing trees of today—that is, until 2019, when a new
fierce competitor for the title of the oldest "tree" emerged.[12]

The findings at Riverside had stimulated further paleobotanical investigation of the rock formations on the East Coast of North America
that eventually led to the discovery of another major ancient fossil in
the region. Again in New York, and again in a quarry (albeit this time
abandoned), scientists uncovered an incredible record of a different
type of ancient arboreal life-form. Dating to 2 million to 3 million years
before Riverside, the new site at Cairo produced the same hollow *Eospermatopteris* and same woody aneurophytalean vines as Gilboa. In addition, the scientists found so-called lycopsids (distant relatives of the still
living *Lycopodium* genus of club mosses, commonly known as "ground
pines") and liverworts. However, most amazingly, these Devonian sediments also preserved the remarkable root and body structure of a member of the *Archaeopteris* genus. Although *Archaeopteris* was known from
previous studies, the Cairo record clearly demonstrated that, unlike
Eospermatopteris, it also already had a clear set of seed plant ("spermatophyte") features much earlier than previously thought. This included
a vast, deep root system indistinguishable from those of modern seed
plants; a large, upright form; thick, woody trunk; varied spore types; and
flat green leaves for the efficient capture of sunlight and absorbance of
CO_2. As a result, it is considered to represent an ancestral lineage from
which all seed-producing plants, which dominate the Earth's forests today, evolved. Furthermore, while *Eospermatopteris* was largely limited to
flooded, lowland swamp habitats, the root systems of *Archaeopteris*, and
the ability of its sophisticated leaves to function in dry systems, enabled
these trees to expand much more widely across the Earth's surface, with
the potential for wider-scale planetary impacts. By approximately 388
million to 359 million years ago, fossil records suggest that *Archaeopteris*
was a dominant member of many forests in not only North America but
also Morocco and China, among other places.[13]

Whichever group of organisms best represents the first trees, the
Gilboa and Cairo fossil sites together provide a remarkable snapshot of

the world's first forest ecosystems. These new leafy kids on the block began to further reconfigure the Earth's surface and atmosphere, and not necessarily to their own advantage. These novel Herculean plants were the world's first "geo-engineers." True, deep, wide root systems converted significant amounts of rock into plentiful, supportive soils. As the roots reached outward, they continued to form ever larger bonds with fungi and microbes that characterize the dynamic floors and soils of modern forests. Calcium, magnesium, sodium, iron, aluminum, and potassium were mobilized from the Earth's crust into new, thick, nutrient- and even clay-rich soil ecosystems and mineral carbonates. Once again, however, not all of these minerals remained on land. As at the end of the Ordovician, they made their way into the Earth's seas, driving an expansion of life. More rapid and wide-scale weathering trapped more CO_2 in solution, further reducing CO_2 in the atmosphere, alongside more efficient photosynthesis by these plants' extensive leaf systems. This new plant-based shock to the planetary system once more sent the world into a drastic phase of cooling, even more significant than before. Polar glaciers pushed right up to the lower latitudes. Numerous extinctions occurred both in the oceans and on land at the end of the Devonian. There was also widespread decline of many of the first forests that had dared to stake their claim on our planet.[14]

Nevertheless, the pioneering biological and geological success of these tree-based configurations meant that the planet was now forced to contend with something of an "inevitability of forests."[15] By the time of the Carboniferous period, 359.3 million to 298.9 million years ago—a time when the supercontinent of Pangaea was forming, glaciers were contracting, and much of the continental landmasses of present-day Europe and America existed around the warm, wet equatorial regions—forests were well and truly back. Woody trees with thick bark, alongside their decking of vines and climbers, expanded in moist, swamp-like habitats across these latitudes. Indeed, the name of this geological period comes from all of the carbon that became trapped in the ground following the deaths of these new, more resistant trees. The thick layers of peat that formed from these extensive forests, in the absence of microbes and fungi that could digest tough wood fibers, were buried deep under the surface. There, millions of years of heat and intense pressure turned them into the Carboniferous coalfields, which geologists around the world could use to visibly recognize a distinct period. In an ironic twist of fate, these same coalfields, formed from forest habitats that had begun to turn the Earth's surface into a lusher and more comfortable setting, were rediscovered as a key fuel source for the Industrial Revolution of the eighteenth and nineteenth centuries. The human-induced release of CO_2 back into the atmosphere has reversed a process that began hundreds of millions of years ago, threatening not only the forest descendants of these Carboniferous ecosystems but also the habitability of various parts of the Earth's surface for plants and animals (including ourselves).[16]

Various ancient trees (known as "lycophytes") lived in these first Carboniferous forest environments, extending between forty and fifty meters high and spanning up to two meters in diameter. Sweeping vines climbed these gigantic arborescent versions of modern-day club mosses, while horsetails added density to the extensive, humid, tropical-swamp "coal forests" that, by approximately 300 million years ago, made up the majority of the now plentiful forest cover visible in fossil records dating to this period. The remaining third consisted of "progymnosperms" and even seed-producing trees, such as the now extinct giant "cordaites" that may have been ancestral to later conifers, ginkgos, and cycads, which broke free of the wetlands and adventurously struck out into well-drained, drier upland conditions. True seed-producing plants ("spermatophytes") emerged in the Devonian, but they expanded across latitudinal zones within the warm conditions of the early Carboniferous. The fossil evidence for vast mats of roots from 305 million years ago shows the "earth-shattering" capabilities of Carboniferous forests—producing ample soils they could anchor into both horizontally and vertically. The resulting active, nutrient-rich soils were teeming with fungi, bacteria, and also now invertebrates (mainly insects), which were highly effective at breaking down organic matter and would not have been out of place on the ground-level of twenty-first-century tropical forests. Photosynthesis continued to absorb and trap significant amounts of CO_2, leading to atmospheric levels less than five times preindustrial levels. At certain times, O_2 was even more abundant in the atmosphere than it is today. The shift in the atmosphere during the Devonian and into the Carboniferous was so significant that even the plants themselves likely had to adapt. While we have already seen that stomata have inferred origins as far back as the Cambrian with the first land plants, there appears to be an increase in the frequency of, and surface area covered by, these pores on plants as the Devonian period wore on. It has been suggested that the proliferation of stomata and a growing function in gas regulation was an adaptation to declining CO_2 and environmental conditions at this time, though some researchers dispute this. Regardless, overall mild global climatic and atmospheric conditions were now ripe for the expansion of life and, as we will see, a proliferation of familiar land-based organisms. Forests growing in the tropical latitudes of the time were the key catalyst for this process, which human hands are only now beginning to undo.[17]

IN CONTRAST TO our tour of the desolate landscapes of the Cambrian, if we strode through the new forests of the Carboniferous, we would feel significantly more at home. Walking across the Earth's surface, we would now see organisms and ecosystems that looked like the plants and forests we know today. Compared to the chatter and squawking of diverse Amazonian rainforests or even the sounds of British woodlands, Devonian and Carboniferous forests would have been eerily silent. No herbivores yet existed that could process the new cellulose- and lignin-rich plants that had begun to populate the planet, leaving them largely unmolested to grow upward and outward. Nevertheless, the rapid formation of soils, thanks to the lofty trees, now provided moist, warm homes for millipedes and centipedes, consuming rotting plant matter or hunting other tiny organisms trying to make their way in this very new world. The earliest insects appeared in Devonian and Carboniferous forests, some flying through the humid swamp forests on wingspans as wide as two feet by the end of the latter period, thanks to soaring atmospheric oxygen. Predatory spiders and scorpions had likewise begun to call these lush forests home, while worms and snails likely slithered along the ground, although their soft body tissues have not been preserved well enough as fossils for us to be certain. Ultimately, if we were to dig away at the ground of the Carboniferous forests, the communities of creepy crawlies we uncovered would not differ significantly from those seen around the world today.[18]

Turning our heads up from the floor, if we looked closely enough at the dense, dark forest subcanopy, we might begin to see how these earliest forests, while relatively quiet, also hosted some of the first major step changes in land-based animal communities. To observe this most clearly, we must travel to Hamstead in Birmingham in the United Kingdom, the "Jewel of the Midlands." Like Wales, it might seem a peculiar location to pick when discussing the origins of tropical forest ecosystems on Earth. However, the coal layers that attracted industrialists to Birmingham in the eighteenth and nineteenth centuries provide an ideal place to look at the forests of the Carboniferous—a time when Birmingham would actually have been located near our planet's equator. In 2016, scientists presented an incredibly significant find in the context of early "tetrapod" (four-legged animal) evolution. Their research concerned twenty red sandstone slabs that had been sitting neglected at the Lapworth Museum of Geology since their discovery by a schoolteacher in the early twentieth century. Using state-of-the-art scanning approaches, scientists took around one hundred photographs at a range of heights and orientations (a methodology known as "photogrammetry") around the specimens to build 3-D models of the marks left behind on the slabs. The lead author of the resulting study, Luke Meade, notes how reexamining these blocks after over a century "opened a window onto Birmingham at the end of the Carboniferous." Analyzing the 3-D models, the researchers found a series of footprints left by increasingly active animal life on the muddy, forested floodplains during this period of Earth's history.[19]

Most numerous were amphibians or amphibian-like creatures. These were the first four-legged animals to emerge from the oceans and inherit the Earth—making short trips onto land as early as the Late Devonian. The warm, humid swamp forests of the Carboniferous period and their regulation of the planet's atmosphere enabled amphibians to diversify in terrestrial habitats. They became the top land predators, some reaching several meters in length, consuming the available creepy crawlies on land and fish in the swamps and rivers. Yet, like frogs and newts today, they were always forced to return to the water to lay their eggs, limiting the degree to which they could successfully colonize the land. The Birmingham footprints, however, reveal two other vertebrate groups that were beginning to walk, reproduce, and form permanent land-based

Figure 1.2. Artistic reconstruction of *Sphenacodon* (an early synapsid, the group from which the ancestor of all mammals would later emerge). *wikimediacommons.org*

communities. Rare tracks of large pelycosaurs, which may have looked a bit like Komodo dragons, represent one of the first members of an ancient lineage that would eventually produce mammals. Meanwhile, smaller tracks show the appearance of the sauropsid reptiles. Initially small and lizard-like, they produced "amniotic" eggs, whose surrounding fluid could resist drying out on land, and initiated a reptilian dominance of the Earth that would include creatures as diverse as dinosaurs, crocodiles, turtles, and, later, birds. Much of what first springs to mind when we think of animal life thus began in these warm, wet equatorial forests.[20]

The earliest forests of the tropics completely redesigned the Earth's surface, providing nutrients, shelter, stable soils, and atmospheric conditions for lively terrestrial ecosystems that contained insects and various animals living at different trophic (predator-prey) positions and exploring different adaptations. Within these new types of environment appeared all of the main terrestrial groups of animals that we now know today, captured in time by the Birmingham footprints. Tropical forests were thus the first complex land-based ecosystems on the planet. They also had a lasting impact on the Earth's atmosphere and climate. Absorbing significant amounts of CO_2, through a now expanded factory line of photosynthesis, and aiding the cycling of water into the atmosphere and through the ground thanks to evaporation from larger leaves and effective root systems, they made themselves critical to life on Earth. In essence they developed an entirely new planetary order that was to host a diversity of life on land from that point onward. Tropical forests were vital to the air breathed in by new, terrestrial organisms and the climates that they experienced, and, as we will see in Chapter 12, they remain crucial to the functioning of the various systems of our planet today. This is something we should perhaps consider more often as they rapidly disappear from its surface. Yet this does not mean that they stood unmoving. In fact, they had only just started evolving. The 3-D scans of the Birmingham sandstone reveal much more than just footprints. They reveal the raindrops that fell in the humid Carboniferous "coal forests" as well as the mud cracks that formed in river basins about 300 million years ago during drier periods that were becoming increasingly common. Tropical forests were going to need to change if they were to survive, as were the creatures that inhabited them. And change they did, impacting and hosting some of the most dramatic periods in planetary history that were yet to come.[21]

Chapter 2

A TROPICAL WORLD

Tropical forests have been on Earth for over 1,000 times longer than we have, so it is easy to act as though they've just always been here—timeless, primordial guardians of our planet. Conservation efforts even often use their great age, contrasting unchanged and uninhabited forests with the dangerous, destructive "progress" of twenty-first-century industry and development. Even how we commonly present or talk about tropical forests in popular culture works to sweep away any ideas of transformation or diversity. Novelists and film producers are all too happy to describe or capture the same tangles of vines, the same dense undergrowth, and the same dark, humid settings to represent a stock perilous "tropical" environment, while, as we have seen, the colloquial word for tropical forests in English, "jungle," simply refers to a generalizable wilderness. If these perceptions of static, homogeneous forests were correct, we would be living in a world still decked in the forests dominated by club moss and ferns of the Devonian and Carboniferous that we met in Chapter 1. Today, however, tropical forests, as a whole, are home to over half of the Earth's plant and animal species. This is not because they are all the same, and it is certainly not because they have remained unaltered for 300 million years. Instead, over the course of their history, they have proven themselves to be just as dynamic, vibrant, and geographically variable as the rest of the Earth's environments.[1]

What are tropical forests? As we have already seen in Chapter 1, there are two broad ways of thinking about them. One focuses on the climatic necessities known to be required for the growth of tropical forest types (e.g., for tropical rainforests, 2,000 millimeters of annual rainfall, minimum temperatures of 18°C, and no significant dry season). These definitions can be incredibly useful in "deep time" when exploring periods in the Earth's history, like the Devonian and Carboniferous, when much of the world was what we would conventionally consider "tropical" in a climatic sense, and forests reliant on warm and wet conditions, once present, could extend well beyond the equator. They also provide us with a way of looking at the surroundings of the first plants and trees to have colonized the Earth in Chapter 1. Definitions focusing on necessary climatic parameters are also useful in acknowledging that the forests of the past may not have had exact modern counterparts. For example, we will see in this chapter how terms such as "megathermal moist forest" can be used to explore how even the most unlikely of regions could have hosted hot, wet, frost-free forest ecosystems of a type that we might expect solely to prevail in the tropics today. The movement of the Earth's surface (and thus the "paleolatitudes" of different continental landmasses at different points in time) and changes in the planet's climate systems add yet another layer of drama to the winding road of "tropical" forest origins and evolution.[2]

The second way of defining "tropical forests," and the one formally used in this book from now on, is "any forest located between the Tropic of Capricorn and the Tropic of Cancer." Although the tropical latitudes were also often the warmest, by not focusing on the climatic requirements of a particular forest "type," this definition allows us to explore the vast variety of the tropical forests we coexist with at present: from the wet, lowland evergreen rainforests we know so well from documentaries on the Amazon and Congo River basins, to the seasonally thirsty, dry, and thorny tropical forests in places such as Central America and Madagascar, to the snow-tipped branches of montane tropical forests that disappear into the heavens in regions such as New Guinea and the Andes. Although rainforests receive a lot of attention for the incredibly diverse insect communities they host, as well as for their great apes and forest elephants, more seasonal tropical forests are home to equally rare, though perhaps less conventionally rousing, specialized species of plants and animals, such as lemurs, fruit bats, and the world's most endangered tortoise. Crucially, this definition also allows us to investigate varied forest ecosystems that inhabited the tropical latitudes at different points in time and how they led to the origins of the tropical forest ecosystems that exist there in the present. We are able to chart a course, from the equatorial Carboniferous "coal forests" we met in Chapter 1 to the first closed-canopy Neotropical rainforests that represent the origins of ecosystems that have made the Amazon Basin one of the most renowned homes to tropical forests today.[3]

Long-term tropical forest evolution and its significance for the emergence and diversification of life on Earth is a story of change. The first complex tropical forests, emerging under warm, moist conditions at the equator, had already played a role in the permanent alteration of the Earth's systems and the appearance of some of the first land-based animal life during the Carboniferous. But this was by no means the final word. To truly understand how—and how much—tropical forests have impacted the various evocative life-forms that have graced our planet in the last half billion years, and to truly determine the degree to which they managed to extend their roots into the functioning of the Earth's atmosphere, climate, and nutrient cycles, we must now investigate how they themselves went on to change. This chronicle is one of ongoing shifts in global climate and, significantly, the crunching process of continental drift and plate tectonics. It is key to the evolution of all plant life and its increasingly permanent seat at the table of atmospheric, geological, and climate change. So, without further ado, let's begin our tour of tropical forest evolution and change. While the Devonian and Carboniferous both represented major ecological accomplishments for arboreal lifeforms, tropical forests, not for the last time, were about to grow headlong into catastrophe.

GIVEN THE VARIETY of life that we met beneath their canopy in Chapter 1, the Carboniferous "coal forests" that covered the continental

landmasses at the equator could already be considered a highly successful evolutionary experiment by c. 300 million years ago. Nevertheless, the
final coalescence of the world's continents into a single supercontinent,
Pangaea, approximately 330 million years ago began a slow shift toward a
warmer and, more significantly, drier planet that would completely overhaul the vegetation communities of the tropics and beyond. So severe
was the resulting "rainforest collapse," as it has become known, that it
arguably led to one of the two mass extinction events for plants revealed
in the fossil record (by comparison, animals have faced five mass extinction events and may soon be heading for their sixth). Tree and treelike
species that had begun to dominate the earliest ancient forests, such
as club mosses, giant horsetails, and the immense cordaites we met in
Chapter 1, gradually disappeared from the fossil record, with the number
of tree families slashed in half by the time of the Permian (298.9 million
to 251.9 million years ago). Not only did species change, but by the start
of the Triassic period (251.9 million years ago), forests in all forms had
almost entirely disappeared from the face of the Earth. Amazingly, this
was the only point in time, since the evolution of the first tree until today,
that forests have been practically absent from our planet's surface.[4]

As these forests had already inserted themselves into the Earth's
systems, knock-on effects were inevitable. Anchoring trees were suddenly missing, causing a thinning of life-giving soils, soil erosion and the
movement of once reliable rivers, and a return to a bleak surface with the
exception of a few isolated "islands" of vegetation. Furthermore, although
the planet was drying anyway, a reduction in forests, greenery, and the
evaporation of water from leaf surfaces almost certainly compounded
this increasing global aridity. Any water that did make its way onto or
into the ground was no longer captured by dense root systems. Instead,
it flowed away, carrying valuable nutrients from the Earth's crust into
rivers and, eventually, thriving Permian marine ecosystems. At the end of
the Permian, explosive volcanic activity in Siberia spat dust and poisonous gas into the atmosphere for hundreds of thousands of years, choking
photosynthesis and the diversity of four-legged creatures that had begun
to expand within and beyond forests. Over half the families of animal
life on land went extinct. However, while these dramatic eruptions at the
end of the Permian rightly get attention in most popular accounts of this

period—for nearly wiping out all life on Earth—the forests had actually already encountered their apocalypse. Although plants faced a further reckoning at the end of the Permian and start of the Triassic, with an additional slashing in number of the remaining plant families, much of the damage to the planet's greenery had been done. Without widespread forests, animal life continued to face frequent flooding and a relatively barren planetary surface, even as it recovered from its brush with volcanic hell. The first real attempts by forests to make their way in the world seemed to have ended in failure.[5]

As is so often the case in evolution, however, adversity can lead to innovation. Out of disaster sprouted the beginnings of all plant and forest life as we know it today. Nowadays we divide seed-producing (spermatophyte) plants into two broad categories, the gymnosperms and the angiosperms. The former can actually trace their origins back to the "progymnosperms" that we met in Chapter 1 and were the first beneficiaries of the Carboniferous "rainforest collapse." Their name comes from the Greek word *gymnospermos*, meaning "naked seed," and refers to the fact that the seeds they rely on for reproduction are unenclosed prior to fertilization. While they had already begun to dominate ecosystems during the Carboniferous, they were often found on forest floors—in and among the earliest cast of tree-forms, such as *Archaeopteris*. However, Pangaean drying and increasing CO_2 concentrations caused problems for other plant life, creaking open a window in the canopy for gymnosperms, such as cycads and ginkgoales (made famous by the ginkgo tree, or *Ginkgo biloba*, which is the only surviving member of this group), to rise to the top. Plants reproducing via seed have the advantage that their seeds can lie dormant, enabling them to outlast dry spells and more seasonal conditions. When forests did finally begin to return to Earth as the Triassic period wore on (251.9 million to 201.4 million years ago), these seed-producing gymnosperms, this time including the familiar conifers, now dominated the forest overstories in the tropics and beyond. The resulting ecosystems are still represented by today's temperate forests, from northern Siberia to their southern outposts in Chile and New Zealand. Meanwhile, spore-bearing plants were left to pick themselves up on the forest floors, and the ancient "progymnosperms," such as *Archaeopteris*, were consigned to the Carboniferous fossil record.[6]

Given the cold temperatures they inhabit today, you might be thinking that the boreal forests of Russia are not really what you would consider
"tropical," and you would be entirely entitled to do so. Indeed, while the
gymnosperm cycads are still represented widely in the tropics, and these
gymnosperm-dominated forests extended across the tropics for much
of the Mesozoic era (Triassic to Cretaceous period), to find the true beginnings of the types of forests that inhabit the tropics in the twenty-first
century, we must turn to the evolutionary history of the other group of
plant seed producers, the angiosperms (or flowering plants). To do so,
there is no better guide than Dr. Carlos Jaramillo, appropriately based at
the Smithsonian Tropical Research Institute in Panama. With his wide-
brimmed hat, geological pick, and mud-covered trousers, Carlos is one
of the finest explorers of ancient flowering plants and tropical forests
around. Undertaking long-term fieldwork and excavations in the Neotropics, he and his team are on the lookout for all types of fossil plant
evidence, from miniscule pollen grains to huge petrified trees, in order
to piece together the deep-time history of tropical ecosystems. While he
cannot resist the occasional analysis of giant turtle and crocodile fossils,
which the team also often find in their Central and South American excavations, he tells me that his heart truly lies with the enchanting and
underappreciated evolution of tropical plants.

The history of tropical forest vegetation is one of radical and fascinating change. While tropical forests might be ancient, they have never
been short of excitement or dynamism, particularly across the last 200
million years of their evolution. As Carlos puts it, "120 million years ago
the entire tropics were home to just a single flowering plant; yet today
they are almost completely composed of some of the greatest concentrations of flowering plant diversity anywhere in the world." In fact, the
origins, expansion, and diversification of the most diverse group of land
plants—the group that yields the plants that today sit in our vases and
manicured gardens, dominate celebration and mourning rites, and make
up the majority of plants exploited by humans for food, medicine, and fibers—are intimately linked to the evolution and emergence of the varied
tropical forest habitats as we know them. So how did these enormous
ecological changes happen? To find out we have to look at the planetary

drama out of which the angiosperms emerged. During the Jurassic period (201.4 million to 143.1 million years ago), the supercontinent of Pangaea began to experience a tectonic divorce, as subsurface tensions bubbled up into the Earth's crust to unleash the irresistible forces of plate tectonics. First, Pangaea broke in two, forming the northern Laurasian and the southern Gondwanan landmasses. By 140 million years ago, Gondwana was itself beginning to split into Africa, South America, India, Antarctica, and Australia.

It is into this fiery Cretaceous (143.1 million to 66.0 million years ago) breakup that the angiosperms entered the frame. After a cooler interlude in the Triassic that favored conifer expansion, this world-making process released masses of CO_2 into the atmosphere as a product of volcanic eruptions, stimulated by the grinding apart of the Earth's crust. Temperatures and humidity rose, creating the ideal "greenhouse" for the first angiosperms that were able to ride their way into a new world order on the backs of free-roaming continents. Based on the majority of existing genetic and fossil evidence, these flowering plants likely first emerged around 140 million to 120 million years ago in the tropical latitudes, although some researchers have also suggested that midlatitudes would have perhaps provided more suitable, stable settings for angiosperm emergence during the Early Cretaceous (143.1 million to 100.5 million years ago) and others still argue that the angiosperms may have begun to emerge as early as the preceding Jurassic. Regardless, analysis of ancient angiosperm lineages certainly makes clear that they first appeared as part of a warm, wet forest undergrowth, lying low under gymnosperm canopies that were gradually losing diversity. Meanwhile, genetic evidence shows that around 100 million years ago, many orders of angiosperm widespread in tropical forests today diverged as Gondwana broke up and global temperatures became warmer and wetter. Between 100 million and 70 million years ago, angiosperms spread across the tropics, experiencing rising species diversity and a growing domination over all other plant groups, particularly in warm, wet land-based environments. One final dramatic proliferation approximately 60 million years ago saw angiosperms spread extensively across the majority of the lower and middle latitudes. This proliferation of angiosperms has also been implicated in alterations to the atmosphere and the water cycle. They have higher numbers of veins in their leaves and can therefore perform water evaporation (transpiration) and photosynthesis more effectively. The release of more water back into the atmosphere from these novel leafy surfaces has been linked to the development of wetter and less seasonal climates.[7]

The scale of the tectonic journeys of the angiosperms is visible when we focus in on the family that includes some of the true tropical rainforest giants that stand today (some as high as eighty to one hundred meters), including major tropical timber trees that are exploited globally, the Dipterocarpaceae. Evidence from fossils, modern genetics, and comparisons of body shapes all indicate an origin for this group before 120 million years ago on the landmass of Gondwana, which included Africa, South America, Antarctica, Australia, and the Indian subcontinent. Some members of this family then rafted on top of the plate comprising India and Sri Lanka as it separated from Gondwana and eventually smashed into Laurasia (a plate including the rest of Asia), completing their global voyage about 45 million years ago. Aided by the forces of geology and perhaps also pollinating insects, the angiosperms accompanied global "greenhouse" conditions that prevailed from the Cretaceous (143.1 million to 66.0 million years ago) right up until the middle Eocene (48.1 million to 37.7 million years ago). As early as the Late Cretaceous (100.5 million to 66.0 million years ago), these flowering plants had also already started to become key members of tropical forest ecosystems, and fossils of broad, woody angiosperm trees and fruit-producing climbers have been found in Nigeria. However, clear fossil records of the earliest angiosperm-dominated forest ecosystems, with a collection of species and a structure (i.e., a number of stratified layers, or "stories," within closed forests, where light reaching the floor is limited) that we expect from tropical forests today, only really began to appear in the Paleocene (66.0 million to 56.0 million years ago), when temperatures started to rise and environments began to recover following the extraterrestrial collision and volcanic winter that drew a dramatic line under the Cretaceous and dinosaurian dominance. To see what these new angiosperm-dominated tropical forests looked like and how they related to the diverse ecosystems we know from the tropics today, we must now follow Carlos and his team into yet another mine in the Americas.[8]

THE CERREJÓN COALMINE in northern Colombia is one of the largest open-pit coalmines in the world. So vast is the expanse of coal in the region that it has been recognized as its own geological entity, the Cerrejón Formation. As we have already seen in Chapter 1, thick beds of coal are something of a "smoking gun" for detectives in search of ancient tropical forests, and these are no exception. The mine has opened an unprecedented window into the communities of plants that formed in the tropics 60 million to 58 million years ago, just after the dramatic Cretaceous-Paleogene boundary (66.0 million years ago) extinction event that wiped nonavian dinosaurs off the face of the planet. The Cerrejón coalmine represents a treasure trove not just for fossil fuel companies but also for paleobotanists, and to date it has produced over 2,000 plant "mega fossils," some nearly twice as big as an average human, which include ancient leaves, flowers, seeds, and fruits. Today, when we think of tropical forests, particularly rainforests, in the Neotropics (i.e., the tropical portions of the Americas), we imagine—or, in the case of Victor and myself, experience—a certain level of plant diversity, the presence of specific angiosperm families, and high numbers of plant-eating insects, high temperatures, and plentiful rainfall. Unsurprisingly, these traits are immensely difficult to find altogether in one spot in the fossil record, making it exceedingly challenging for us to properly identify the origins of this evocative twenty-first-century tropical forest community. The Cerrejón coalmine is perhaps *the* major exception in this regard and represents a paradise for researchers like Carlos.

Carlos and his team have discovered that the plants found at Cerrejón represent many of the families that dominate tropical rainforests in this part of the world today. When analyzing the fossils under the microscope, the team could identify characteristic plants, including palms, legumes, and even the ancient relatives of the avocado, and show that angiosperms had truly come to dominate these new, post-dinosaurian forests, with multilayered rainforest canopies providing the stage for angiosperms to diversify and experiment with the astoundingly different forms we see in contemporary tropical rainforests. Amazingly, study of the microscopic (e.g., fossil pollen) and macroscopic (e.g., fossils of leaves) remains preserved at Cerrejón can find about 60 to 80 percent of all of the main groups of plants found in twenty-first-century Neotropical

Figure 2.1.
Image of an
angiosperm leaf
recovered from
the Cerrejón
Formation in
northern Columbia
by Carlos and his
team. The density
of veins in the leaf
surface remains
clearly visible after
60 million years.
Carlos Jaramillo

rainforests. While this ancient forest was slightly less diverse, its likeness to those seen in Central and South America today is remarkable given the subsequent geological processes (e.g., uplift of the Andes between 55 million and 40 million years ago) and changes in climate (fluctuating between warmer and cooler than present). Effectively, at Cerrejón we can start to see the origins of some of the modern tropical forests we can still encounter around the world today, 50 million years before our first hominin ancestors would appear. These plant communities, while changing subtly and waxing and waning in spread, were to stand tall

throughout major upheavals and remain around us, spanning a significant portion of the Americas, including, of course, the Amazon Basin.
After experiments of varying degrees of success, a new type of tropical
forest had now established itself on Earth, and it was here to stay.[9]

Carlos and his team did not stop there, however. The remarkable
preservation within the Cerrejón Formation even allowed them to analyze the cell and leaf structures of these ancient forests. The leaves had
the same forms as those found in modern Neotropical rainforests: big
with smooth edges, indicating similarly hot and wet conditions. Indeed,
comparison with modern leaf shapes suggests a past rainfall greater than
2,500 millimeters per year and an average annual temperature of greater
than 29°C, about 2°C higher than that of today. In fact, the leaves were
so complete, even after a remarkable 60 million years, that Carlos and
his colleagues could analyze the bite marks of insects that had feasted
on these tropical bounties. This damage showed that a variety of plant-
eating insects now bustled around tropical forests, though, like the
plants, they were less diverse than the menagerie of organisms found on
the tree trunks and floors of the Amazon rainforests today. Finally, the
available fossils also show the beginnings of a classic Neotropical animal community, including the remains of a now extinct giant snake, the
terrifyingly named "Titanoboa." Related to today's boa constrictors and
anacondas, this monster would have been right at home in Luis Llosa's
1997 Amazon-based horror film *Anaconda*. As in the case of the large
leaves, the significant size of this snake (nearly thirteen meters in length)
suggests that the Cerrejón forest was warmer than the tropical rainforests of modern South America. This is because snakes are "ectothermic,"
relying primarily on environmental sources of heat to regulate their body
temperature.[10]

Cerrejón was not the only tropical forest in town. By the Paleocene-
Eocene boundary (56.0 million years ago), tropical forests, with
characteristic plant families and structures that we still see in these forests today, spanned the tropics. Not only that, but similar "megathermal"
forests reached the greatest extent they have ever managed in the history
of our planet. Although incredibly diverse and not always "tropical" in the
context of the definition I provided at the start of this chapter, whether
we set down in North America, Africa, South America, Australia, Asia,

or Europe, warm megathermal forest ecosystems would have greeted us, reinforcing a global move toward warmer and wetter conditions as they further inserted themselves into the Earth's systems. Mangrove-like forests even adorned the now frosty coasts of the United Kingdom, while so-called boreotropical forests (the name for forest communities with similarities to those seen in the tropics today that extended well into northern and southern latitudes) reached as far south as Tasmania and as far north as Alaska. Importantly, unlike those of the Carboniferous forests, the structure and species of the new forests in and around the equator, like that at Cerrejón, would have appeared much more familiar to those of you who live in or visit the tropics today. The distributions and evolution of various tropical angiosperm plants continued apace, and certain plant families were able to move rapidly between the tropical continents via land bridges. However, the ongoing processes of tectonic separation throughout the Eocene (56.0 million to 33.9 million years ago) meant that tropical forest communities began to take on their own regional "flavors." By the end of this period and the final division of Antarctica, South America, and Australasia, the Earth's continental makeup was pretty much decided. Species that had previously surfed on moving geological plates now became largely trapped. The result was the gradual, independent evolutionary formation of the unique forest ecosystems we know from different parts of the tropics today.[11]

The process of continental separation, as well as isolating tropical forests in different parts of the globe, also stimulated the development of new climate systems. The moving apart of landmasses opened up new spaces for vast ocean circulation systems that extended across the planet. Cold water flowed from the north polar region into the Atlantic, cooling and drying the Northern Hemisphere thanks to a reduction in evaporation, while Antarctica became separated from warmer equatorial waters. From about 47 million years ago, the formation and rise of many the world's great mountain ranges exacerbated this trend. They began to force the megathermal forests back toward their current distribution as a belt around the Earth's equator. While trees retreated, new players began to expand from approximately 20 million years ago: players that had crept onto the vegetation scene during the time of the dinosaurs. With a shape, root system, and physiology adapted to resisting droughts and cold snaps, grasses provided a small but collectively mighty challenge to forests during times of disturbance. And thus, present-day ecological battlelines were formed within the tropics: on one side, tropical and subtropical forest ecosystems, with their vast amounts of angiosperm diversity; on the other, novel grassland ecosystems. These two formations have been locked in an eternal skirmish ever since, waxing and waning in the face of periodic global climate swings.[12]

BY 10 MILLION years ago the modern tropical forest toolbox was largely complete, with characteristic plant species, continental location, and the broad global conditions we associate with them today all in place. This did not mean that they remained idle, however. Although their exact impacts remain hotly debated, changes in precipitation, temperature, and global CO_2 continued to drastically impact the extent and structure of tropical forests. During the Quaternary (2.58 million years ago until present), these shifts have been driven primarily by cycles in the distribution of the Sun's energy around the Earth, linked to our planet's changing orbit and tilt. Broadly, over the last 1 million years these "Milankovitch" cycles (named for the Serbian scientist who discovered them) have resulted in an ice age and temporary dramatic expansion of the polar ice caps approximately every 100,000 years. Under ice age conditions, sea ice advances in the North Atlantic, which, through circulation systems in the ocean and air, can also cool the tropics. This cooling in turn leads to movements and weakening in the operation of key climate systems, such as the Intertropical Convergence Zone (where northeastern and southeastern trade winds meet around the equator) and the Indian summer monsoon system, which bring rainfall to many parts of the tropics today, resulting in drier conditions around the equator. Nevertheless, these broad solar models do not always explain all of the variation in the Earth's climate over the last 10 million years, both in terms of extremes and frequency. In fact, temperature, precipitation, and CO_2 at a given time and place also remained dependent on changing relationships between how heat and water circulated between the oceans, the atmosphere, and, crucially, land cover, including the greatest of earth systems engineers, the plants and forests themselves.[13]

Taking a broad-brush approach, we can identify some major climate- and atmosphere-driven changes in the varying fortunes of tropical forests and their grassland competitors over the past 10 million years. Starting in the Miocene (23 million to 5.3 million years ago), warm global temperatures and high CO_2 concentrations had seen megathermal forests once again expand across much of Africa and Eurasia, reaching a maximum extent between around 21 million and 14 million years ago. From 10 million years ago, however, as CO_2 concentrations declined, profound changes in vegetation occurred. Grasses practicing a novel form of photosynthesis (named "C_4" after the number of carbons in the first sugar produced by the plant during the process), which was more efficient at absorbing CO_2 and photosynthesizing in drier conditions and which dominates tropical savannah habitats as we imagine them today, joined existing "C_3" grasses in the fight against warm, wet forests that saw the latter removed from Europe and Central Asia. Overall cooling and drying that had begun in the Eocene continued into the Pliocene (5.33 million to 2.58 million years ago), dealing another gradual but lasting blow to nontropical megathermal forests. As vegetation that had adapted to cooler, "temperate" climates expanded in the Northern and Southern Hemispheres, they retreated, regrouping near the equator. This was a new dawn. The age of tropical grasslands had begun, while other types of seasonally dry tropical forests (including those containing "deciduous" plants, which shed their leaves for part of the year) also began to emerge and diversify across the subtropics and tropics. Continued drying also saw deserts form in the most water-deprived areas of Africa, South America, and Asia, resulting in the layer cake of environmental variation we see as we travel from north to south across the globe today.[14]

While overall planetary cooling continued, from a mean annual surface temperature around 2°C warmer than today at the start of the Pliocene to as low as −6°C cooler than today during parts of the Pleistocene, the greatest characteristic of the Pleistocene epoch (2.58 million to 0.0117 million years ago) was actually the intensification of the impact of Milankovitch cycles. In the early Pleistocene (2.58 million to 0.77 million years ago) glacial cycles were more frequent (around every 41,000 years) but less extreme, while the middle and late Pleistocene saw the less frequent but more significant 100,000-year swings in global climatic states. In a tropical context, Pleistocene glacial cycles and a drying of the tropics have been argued to have led to an expansion of "savannah corridors" across different parts of Africa and Asia, perhaps even reaching between the two continents. Growing larger as glacial shifts became more extreme into the middle and late Pleistocene, these corridors would have provided ideal grassland conveyor belts for medium- to large-sized animals, which, unlike the usual dense tropical forest that had covered these regions, enabled them to traverse vast areas. In Southeast Asia, for example, some argue that such a "corridor" extended from southern China in the north down beyond the Wallace Line that runs through Indonesia, aided by the formation of a land bridge between mainland and island Southeast Asia, which appeared as growing ice caps sucked up water from the oceans. The end of wide-ranging savannah ecosystems was nonetheless inevitable, due to the fact that glacial ice ages, though longer and more extreme in this period, were not indefinite. The return of warmer, wetter, "interglacial" conditions broke up savannah extents, replacing them with classic tropical rainforest environments—hostile to many of the large-bodied grazing animals that had moved in during their absence.[15]

The last great glacial period, or Last Glacial Maximum (LGM), occurred between 26,000 and 21,000 years ago. Greater numbers of ancient archives of past climate and environment that relate to this more recent period allow us to determine the exact effect such a period might have had on tropical forests around the world in more specific detail than we can for earlier glacial periods. Although changes were not as extreme as in temperate areas, staggering drops in temperature of as much as 4°C to 8°C have been suggested for different parts of the tropics during the LGM. Given that our current rates of emissions predict a rise in tropical temperatures of anywhere between 2°C and 7°C between 1980–1999 and 2080–2099, this should highlight both the extreme nature of the LGM and also the disastrous biological consequences potentially awaiting us if we do not act. The LGM also made the tropics more arid. In South America and Africa, for example, it may have resulted in a rapid retreat of tropical rainforest to just a few small *refugia* within the usually vast

Amazon and Congo Basin forests. Glacial declines in atmospheric CO_2 may also have changed the structure of tropical forests, leading to smaller, more efficient leaves, open canopies, and increasing forest floor growth, as well as the increasing presence of mosaics of forests, woodlands, and grasslands. Whatever the cause, subtropical and tropical forest retreat between 26,000 and 21,000 years ago has been recorded in Africa, Southeast Asia, South Asia, New Guinea, Australia, and South America. The rate of forest re-expansion varied by region and was often complicated by other factors, including, in Southeast Asia, the impacts of rising sea levels on rainfall patterns.[16]

Better-dated and finer-grained records of climatic and environmental change for our current geological period, the Holocene (11,700 years ago to the present), show us that the broad changes in tropical forest structure and extent noted above for time periods where resolution is limited to 1,000 years or even more miss a variety of finer-scale variabilities that would have confronted tropical forest environments. Study of the changing operation of climatic systems today and in the recent historically recorded past reveals that parts of the tropics can rapidly switch from cold and dry to warm and wet and vice versa. For example, variations in how winds and sea surface temperatures interact over the Pacific Ocean (as part of the El Niño Southern Oscillation climate system) have led to extreme and sometimes unanticipated droughts and flooding in the recent past. Although the broader planetary changes seen on geological timescales undoubtedly influenced evolutionary processes, these short-term shifts would have shaped the experiences of individual organisms and communities. Feedbacks between vegetation and climate systems are also important for these high-intensity processes. This is acutely visible when we take the dramatic example of the "Green Sahara." Stretching from the Red Sea in the east to the Atlantic Ocean in the west, the Sahara is today the world's largest nonpolar desert, encompassing much of North Africa. Between, 15,000 and 5,000 years ago, however, conditions were very different, and this gigantic field of dunes was covered in vegetation and water. Climate models have demonstrated that alterations to the Earth's orbit are not enough to produce this effect alone. In fact, only when earth scientists factor increasing precipitation *plus* feedbacks from expanding forest and plant life into their computer software can they reproduce a vast oasis, with vegetation likely increasing water in the atmosphere through evaporation from its green surfaces, while also stopping the usual buildup of dust. Certainly the Holocene subtropical and tropical forests—which enable increases in regional air moisture, stabilize landscapes to avoid runoff, act as a buffer against flooding, and absorb solar energy—were critical players in the translation of climate system changes into on-the-ground environmental dynamics. This is something we will see later in the book, as our own impacts on these remarkable environments, and thus the planet's climate, come to the fore.[17]

AS WE HAVE seen, the history of tropical forests on Earth is one of upheaval, not timelessness. Extreme geological processes saw the first tropical trial runs begin to disappear from c. 300 million years ago to a point, c. 250 million years ago, when there were basically no forests on Earth. This seeming calamity paved the way for new forests, eventually composed of a new form of pioneering, flowering plant life that came to dominate global vegetation between 140 million and 60 million years ago. By around 60 million years ago, tropical forests closely resembling their twenty-first-century descendants had appeared in the tropics (and even pushed out beyond into a warm world as megathermal forests). But this was not the end. From 10 million years ago they were pushed back by new grassy adversaries to roughly their current equatorial distribution. Still the world was not done with them. Ongoing cycles in the availability of the Sun's energy and changes in climate circulation systems have buffeted these subtropical and tropical forests, causing them to wax and wane, open and close, ever since. However, tropical forests were never just victims of external change. As geology and climate shaped and moved them, they slowly but surely inserted their green tendrils into the Earth's systems. While the Sun dominated planetary climate change, this new vegetation cover decided how this solar energy manifested on the ground, whether it made local temperature and precipitation changes more dramatic or buffered local landscapes against them. Furthermore, tropical forests controlled the amount of CO_2 in the atmosphere and its distribution, providing some of the most significant land-based storage areas, or "sinks," of carbon on Earth.

This dynamism, both in terms of evolutionary change and the increasing role played by tropical forests in climate systems, gives us a new appreciation of their significance for life on Earth. Broadly considered, they were the sites of the most significant evolutionary events in the evolution of plants: the rise and diversification of the gymnosperms and angiosperms. They were also changing their distribution and structure as some of the most famous cast members of the animal kingdom arrived, evolved, and left the planetary stage. At the beginning, this involved insects and amphibians, but by the end it included the appearance of our species, its migration across the global tropics, and its formation of a vast variety of social, economic, and political systems that have varied radically until the twenty-first century. Throughout the rest of this book, we will meet the hugely diverse types of tropical forest—from the seasonally dry forests of Mexico and Guatemala to the often frosty forests of New Guinea, from the isolated island forests of Wallacea and the Pacific Islands to African tropical forest-savannah mosaics—encountered by our species, which are the ultimate products of this vibrant evolutionary history. Before we examine *Homo sapiens* and its closest relatives specifically, however, the significance of tropical forest change to our world as we know it is vividly visible when we look at two of the most evocative groups of animal life to walk the Earth. The next two chapters explore how the dynamics of tropical forests, from the Permian to the Pleistocene, described above, influenced the evolution and fates of, first, the dinosaurs (Chapter 3) and, later, their eventual inheritors, the mammals. Captivated by the fossils of these fascinating creatures, we have perhaps too rarely explored or given credit to the leafy, canopied companions that escorted them on their biological journeys. It is to these stories that we now turn.

Chapter 3

"GONDWANAN" FORESTS
AND THE DINOSAURS

I will always remember visiting the National Museum in Cardiff, the capital city of Wales, as a young boy with my family. Unlike many museums, which value academic silence and regular processions around their secure glass cases, this museum created a cinematic experience for its visitors—dramatically using sound and atmospheric visual displays to form the most vivid picture possible of our planet's history. The continental plate exhibit featured a darkness that focused your eyes onto the flowing red-orange lava of volcanoes and your ears onto the loud, crunching sounds of earthquakes. The exhibit of some of the earliest humans to reach a then icy Wales involved a trip into a cave-like chamber to meet moving, trumpeting mammoths. Most stunning of all, however, was the hall of the dinosaurs. Here, fossil casts of giant reptiles emerged out of shadows, illuminated by well-placed lighting and a chorus of roars and brays. This room has stayed with me as one of my earliest, and perhaps also scariest, encounters with the evolution of life on Earth. However, looking back now, something major was missing from this drama. In fact, it has been missing from almost all of the museum exhibits of dinosaurs I have encountered since. While we are staring at monstrous jaws, reconstructions of colorful feathered or hairy skin, or artistic impressions of vicious hunting scenes, we always seem to forget one of the most key parts of the dinosaurs' existence: their environments.[1]

Approximately 240 million years ago, the dinosaurs began as a predominantly carnivorous group of initially small "dinosaurmorph" reptiles. They bucked the trend of other sprawling and shuffling Triassic life and stood upright—adopting a very different way of living and moving. Indeed, when many of you think of dinosaurs, your thoughts will quickly turn to the hulking bipedal carnivores such as *Spinosaurus* and *Tyrannosaurus* that have haunted dreams since Hollywood pushed them onto our screens in the early 1990s. Yet these gigantic killers ultimately needed something to eat. From the Jurassic (201.4 million to 143.1 million years ago) into the Cretaceous (143.1 million to 66.0 million years ago), the dinosaurs began to form some of the largest land-based food chains by mass that have ever existed. Herbivorous dinosaurs accounted for around 95 percent of vertebrate biomass in some ecosystems and included the long-necked, long-tailed quadrupedal "sauropod" dinosaurs, such as *Argentinosaurus*, *Diplodocus*, and the aptly named *Supersaurus* and *Patagotitan*, which were the largest organisms to ever walk on solid ground: with a weight maximum of seventy tons, these sauropods would have dwarfed even the largest of African elephants recorded at ten tons. Researchers estimate that the biggest sauropods would have needed to consume a staggering two hundred kilograms of vegetation every day. For reference, that's nearly seven hundred cans of baked beans. Amazingly, we still rarely consider the key role that plants must have played in the evolution and sustenance of dinosaur ecosystems or the combined impacts that herds of herbivorous dinosaurs must have had on ancient vegetation—with plants portrayed as a convenient camouflage or as passive backdrops rather than a crucial part of life.[2]

This is perhaps even more remarkable when we think about the spectacular geological and environmental processes that happened from the Permian (298.9 million to 251.9 million years ago) to the Cretaceous (143.1 million to 66.0 million years ago). Major changes in vegetation accompanied the collapse of Carboniferous rainforests, the rise of the coniferous and cycad gymnosperms, and, later, the flowering angiosperms. Warm, wet forests at times blanketed the Earth before shrinking

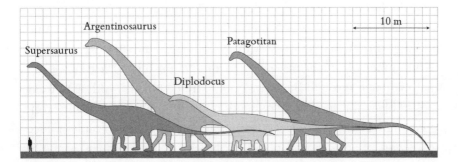

Figure 3.1. Comparison of selected giant sauropods. Several length estimates have been proposed for different taxa, with varying degrees of accuracy. *Adapted from Wikipedia / KoprX*

to more restricted distributions. Meanwhile, the plates of the Earth kept drifting as first Pangaea and then Gondwanaland were rent asunder to form the current continental layout of the tropics. Throughout, new forms of plants were emerging, moving, and evolving in novel, increasingly isolated ecosystems in the tropics. Today we accept that climate and environmental change associated with such significant geological and biological shifts would have massive implications for animal life (and in this chapter, we will take a look at the science that enables us to be more confident in thinking that). Nevertheless, while frequently used for dramatic effect in films, the vegetation of the dinosaurs has often been cut out of their evolutionary story, from their emergence and diversification through cataclysmic disaster to their persistence as birds today. Is this due to the poor preservation of fossil plants or simply a result of plants lacking the media appeal of the reptiles they supported? Either way, it's high time we inserted plants into the history of these "terrible lizards." Today leading researchers are using state-of-the-art investigation of reptile (including dinosaur) fossils, fossil pollen and wood records of past environments, and, incredibly, even the occasional preserved last meals of particular dinosaurs to do precisely that.

IT MIGHT BE hard to imagine, but the first "true" dinosaurs, defined by a series of particular features seen in their fossilized skulls, hips, and limbs, did not immediately dominate the Earth when they appeared in the Middle-Late Triassic approximately 240 million to 230 million years ago. Scampering alongside large amphibians, other diversifying reptiles, and tiny primitive ancestors of mammals, the dinosaurs were just one of a number of organisms trying to survive on the surface of a planet that had been through the wringer of extreme change. Although things looked bleak for animal life on land following the massive plant extinction event at the end of the Carboniferous and the volcanic-induced extinction of animals and plants at the end of the Permian, a number of researchers, including Dr. Emma Dunne, believe that in these cataclysmic shifts the dinosaurs grasped their chance to rise to the top of the animal kingdom. The changes to the Earth's surface that began with the collapse of the Carboniferous rainforest, whose canopies had coated practically all exposed equatorial land approximately 310 million years ago, would gradually, but fundamentally, alter the course of evolution. In time, they were to pave the way for all of the main groups of land-based animals we know today. The first key beneficiaries were, however, the reptiles and, more specifically, the dinosaurs.

Emma, a vertebrate paleobiologist at the University of Birmingham, has been interested in fossils for as long as she can remember, though she sees them in a different way to many people. Perhaps unsurprisingly, there is often a fascination with individual fossils, with people wanting to find the first type of a certain animal or particular body shape. However, a lot of information can be missed if these remarkable traces of life are treated as single objects. Emma prefers to think of them together as "parts of a living, breathing ecosystem." In this way she can try to reconstruct not just what they looked like but also the type of environment they might have lived in. This is essential if we are to go from looking at a stationary bone to determining what these animals might have eaten, what climatic and environmental conditions might have surrounded them, and what impacts they themselves might have had on the world around them. Emma is particularly interested in how land-animal communities changed between the Carboniferous and the early Mesozoic—an age that includes the Triassic, Jurassic, and Cretaceous periods—taking us right up to the dawn of the dinosaurs. To look at this change, she and her colleagues have tried to find a way to properly measure the number of different types (or "diversity") of animals living in different parts of the world, as well as across the world as a whole, at a given point in time. This, in turn, allows them to investigate how and why this diversity might have changed across space and time. This is not as easy as it might sound, however. The fossil record from this time period reflects huge biases. Most of our knowledge, like that about the ancient forests we have already encountered, comes from particular sites, mainly in North America and Europe. This is a product of the longevity of geological and paleontological study in Euro-American societies, as well as the availability of funding for such enterprises in different social, political, and economic situations—something, as we will see in Chapters 10 and 11, often linked to historical global imbalances in wealth distributions.

To try and get around this problem, Emma and her team made use of a remarkable tool that is actually available to every single one of us: the Paleobiology Database, a free-to-use record of fossil discoveries around the world, from the Proterozoic (2,500 million to 538.8 million years ago) to the Holocene. I can guarantee you that the beautiful software and easy-to-access data will give you hours of distraction, should you let it—allowing you to discover long-extinct creatures and the locations in which they have been found. Playing around with its functions, I went on an evening-long search for the find-spots of some of the most iconic creatures we commonly hear about from paleontologists, from the gigantic Megalodon shark to *Triceratops*. Most importantly, however, it represents part of a growing push to make scientific data freely available, or "open access," so that they can be used and combined to answer larger, planetary-scale questions. In this case, as paleontologists repeatedly update the database, adding the location and age of all finds from across the world, some of the more glaring geographical biases can be increasingly overcome. Instead of simply counting species for which there are fossils, larger databases such as this enable the application of novel statistical approaches so that we can get as complete a picture as possible of the *entire* animal fossil record, spanning the earliest tetrapods (four-legged land animals) to the first dinosaurs and beyond.[3]

The results of Emma and her team's research are fascinating. They demonstrate that the collapse of the tropical Carboniferous "coal forests" had a dramatic impact on land-animal diversity. Those affected most were the large, diverse amphibian communities, whose footprints were preserved in the rainforests of Birmingham approximately 310 million years ago. Although early dark, wet, and warm forest ecosystems had provided a haven for animals that needed water to lay eggs vulnerable to drying out, more arid conditions initiated the dramatic decline in these habitats. What happened next is critical. The small reptilian footprints, which were very much in the minority in Carboniferous Birmingham, suddenly proliferated, growing larger. Those tetrapod species that did survive the rainforest collapse began to disperse more freely across the planet, colonizing a number of new environments and moving further away from the tropics. Their laying of eggs resistant to drying unleashed a massive potential for expansion and gave them a number of advantages over the amphibians as forests continued to decline through the Permian and into the Triassic. In summary, then, the disappearance of tropical forests evidently, at first, actually aided the rise of the reptiles and, in turn, the dinosaurs. The changes in these ecosystems reshuffled the hands of the different animal groups and set out a new field of ecological competition.[4]

Dinosaurs have often been portrayed as "tropical" animals, existing within a framework of stable, warm climates from the Triassic to the Cretaceous. Yet, from the above, it seems that they could both be affected by and benefit from climatic and environmental instability. This is certainly true when we look at the early evolution and expansion of the first "true" dinosaurs from 230 million years ago. Although dinosaurs rapidly divided into three distinct successful lineages (Theropoda, Sauropodomorpha, and Ornithischia) and had achieved a remarkable global distribution by 206 million years ago, why it took nearly 30 million years from their first appearance for them to enter the tropics and complete their global conquest has remained something of a mystery. Furthermore, even when they did arrive, they did so in a tentative rather than a sweeping manner. The early theropod dinosaurs, which were to later include the infamous velociraptor, made it to the tropics and subtropics by the Late Triassic. This clade, characterized by animals with hollow bones and three-toed limbs, surviving in the form of modern birds, was initially dominated by predatory carnivores hunting small animals scurrying along the ground. By contrast, the other two main dinosaur clades, the increasingly large herbivorous sauropodomorphs and ornithischians, were restricted to the higher latitudes, being somehow "kept out" of the tropics until the Jurassic (201.4 million years ago). Although this division has often been attributed to bad fossil preservation caused by high humidity in the tropics or to potential fights for survival with crocodile-like reptiles that already inhabited the equatorial regions, a growing collection of fossils and the uncanny patterning of dinosaur types by latitude suggests that a deeper underlying geographical trend was at play.[5]

To try and unravel this problem, researchers with an interest in past climates have studied preserved woody charcoal, pollen remains, and the geochemistry of river sediments in New Mexico that have also preserved a variety of early dinosaur fossils. The results demonstrate that although the world was still generally warm and dry during the Late Triassic, it was increasingly affected by frequent, rapid, and extreme shifts in atmospheric CO_2 that led to sudden temperature spikes, increasing fluctuations in rainfall, and, most significantly, persistent wildfires that scorched wide areas of ground. The resulting climatic unpredictability in the tropics that we touched upon in Chapter 2 may have placed particularly strong stressors on animal communities living at the equator. As a result, these regions remained under the rule of other giant reptiles (pseudosuchian archosaur communities that bear greater resemblance to crocodiles and alligators), with some theropod dinosaurs eventually arriving and going about their business capturing prey, fishing, or foraging. Meanwhile, the increasingly large-bodied, fast-growing sauropodomorph herbivores, requiring much more reliable food sources, could not break into lower latitudes until conditions stabilized during the Jurassic. The state of tropical environments thus shaped even the ferocious dinosaurs from their earliest appearance and evolution. These habitats at once locked out some of the newer, more experimental forms while also providing a productive home to some of the earliest types of carnivorous dinosaur that picked their way through the bleak world of the Triassic.[6]

IN 2009 ISABEL Valdivia Berry and her husband, Erico Otilio Berry, were exploring the locality of Los Molles, Neuquén Province, Argentina. Many tourists come to Los Molles for its spectacular mountain views, hiking, and hot springs. It is also exploited by multinational companies as one of the largest shale and gas reserves in South America. However, Isabel was more interested in the other treasures that the surrounding rocks are known to offer, at least to those who know where to look. The Los Molles Formation dates to the Early to Middle Jurassic (201.4 million to 161.5 million years ago), and its rocky outcrops provide vivid snapshots of the marine and river delta ecosystems that were forming on the coasts of the Gondwanan landmass, a considerable portion of which, including much of South America, was primarily located within the tropics during this period. Ichthyosaurs, which resemble a cross between a dolphin and a crocodile and whose name aptly means "fish lizard," ruled the Early Jurassic seas and are found in abundance at Los Molles. However, the Berrys, on their winter expedition to the region, happened upon a very different type of fossil. Poking out of the rock face was a near-complete skull and significant portion of the body of a dinosaur, which, as they would discover, had never been seen or described before.

After Isabel brought her startling find to Museo Provincial de Ciencias Naturales "Professor Dr. Juan A. Olsacher" of Zapala, a team of paleontologists, headed by Dr. Leonardo Salgado of the National University of the Río Negro–Conicet region, set off to perform detailed excavation at the find site, recovering further fragmentary material. Following years of painstaking research, Leonardo and his team eventually published the find as *Isaberrysaura mollensis* in 2017. More recently, this dinosaur has even been recognized as one of the earliest stegosaurs, large, heavily built herbivores clad with rows of plates and bearing spike-tipped tails. Beyond its evolutionary position, however, *Isaberrysaura* is significant for two main reasons. First, it represents overdue recognition of the remarkable women, from the famous nineteenth-century paleontologist Mary Anning onward, who have made important contributions to paleontology, with "Isaberry" dedicating the find to Isabel herself. Second, when the paleontologists analyzed the specimen, they realized that within the skeleton, where the stomach should have been, there was, astonishingly, a mass of mineralized food. As Leonardo puts it, "We always ask ourselves what dinosaurs ate. Usually we have to rely on assumptions based on the shape of teeth. However, here was our chance to go one better."[7]

The team's findings provide a new, more intimate picture of increasingly large dinosaurs and their increasingly green surroundings in the tropical latitudes of the Jurassic. The gut of *Isaberrysaura* contained two types of seed. One set could not be identified, but the larger seeds were clearly members of the cycad group. The seeds were almost entirely intact, suggesting that they were gobbled down, in a manner similar to that used by modern reptiles to eat their food, instead of being chewed. Furthermore, as cycads are toxic to most animals, if *Isaberrysaura* survived this culinary experience, then the microbes inside its stomach must have been well adapted to digesting these plants. This one-in-a-million find— while providing a vanishingly brief single-meal snapshot of the dietary behaviors of dinosaurs—gives us a new appreciation of the close relationship that diversifying dinosaurs had with novel emerging plant life in the Late Triassic and Early Jurassic. Not only were herbivores consuming gymnosperms, the new green arrivals on Earth, but, Leonardo argues, they may also have played a role in their increasing success. As mentioned, he found that the seeds left in the stomach were almost complete. This means that when they passed out of the *Isaberrysaura*, they could go on to be fertilized and grow. In essence, these early stegosaurs were potentially some of the first wide-ranging seed dispersers, perhaps even contributing to the health of these rapidly expanding gymnosperm plant communities.

At the end of the Triassic 201.4 million years ago, there was yet another mass extinction event on land. Global drying, an increasing instability and seasonality of climate, and the deadly accompaniment of volcanic eruptions killed many large amphibians and all reptiles with the exception of the dinosaurs and the ancestors of those that are alive today, including crocodiles, turtles, and lizards. It also caused a major turnover in seed-producing plants. With the rise of the Jurassic, however, the Earth's climate stabilized. The arid conditions of the Triassic gave way to wetter and warmer (approximately 3.5°C warmer than today) conditions that encouraged the expansion of lush vegetation, including tropical forests. While conifers continued to dominate, as they had done from the Middle-Late Triassic, these new conditions enabled the diversification of gymnosperm plants. Study of the fossil pollen found from the same geological formation in which *Isaberrysaura* was found, under a microscope, shows that it belonged to a mix of conifers (such as the family Araucariaceae that now extended across the entire planet), cycads, and non-seed-producing ferns. Together these plants would have provided a combination of rich open and forest environments for increasingly diverse dinosaur species, not just in the tropics but also beyond. Ginkgo trees began to dominate at warmer temperate latitudes, with pine-based boreal forests extending into the cooler environments. Although atmospheric CO_2 levels stabilized in the Jurassic, as was seen for the end of the Triassic, generally high CO_2 could still lead to frequent, regionally variable high-temperature events and rainfall variability. Nevertheless, overall increasing stability and a lush gymnosperm vegetation relied upon by ever-growing herbivorous dinosaurs provided some semblance of steadiness, and dinosaurs rapidly filled every single ecosystem and niche in an increasingly mixed landscape that was full of vibrant green life. The Jurassic was thus a time when not just dinosaurs but also their gymnosperm food sources came to rule the Earth.[8]

The relationship between dinosaurs and gymnosperms may even have been so intimate as to represent an example of "coevolution." The sauropods—dinosaurs with massive necks, huge tails, and an overall tendency toward gigantism—flourished by the end of the Jurassic. Studies of the shape and size of their teeth, as well as reconstructions of their jaw movements and skull dimensions, show that they were herbivores. The vast size of these animals and their rapid growth rate, when compared to known modern large herbivores, shows that they must have been bulk feeders, with a paraphernalia of gut bacteria that helped them to digest their leafy meals through fermentation. Like large stegosaurs (later relatives of *Isaberrysaura*), they did not chew their food; instead, they swallowed it whole. Study of fossil plant remains found in the same contexts as sauropods has provided some insights into the kind of plants that were likely on the menu of these giant vegans. When all of this information is combined, it seems clear that these proliferating long-necked dinosaurs fed in bulk on gymnosperms, including conifers such as *Araucaria* and *Ginkgo* trees, and ferns in a variety of different settings, ranging from humid forest edges to more open areas. Subtle variations in the shape of their teeth suggest that they diversified so that while some may have focused on treetops, others swept more mobile necks around ground level to feed. This was not just a one-way street, however, and like the large herbivores of the modern tropics, the sauropods likely "gave back" to their ecological settings. Their wide-ranging, crashing movements would have benefited gymnosperms as their seeds were dispersed without damage, and plants such as cycads may have attracted color-seeing dinosaurs with their brightly colored seeds. It has even been suggested that the cones on conifer trees, particularly the so-called monkey puzzle Araucariaceae, evolved their shape as a way to survive and benefit from dinosaur herbivory! Although the strength of this relationship has been hotly debated and an exact link between gymnosperm success and dinosaur abundance seems unlikely, the varied, enormous herbivorous dinosaurs of the Jurassic, as well as their predators, undeniably benefited from an expansion and diversification of lush gymnosperm vegetation. It is also intriguing that, in North America at least, sauropod diversity eventually decreased, while stegosaurs all but disappeared, at a time of gymnosperm decline in the Early-middle Cretaceous (143.1 million to c. 100 million years ago). In fact, it is around this time that an even more dramatic link between dinosaurian green "fingers" and plant evolution has been proposed—this time centering particularly on the tropics.[9]

PROFESSOR PAUL BARRETT has the job many of us would have wanted as children (and, let's face it, probably still want). His office in the Natural History Museum in London is just down the hall from one of the largest and most famous collections of dinosaurs anywhere in the world. Even better, every day he gets paid to walk down the hall to study these bones, which include the first skeleton of an *Iguanodon* known to science, used as part of some of the first academic descriptions of dinosaurs as a biological group, and a part of the first *Tyrannosaurus rex* skeleton ever discovered. Nevertheless, like the other researchers in this chapter, Paul is not just interested in dinosaurs for dinosaurs' sake. He wants to know how they influenced life on Earth and how they were a part of broad processes that have shaped the environmental and evolutionary history of our planet with a lasting legacy to this day. As many of us have only ever seen a dinosaur as a collection of bones, it can be hard to think about them as living, active beings, particularly when we often imagine them as ultimately unsuccessful and consigned to the static fossil record by an as-teroid. Furthermore, in museum collections that include dinosaurs with machete-like teeth, the gentler herbivorous beasts of this earlier time are not often thought of as major players. Yet Paul's work over the last twenty years has looked into some of the most intriguing hypotheses for how populations of herbivorous dinosaurs impacted changing plant evolution and how, in turn, changing climates, environments, and plant communi-ties affected the dinosaurs.

In the 1970s, the American paleontologist Robert Bakker presented the intriguing idea, in the world-leading scientific journal *Nature*, that "dinosaurs invented flowers." Looking at broad changes in the fos-sil record, he noticed an apparent link between a significant change in dinosaur feeding behavior and the origin and spread of the flow-ering angiosperm plants, at the expense of gymnosperms, during the Cretaceous (143.1 million to 66.0 million years ago). In the Jurassic, a dominance of long-necked sauropods feeding high in the canopy al-lowed slow-growing gymnosperms to proliferate from the floor up. By the Early Cretaceous, ornithischian browsers that preferred to remain lower down for their meals, such as the tank-like anklyosaurs, which had wrecking balls for tails, and the *Iguanodon*, one of the most numerous dinosaurs across Europe, massively increased pressure on plants grow-ing on the ground floor of vegetation communities. The resulting high disturbance was hugely problematic for the lethargic gymnosperms, leaving the fast-growing angiosperms to radiate out from the tropics and dominate the plant world. Whether due to a change in behavior, to new, sophisticated jaw types developed by certain herbivorous dinosaurs to specialize in angiosperm consumption, or simply to the scale of distur-bance caused by hulking Cretaceous dinosaur communities dominated by animals larger than one ton, the global spread of angiosperms, and flowers, from the equator to the poles may, remarkably, have been a di-nosaurian achievement.[10]

"While it is a wonderful idea, I will unfortunately have to stop your readers there," says Paul. He tells me that when we review fossils, dino-saur feeding behavior, and plant evolution in detail, it becomes clear that the dinosaurs did not "invent" the angiosperms. Nor can they be given sole credit for their global expansion. First, when we take into account preservation and geographical biases in the dinosaur fossil record and limited understandings as to how different dinosaur fossils relate to each other, we see that the sauropods did not decline everywhere in the Cretaceous. In fact, they even flourished in areas such as tropical Gondwana, where the giant, sixty-ton, long-necked and long-tailed *Patagotitan* roamed the forests of Argentina late into this period. Nor did the ornithischians and their "low-browsing" tendency succeed everywhere, remaining largely absent from Gondwana and the Southeast Asian portion of Laurasia. Second, there is actually little difference in the spatial distribution of major herbivore types between the Jurassic and Early Cretaceous. Third, as noted above, not all sauropods browsed in the canopy, and alongside stegosaurs and other herbivores, many long-necked giants would have actually already placed major pressures on vegetation growing under one meter tall in the Jurassic, prior to the advent of angiosperms. Finally, the major expansion of dinosaurs with special jaws, such as the iguanodonts, occurred a considerable time *after* the origins of flowering plants.[11]

Nevertheless, all is perhaps not lost in the search for dinosaur "gardening," and there may be some indications that a more "diffuse" coevolution, as Paul puts it, existed. The disturbance of ground vegetation by increasingly large dinosaur ecosystems, or "dinoturbation," evidently increased across America, Africa, and Asia from the Early Jurassic, reaching its maximum during the Cretaceous. This undoubtedly provided some assistance to fast-growing angiosperms and potentially drove their diversification by the Late Cretaceous. It also remains possible that one major group of dinosaurs, the hadrosaurs, or "duck-billed" dinosaurs, which originated in the Early Cretaceous and exploded in number in the Late Cretaceous, both helped and depended on the planetary colonization of angiosperms. In particular, novel innovations in hadrosaur teeth may have differentiated them markedly from other reptiles, giving them a tooth complexity rivaling that of modern herbivorous mammals and allowing them to eat a diversity of plant matter. Perhaps the starkest evidence of a direct relationship between dinosaurs and angiosperms exists in the preserved last meal of a *Kunbarrasaurus* individual, an ankylosaur from the Early Cretaceous of Australia, which included a variety of angiosperm fruits, meaning that we know dinosaurs did eat these plants. Indeed, it has been suggested that the lack of mechanical defenses against herbivores, such as thorns and spines, among the earliest angiosperms shows that these plants actually wanted to be eaten and dispersed by bulk-feeding dinosaurs.[12]

In fact, our record of dinosaur fossils and angiosperm origins and diversification is simply not yet complete enough to rule out a close relationship between dinosaurs and angiosperms. The origins and expansion of this new type of plant certainly benefited from increased levels of Cretaceous CO_2 as well as the expansion of insect pollinators. It would not be surprising if dinosaurs also added their considerable weight to this process, especially given evidence for the role of large animals in seed dispersal and habitat modifications across the tropics and beyond today. There also seems no doubt that dinosaurs benefited from the varied ecosystems that angiosperms were to dominate by the end of the Cretaceous. Spreading out from tropical forests, angiosperms accompanied dinosaurs all over the planet, into more arid, riverine environments and even as far north as the Arctic Circle in Russia, potentially providing dinosaurs with more varied ecosystems in which to lay their eggs, reproduce, and feed. Larger and larger food chains that supported the giant carnivorous tyrannosaurs by the end of the Cretaceous suggest that the rise of the angiosperms and the loss of dominance of gymnosperms did not hinder the dominant land animals of the time but rather drove them to bigger, wider, and greater success. Popular imagery and even conservation presentations tend to link dinosaurs to "ancient" Gondwanan cycads and conifer trees. Yet, the angiosperms, which diversified rapidly in Cretaceous megathermal forests and went on to dominate the majority of modern environments, clearly grew alongside dinosaurs and also sustained them right up until the end. Then about 66 million years ago an asteroid hit Mexico and brought their green relationship to an end. Or did it?[13]

PORTRAYALS OF A dramatic "extinction" of dinosaurs often lead us to associate them with failure. Not only do we forget that by existing between 230 million and 66 million years ago, they are one of the most

Figure 3.2. Hadrosaurids developed highly complex jaw and teeth morphologies that would have allowed them to consume a diversity of plant matter, including, perhaps, expanding angiosperms. *De Agostini via Getty Images*

successful animal groups ever to have roamed the planet (around 23 times longer than our own hominin clade and 547 times longer than our own species), but we also commonly neglect the fact that they actually still live, or rather fly, around us today. Birds are dinosaurs. They share a huge number of clear, unique anatomical features with the theropod dinosaurs that we now know are their closest ancient relatives. We have the fossils to show that dinosaurs had feathers, or at least feather-like structures, and thus that the ancestors of birds metaphorically flew out of the end-of-Cretaceous doomsday brought on by an extraterrestrial impact. And, as we all know, birds are the most successful vertebrates on Earth, with a staggering 18,000 species distributed over practically the entire planet, according to some estimates. Birds also maintain an incredibly close relationship with seed-producing plants, including within the tropics. Tropical forests are home to a vast variety of birds today, from the fantastic bowerbirds of Australia and Papua New Guinea, which clear themselves a stage within the forest to dance and prance for potential mates, to the harpy eagle of the Amazon Basin, which is one of the largest eagles worldwide and terrorizes monkey communities in the canopy.

Across the world, tropical forests, including both gymnosperms and angiosperms, continue to provide green shelters for some of the greatest
diversity of living dinosaurs.[14]

Whether or not Jurassic or Cretaceous dinosaurs coevolved with
gymnosperms or angiosperms, birds undoubtedly play some of the most
important roles in keeping flowering plant populations alive and kicking
around the world today. Plants actively seek to attract the color vision
of birds by producing bright fruits so that these wide-ranging fliers will
eat them and spread their offspring to new horizons. So important are
birds to the spreading of seeds that, in the tropics, there is hardly a forest
ecosystem that could remain healthy without them. Again and again,
biologists find that birds ensure the reproduction and genetic diversity
of tree populations, from the tropical montane forests of the Andes to
the lowland evergreen rainforests of the Amazon Basin, from the dry
evergreen forests of India to the island tropical forests of the Pacific.
Not only that, but birds are some of the most important agents in the
recovery of disturbed tropical forests. Modern agriculture, ranching, and
infrastructure not only cause dramatic deforestation but, by disturbing
resident birds, also threaten any hope of a new beginning. It seems likely
that nonbird dinosaurs had significant relationships with plants and
forest communities that were dynamically changing on an increasingly
fractious tropical "Gondwanan" surface. Their descendants certainly did
and still do. However, the dinosaurs, while spectacular, are not the only
major type of animal to have both been buffeted by forests and to have
gnashed them back. We now turn to another key group of animal life
with a deep-rooted, symbiotic relationship with tropical forests—one
that is perhaps a little closer to home . . .[15]

Chapter 4

"TREE HOUSES"
FOR THE FIRST MAMMALS

Mammals are everywhere. They stock our supermarkets with meat and dairy products. They fill our houses with barking, meowing, and squeaking. They carry us trekking over rough landscapes or on a day out at the beach. They are some of the most frequently used symbols for brands, making appearances on logos ranging from the badges of sports teams (the bulls of Chicago) to boxes of cereal (Tony the Tiger). They dominate some of the most iconic ecosystems presented to us in documentaries or recreated for us in zoos. And in the form of the trumpeting African elephant and the hulking blue whale, members of the class *Mammalia* represent the most giant living animals on both land and sea. They are found in every single land-based environment on Earth. They are also, of course, us, comprising the approximately 7.8 billion of the world's still-growing human population. In this bustling "Age of Mammals," we rarely take pause to ask, "How did we get here?" In fact, while we take our current mammalian success for granted, for two-thirds of the existence of mammals on the planet, from c. 210 million to 66 million years ago, they were never larger than a badger, cowering beneath the mighty dinosaurs. Not only that, but it is mind-boggling to think that *all* modern mammals—from the Etruscan pygmy shrew, which weighs in at two to three grams, to the blue whale, at 190,000 kilograms—are actually related to just a single common and tiny ancestor from a time when the planet was ruled by reptiles.[1]

Approximately 66 million years ago a giant asteroid crashed into the coast of Yucatán in Mexico, leaving a crater with a diameter of 180 kilometers—the center of which lies directly under the modern-day town of Chicxulub. The impression left on our planet by this vast extraterrestrial object was not just geological. Its impact incinerated much of North America, threw soot and ash into the atmosphere, and likely triggered massive volcanic eruptions in the Deccan Plateau of India that induced a global winter. With sunlight reduced to around half previous levels and lashings of acid rain caused by the released volcanic gases mixing with water in the air, the Earth's surface and the life it sustained spiraled into disarray. Many species of plants, birds, insects, and marine life and, perhaps most significantly, *all* of the nonbird dinosaurs went extinct—more than 75 percent of all species on Earth at the time. Although some mammal species were also wiped out in this "end of days" event, traditionally the fire and brimstone at the end of the Cretaceous period has been seen as a window of opportunity for our evolutionary line. With no dinosaurs left to molest us, we could fill landscapes left empty by disaster, take our pick of remaining food sources, and expand and diversify in a march of seemingly inevitable progress until the present day. However, while attractively simple, such a narrative certainly does not tell the whole story.[2]

Fantastic fossil finds are now beginning to show us that mammals and their closest, but now extinct, "mammaliaform" (literally meaning "mammal-shaped") relatives were actually already experimenting and expanding *before* the dinosaurs went extinct. These little but increasingly varied critters did not just scurry below these fearsome reptiles; they also climbed above and even flew around them—first in Jurassic gymnosperm-dominated forests and later in Cretaceous forests that were increasingly invaded by flowering (angiosperm) plants. Furthermore, the asteroid-induced mass extinction at the Cretaceous-Paleogene (K-Pg) boundary was, for mammals, merely the beginning of our brushes with climatic and environmental change. During the following Paleocene (66.0 million to 56.0 million years ago) mammals accompanied the expansion of angiosperm-dominated megathermal forests across an increasingly fragmented planet by the end of the period. A climatic record from the Arctic has shown that mean atmospheric temperatures rose from between 16°C and 18°C to a balmy 25°C at the Paleocene-Eocene boundary (56.0 million years ago). As we saw in Chapter 2, tectonic shifts continued to rearrange the world's continents toward their current orientation. In particular, in the middle Eocene, Australia and Antarctica split apart, creating a deep-water passage for the flow of cold polar water toward the equator. The new patterns of ocean circulation and their relationship to the atmosphere saw swings between warmer and cooler states in the Eocene (56.0 million to 33.9 million years ago) and Oligocene (33.9 million to 23.04 million years ago). They also resulted in the global expansion of grasslands and retreat of megathermal forests as the planet broadly became cooler, drier, and ever more seasonal heading through the Miocene (23.04 million to 5.33 million years ago). The latest science shows that the geology- and climate-induced ebbs and flows of warm, wet forests over this span of 45 million years played a major role in shaping mammalian distribution, diversity, and evolution, leaving us with the modern ecosystems we gaze in awe at today. Not just that, but without forests growing in the tropics of the Jurassic, the "Age of Mammals" may never have even begun.[3]

IN NORTHEASTERN CHINA, the provinces of Liaoning, Inner Mongolia, and Hubei meet in one of the most important regions for paleontologists trying to track the world-changing transition of life on Earth as it swung from the realm of the dinosaurs to the dominion of mammals. Here, fine particles of ash and waterborne mud generated by past volcanic activity and flooding sealed soft body parts and entire skeletons across millions of years into what have become some of the world's most-celebrated and best-preserved fossil beds. This part of northeastern China has yielded *Sinosauropteryx*, one of the first dinosaurs found with intact feathers, so complete that paleontologists could even identify cells that give the feathers of birds today their color. While not part of the direct lineage that eventually led to modern birds, *Sinosauropteryx* nonetheless highlights the biological innovations of some avian dinosaurs in the Cretaceous period that allowed them to hang on beyond the K-Pg boundary. Even more importantly, this region has also preserved one of the most significant fossils in the evolutionary history of the group that was to replace the dinosaurs as the new heavyweights of the animal kingdom. In 2011, Professor Zhe-Xi Luo of the University of Chicago and his colleagues uncovered the remains of an animal they named *Juramaia sinensis*, which they dated, using radiometric dating methods, to 165 million to 160 million years ago and the Middle-Late Jurassic period, a time when this part of the world would have been located at the boundary of the tropics. These methods compare the amount of naturally abundant radioactive isotopes (in this case Uranium[235] or [238]) to the elements they are known to decay into (Lead[207] and [206]) within minerals such as zircon trapped in the fossil layers of interest. As the rate of decay can be calculated, these measurements can be used to estimate how long ago the fossil was buried. Based on its age and the characteristics of the fossil bones, the team gave it a name that denotes it as the "Jurassic mother" of all true placental mammals. "Despite its minute stature, it gave birth to almost all of the mammalian diversity we see around us today," says Luo.[4]

The word "mammal" literally means "breast" in Latin, giving immediate insight into the new biological features that separated these animals from their dinosaurian overlords. Although monotremes (like the duck-billed platypus) lay eggs that are later hatched and raised in a

Figure 4.1. Artistic reconstruction of *Juramaia sinensis*, the "Jurassic mother" of all placental mammals. *Carnegie Museum of Natural History / M. A. Klinger*

pouch, the majority of mammals give birth to live offspring, which they feed with milk. Back in the Jurassic and Cretaceous periods, this would have represented a major evolutionary adaptation, allowing mammals to develop their young in the comfort of safe burrows. Like their extant relatives, these early mammals would also have been warm-blooded and had fur, enabling them to hunt at night when temperatures were low and most predators were asleep. Unsurprisingly, however, spotting the first appearance of these traits in the fossil record can be incredibly difficult. Soft tissues rarely, if ever, preserve over these vast timescales. This has meant that scientists have looked to sturdier, more resistant skulls for clues. Comparison of fossils with modern mammals demonstrates what sets the earliest mammals apart from their reptilian counterparts: the separation of a newly fused jawbone from ear bones that moved backward in the skull. A more stable jaw supported teeth that interlocked like a jigsaw, slicing and dicing through food. While reptiles tend to swallow their food whole, using their teeth simply to lock onto and tear prey, this new adaptation enabled a "preprocessing" of food before it reached the stomach, ultimately releasing more nutrients. Meanwhile, separate ear bones, located away from the gnashing jaw, meant better hearing (essential when active at night) and also provided space for mammalian brains to grow outward and backward. The earliest mammaliaform creatures in which these features have been found are known as morganucodonts. Fossils uncovered from the Late Triassic across Europe, Africa, America, and Asia show this group was starting to scamper around just after the arrival of the first dinosaurs.[5]

New and varied mammaliaform creatures continued to make an appearance throughout the Jurassic and Cretaceous periods. Paleontologists have been frantically searching for when these mammal-like creatures became the "true" ancestors of most mammals alive today. In *Juramaia*, Luo thinks we have it. *Juramaia* had a miniscule body length of between seven and ten centimeters. We can actually see its soft tissues, including fur, thanks to impressions left in its imprisoning rock. Importantly, detailed comparison of *Juramaia*'s paw bones and teeth with other extinct and living mammals demonstrates that it was alive shortly after the evolutionary division between all of today's placental mammals (mammals that protect and feed their young in a womb using a placenta until they are fully developed) and marsupial mammals (mammals like kangaroos that release their embryos early to be developed in a pouch). *Juramaia*, which, like us, is part of the placental line, shows that this split must have occurred approximately 170 million years ago. This is in line with molecular "clock" estimates that compare genetic diversity in modern mammals, as well as assumed rates of mutation, to estimate the age of their last common ancestor, although more recent finds by members of the same team may ultimately push the beginnings of this process back still earlier. More incredible fossils, again from northeastern China, show that the split between placental and marsupial mammals was certainly complete by the Early Cretaceous, with the earliest marsupial, the tiny pointed-nosed *Sinodelphys*, and, until *Juramaia*, the earliest placental mammal, *Eomaia* ("dawn mother"), dating to 125 million years ago. From *Juramaia* and from northeastern China, we can therefore see the building blocks needed to reconstruct the evolutionary history of the majority of mammals that exist on the planet today.[6]

Significantly, the rich fossil beds of northeastern China have also provided insights into the early interaction between some of our earliest mammalian ancestors, their closest "mammal-shaped" relatives, and diverse subtropical and tropical forest ecologies. Until recently, scientists had assumed that the earliest mammals were all small, all ground dwelling, and all nocturnal, with variation in mammals occurring only *after* most dinosaur communities were blasted into oblivion. Nevertheless, studies of plant fossils found alongside Jurassic and Cretaceous mammals, as well as analyses of the shape and size of their teeth and limbs, which provide insights into the stresses and uses they were put to, have revealed that early mammals and mammaliaforms went through two major periods of ecological of "radiation" *prior* to the K-Pg boundary: one in the Jurassic and one in the Cretaceous. In the Jurassic period (201.4 million to 143.1 million years ago), weighing in at fifteen to seventeen grams (just a little more than a "AAA" battery), *Juramaia* would not have been particularly conspicuous, and study of its teeth reveal that it mainly ate insects. Nevertheless, its strong forearms suggest that, rather than cowering in fear on forest floors, it ascended trees to live up at eye level with the dinosaurs. Indeed, living in trees significantly benefits mammals in terms of life expectancy and protection from ground-based

Figure 4.2. A reconstruction of *Repenomamus* snacking on a young *Psittacosaurus*. *Wikipedia / Nobu Tamura*

threats. Moving away from *Juramaia*, fossil finds of Jurassic mammals reveal they practiced a variety of arboreal, ground-based, burrowing, and even aquatic modes of movement. Another Jurassic fossil from northeastern China has even revealed the existence of "gliding" mammaliaforms in the Jurassic approximately 160 million years ago. Remarkable preservation has enabled scientists to observe the winglike membrane that would have enabled these animals to fly between trees. Not only that, but through comparison with modern gliders, which are all herbivores (with some supplementing their diet with insects), and a detailed analysis of the shape of their teeth, Luo and his team argue that these Jurassic mammaliaforms had developed a new dietary adaptation—one involving consumption of the young leaves, cones, and tender tissues of seed ferns and various lush gymnosperm forest vegetation that dominated Jurassic landscapes. As Luo puts it, "The forests of the Jurassic provided the perfect 'tree houses' for the earliest mammals, not just for protection, but also for food as part of increasingly diverse mammalian ecological niches."[7]

This diversity continued into the Cretaceous period. In the Early Cretaceous, findings of fossils of the badger-like *Repenomamus*, again in northeastern China and dating to 125 million to 123 million years ago, overturned ideas of what early mammals were capable of. Not only were this mammal's teeth adapted to carnivory, but it has even been found with a dinosaur preserved in its stomach! The Late Cretaceous saw a further, second major radiation in mammal and mammaliaform diversity. A number of the early mammalian groups went extinct between 125 million and 80 million years ago, with those surviving being primarily small and adapted to eating insects. However, from about 80 million years ago, changes in the teeth, jaws, and body sizes of eutherian ancestors of placental mammals, metatherian marsupial ancestors, and their close relatives indicate the development of varied dietary specializations. So great was this diversity that the numbers of different Late Cretaceous mammaliaform species are similar to the renowned diversity of small-bodied mammal communities that live in tropical forests today, and mammals were thriving at the time of the Chicxulub impact. This mammalian ecological radiation once again seems to have links to the equatorial forests of the time. This time, forests that were being encroached on by flowering angiosperms, with their fleshy fruits, nutrient-rich leaves, seeds, and tubers. Indeed, one of the earliest-dated fossil angiosperms (called *Archaeofructaceae*, or "ancient fruit") comes from Early Cretaceous deposits in same part of northeastern China that we have explored throughout this chapter. The teeth and jaws of our placental ancestors, with their dual-purpose crushing and shearing capacity, were well set up to access the new nutrients provided by novel vegetation communities, as well as insect populations that blossomed as the pollen couriers for the new flowers. They were not alone, however, and a now extinct mammalian lineage, the rodent-like "multituberculates" (so named because of the many tubercles, or "cusp bumps," on their teeth) also developed teeth specifically adapted to the consumption of complex angiosperm vegetation, with sharper, bladelike teeth for slicing at the front and bumpy teeth for crushing at the back. Although debate continues as to the significance of the role of mammals in the initial Cretaceous origins and expansion of these new types of plants, and many of the more specialized mammal groups, like the multituberculates, eventually disappeared, creating an intimate relationship with these innovating plant communities was perhaps the best thing the mammals ever did. Not only did it give them a number of sturdy companions with which to survive one of the worst extinction events ever seen on Earth, but, coming out on the other side, it also guided them into a new era of global domination.[8]

THE AMAZING FOSSIL finds from the Jurassic and Cretaceous show that diverse mammalian pioneers were flying around the heads of dinosaurs, scurrying between their feet, and swimming in their rivers; some were even beginning to turn the traditional predator-prey relationship on its head. This growing roster of mammals and their relatives seems, quite literally, to have been energized by gymnosperm-dominated Jurassic forests as well as Cretaceous forests with an increasing representation of angiosperm communities. Nevertheless, while our earliest mammalian forerunners were certainly more vibrant than is often appreciated, their main break still clearly came with the fall of the dinosaurs. To watch how mammalian understudies became the stars of the show, we must leave China and travel to North America. Here, at the Corral Bluffs in Colorado's Denver Basin, we can find one of just a handful of localities in the world with a fossil record that brackets the disastrous K-Pg impact. The Corral Bluffs are a single, continuous, exposed outcrop that runs for 27 km^2 on either side of Highway 94. Thousands of motorists fly past this part of the windswept plains of western North America without ever realizing its significance. Yet approximately 66 million years ago, this location was in the direct firing line of the pulse of heat emanating from the massive Chicxulub impact. Frozen in time, fossils found within the Bluffs span ecosystems that were heading obliviously into disaster during the last 100,000 years of the Cretaceous and, crucially, those that emerged from the rubble in the first 1 million years of a new, green Paleocene era.

"The fossil bonanza that has emerged from the Bluffs makes it one of the best sites, if not the best site, for studying the 'comeback' tour of life after the lethal asteroid collision," says Dr. Tyler Lyson, a paleontologist at the Denver Museum of Nature and Science. Although ancient-fossil detectives are usually on the lookout for skeletons poking out of sediments, this time the treasure could only be found by looking inside inconspicuous "concretions." To the untrained eye, these might look, for all intents and purposes, like solid rocks. However, as the team began to split these concretions open back in the lab, they discovered the biological secrets that lay within. Specifically, they found the incredibly well-preserved remains of turtles, crocodiles, plants, and, most significantly, a rapid (at least in the context of geological time) succession of mammal forms that

show the way in which these animals strutted to the top of the Paleocene. Mammals were undoubtedly also massively impacted by the K-Pg
extinction event, and the largest of this lineage that scrambled through
this apocalypse weighed no more than half a kilogram. Nevertheless, just
100,000 years down the line, their descendants weighed six kilograms;
200,000 years later, mammals had reached twenty kilograms—far heavier
than any pre-K-Pg mammals. By 700,000 years after the disaster, mammals weighed over forty-five kilograms—more than an adult hyaena or
chimpanzee. Traditionally, the unrelenting arrival of ever larger mammals so soon after the global apocalypse at the end of the Cretaceous
has been put down to a simple absence of dinosaurs. Basically, without
the "terrible lizards," our ancestors would no longer have had any reason
to hide in the shadows. However, to settle for such a simple explanation
would be to do the dynamic early mammals and their vibrant leafy companions a disservice.[9]

The asteroid impact and the reduction in access to the Sun's energy
that followed undoubtedly caused ecological disruption and localized
loss of vegetation diversity. However, many plant families found as Cretaceous fossils exist today and seem to have had a special capacity to
adapt to rapid, global environmental change. The angiosperms were
here to stay and were not about to stop expanding. Fossil pollen from
the Corral Bluffs record shows us that the increase in mammal body
size occurred at the same time as plants were getting more diverse and
more nutritious in a warming world where tropical climatic conditions
prevailed across much of the planet. While ferns were the first to emerge
out of the North American ashes, the angiosperms, growing taller and
more nutritious than ever before under the warming Paleocene conditions, soon dominated the landscape. Three hundred thousand years
after impact, the largest mammal around was *Carsioptychus*, a distant
ancestor of all modern hoofed mammals, or "ungulates." This animal
had large, flat premolar teeth, with bizarre fold-like shapes, that were
ideally adapted to the munching of hard objects, like the nuts of trees
from the walnut family, which appears in the fossil plant record at the
same time. After another 400,000 years, the largest mammal in the Corral Bluffs record occurs alongside the sweeping entrance of beans into
the fossil plant records of North and South America. Their energy-filled

leaves and protein-rich seed pods provided ideal sustenance for growing mammal bodies. The close relationship between increasingly large mammals and their tropical angiosperm companions can be seen in the Corral Bluffs fossil record through the ever more popular trend toward vegetarianism. The close bond between mammals and tropical plants seen in Colorado can be found across the global fossil record of the warm, dense forests of the Paleocene.[10]

The only way was up for mammals and their angiosperm partners as the Eocene (literally, "new dawn") broke 56.0 million years ago. After a cooler blip in the late Paleocene, the planet warmed rapidly, spurred on by tectonic plate movements, and megathermal forests expanded all the way toward the polar regions. More forests meant more vegetation and more opportunities for mammals. The modern ungulates, which include the majority of large land mammals still around today, expanded. One group, the perissodactyls (which bear their weight on one of their five toes), including the ancestors of horses, tapirs, and rhinoceroses, were particularly successful, with their special gut fermentation system allowing them to process the fibrous leaves of dense forests. Another group, the artiodactyls (which bear their weight equally on two of their five toes), also emerged, including the ancestors of pigs, goats, hippos, and cattle in Africa and the ancestors of the camel in North America. This group remained at the fringes of the largely forested landscapes until the middle-late Eocene. Some even took their chances on the high seas, and evidence from a hippo-like artiodactyl found in Pakistan, dating to approximately 50 million years ago, has been used to show that modern whales and dolphins derive from early land-dwelling artiodactyls that gradually went on to adapt to ocean life through the Eocene. In South America, marsupials, armadillos, and now extinct ungulate forms diversified. Early forms of carnivores with paws, rodents with claws, elephant-related "proboscideans," bats, and arboreal lemur-like primates had also all appeared on Earth by the middle Eocene, completing the sweep of most of the modern mammal groups across ecosystems that would have seemed remarkably similar to those of the modern world. In fact, the main difference to today would have been the size of these mammals, and many of these early forms would have looked like miniaturized models of their twenty-first-century relatives, with small sizes serving as a coping mechanism against the heat and dense forests that prevailed up to the middle Eocene.[11]

The Messel Pit, a disused quarry previously mined for oily shale, just southeast of Frankfurt am Main in Germany, provides a remarkably vivid picture of these flourishing, cosmopolitan middle Eocene mammal ecosystems. Here, slap bang in the middle of modern Europe, a fair distance from the tropics, paleontologists have discovered a wet, warm forest ecosystem where early hyaena relatives menaced early rodents, hedgehogs, and even marsupials, where a semiaquatic otter-like mammal took dips in swampy wetlands, and where early relatives of horses and an ancestor of rhinoceroses made increasingly larger strides. In the trees themselves, early primates consumed fruits, and the small catlike carnivore, *Paroodectes*, stalked along branches of tropical trees like *Canarium*, cashew, and nutmeg that covered central Germany at this time. Taking leave of the ground, early bats flew around the heads of these other flourishing mammals. Study of the small mammals at Messel show that a "rich" community existed that would not be out of place in the canopy of a multilayered tropical rainforest today, sustained by angiosperm fruits and seeds. Undoubtedly, the spread of angiosperms to almost every corner of the globe encouraged and supported this cornucopia of early mammals found across the major continental landmasses at the time. But this isn't simply a story of take, take, take on the part of the new and increasingly diverse mammals: it may be that their green companions also benefited.[12]

Indeed, not just mammals, but also the angiosperm plants themselves, underwent major changes in shape and diversity between the Cretaceous and middle Eocene. First and foremost among these changes was the development of the fruits that many modern plants use to attract primates, bats, and birds so that they will disperse their seeds, carrying their potential offspring to new parts of the landscape often many kilometers away. This strategy is particularly important in tropical forests, where around 70 to 94 percent of woody species produce fleshy fruits. Fruits have been preserved in the angiosperm fossil record back to the Late Cretaceous, but, significantly, angiosperm creativity in terms of seed sizes and fruit types gradually increased, peaking in the early Eocene 56.0 million to 48.1 million years ago, just at a time when most modern mammalian orders were arriving on the scene. Some trees, like the walnut family mentioned at Corral Bluffs, even shifted from winged seeds that were good for flying to more robust seeds that provided better packaging for being couriered by animals. Despite dispute about the extent of the earliest mammals' involvement in tropical angiosperm evolution and dispersal, it remains highly likely that at least some of the earliest mammals and their relatives, like the multituberculates, had a symbiotic relationship with the plants that were sustaining ever-increasing mammalian dominance.[13]

THE PERIOD FROM the late Eocene (37.7 million to 33.9 million years ago) through the Oligocene (33.9 million to 23.04 million years ago) to the Miocene (23.04 million to 5.33 million years ago) saw the beginning of significant global swings between cooling and drying and warming and wetting as the continents and increasingly active flows of water between the world's oceans and hemispheres began to assume their present-day positions. The late Eocene heralded a dawning of seasonality in planetary climate conditions and the gradual retreat of warm, wet forests back toward their current tropical positions. To some extent, this perhaps foreshadowed a decline in the role of these environments in mammalian evolution beyond the tropics. However, these broad vegetation changes forced mammals who had, to this point, been flourishing in megathermal forests to adapt. This process left us with some of the most iconic mammals around us today in both tropical and temperate regions. To take the first example, horses are important not just for transport, or for their cultural significance ranging from western films to a day at the races, but also because their evolution provides one of the clearest perspectives on how the role of megathermal forests has changed in mammalian evolution over the last 35 million years. As paleoecologist Professor Gina Semprebon, of Bay Path University in Massachusetts, an expert in early horse evolution, states, "We might not recognise it today, but our hooved friends, from their appearance in the North American Eocene, through to the present, have actually undergone radical changes in their body shape, size, and preferred diet that track global shifts in environments across much of the Earth."[14]

The teeth of animals, including our own, have much to say about their lifestyle and environment. For herbivores, specialized grazers can have high, ridged, "hypsodont" teeth, which extend past the gums and provide greater resistance to gritty diets of coarse, fiber-filled material like grass. Meanwhile, browsers, such as wild goats and deer, tend to have flatter, lower-crowned, and shorter teeth for crushing and grinding. Not only that but the types of food eaten leave characteristic scratches, or "microwear," on the tooth surface that can be observed through a microscope. Gina is a specialist in using these approaches to study how the ecology of the horse has changed over time. The first horses, represented by the short-faced *Hyracotherium* and, later, *Eohippus* (or "dawn horse"), were no bigger than a relatively large dog, such as a Labrador—a far cry from the long-legged, galloping beasts we admire today. Horses are also generally thought of as grazers. However, the teeth of these early horses show only the faintest traces of ridges that were to become so important later on in their evolution. Otherwise, they were generally low and flat with a rounded chewing surface—well suited for eating soft, lush leaves, fruits, nuts, and plant shoots. You would not know it today, but the first horses were right at home in the dense megathermal forests of the early-middle Eocene. Small bodied, and focused on picking off bits of nutritious angiosperm vegetation, they made the most of expanding forest floors, from North America to Europe.[15]

In the late Eocene and the early stages of the Oligocene (33.9 million to 23.04 million years ago), things got drier, and sandy grasslands began to expand more widely. The new types of horse, or "equids," changed in response. About 40 million years ago, the so-called *Mesohippus* (or "middle horse") appeared in North America. Now that forests could not always be relied on as places to hide away from growing predator threats, the longer and leggier *Mesohippus* could run faster and more widely in the expanding open areas. Still relatively small (only sixty centimeters at shoulder height), this new horse form had grinding cheek teeth with sharp crests that allowed it to break down some of the earliest gritty grasses that were emerging at this time, something also visible in Gina's microwear work. The stage was now set for the rise of modern horses, and from the Miocene to the Pleistocene appearance of the modern genus *Equus*, horses got bigger. Their feet became increasingly focused on the single toe formation used by horses today, and their teeth increasingly tended toward high, ridged, "hypsodont" cheek teeth ideal for dealing with the expanding C_4 grasslands of this period. While horses were clearly able to switch between both leaves and grasses right up to the Pleistocene, and indeed still do on the Central Asian steppes, the overall trend toward sprinting, grassland-adapted animals is evident in the mouths of these animals. Continental land bridges played a major role in this "Great Transformation," enabling new horse forms to repeatedly pass between America and Europe, and later Asia, to make the most of expanding grasslands.[16]

However, these changes did not just influence the mammal communities left outside the tropics. Many of us, all over the world, flock to zoos to see the peculiarly shaped giraffe. Appearing in the Miocene, a number of giraffids that roamed Africa and Eurasia, including the diminutive *Giraffokeryx* in India, show a pretty speedy tendency toward longer-neck evolution by 14 million to 10 million years ago. Land bridges between continents provided an ideal means to range widely as well as a safety net to move to new areas should food run scarce in one region. The modern genus *Giraffa* seems to have evolved somewhere in East and South Asia before entering Africa approximately 7 million years ago. Although ongoing climate change then caused it to be extirpated in Asia, *Giraffa* found its long-term home in the subtropics and tropics of Africa where the modern-day giraffe eventually emerged 1 million years ago. The main evolutionary driver behind the unique giraffe features we know and love today is thought to be another retreat of megathermal forests rapidly following the expansion of arid-adapted C_4 grasslands in India and Africa from 12 million years ago. However, unlike other previously forest-adapted animals, such as horses, that shifted to grasses, the giraffes stubbornly persisted in browsing whatever islands of trees were left. They developed long necks, like the dinosaurian sauropods, to access nutritious evergreen leaves in dry, deciduous woodland and grassland ecosystems in the tropics, as well as teeth and digestive systems that could cope with the tough foliage and toxins produced by some of these trees. One of the most key members of sweeping eastern African tropical savannah ecosystems known today formed among the chaos of the retreat of warm, wet forests to their current tropical realm.[17]

Bats provide another example. Today bats make up one-fifth of all living mammal species and are one of the most diverse of the mammal orders. The earliest bats of the Paleocene and Eocene evolved powered flight and their amazing echolocation senses in order to hunt the blooming insects in a warming, flower-dominated world. Yet only between the late Eocene and Miocene did some bats, including the group of "megabats" (*Pteropodidae*) that still fill the skies and flock around tasty fruit trees in tropical Asia, start along an evolutionary path toward specialized fruit consumption. Changes to the skull, teeth, and jaw occurred among different diversifying lineages of increasingly specialized bats that fed on first soft and then harder and more fiber-rich fruits. This shift occurred at a time when megathermal forests were contracting, leaving behind woodland and forest patches that were becoming less reliable for larger fruit-eating mammals. Flying creatures, such as birds and bats, however, were ideally placed to link up these patches. The angiosperm plants themselves seem to have recognized this, and the development of smaller seeds and fruits among many tree families during the Oligocene may be associated with the attraction of these smaller-bodied but more mobile consumers. Today, these adaptations have left fruit bats, giant or otherwise, as crucial players in seed dispersal in many tropical forests.[18]

WATCHING COWS GRAZE on pastures, wildebeest flow across savannahs, and whales dive among the deep blue, it is hard for us to imagine that our earliest mammalian forebears primarily lived and flourished in dense, warm, wet forest environments. From the first experiments among the dinosaurs to their rise and domination following the K-Pg global disaster, mammals were tropical creatures, living in trees, gliding between branches, and eating insects or increasingly lush vegetation. The close relationship between mammals and the expanding angiosperms seems to have been a key part of their survival as well as their spread and diversification around the planet. It may even be that the mammals, in turn, helped these new, innovative, nutritious plants to succeed. From the peak warm period in the middle Eocene, however, and despite some brief outward flourishes, megathermal forests were on the back foot, retreating in stages, in the face of dry, seasonally adapted grasslands and woodlands, toward the equator. Horses, giraffes, and bats are but three of many more groups of mammals affected by these global environmental changes between the Eocene and Miocene, showing how the dynamism of warm, wet forests and their retreat toward their modern tropical distribution influenced some of the more characteristic mammalian lineages we know and love today—including our own lineage, the primates.

The earliest true primates to appear in the fossil record were those of the *Strepsirrhini* "wet-nosed" group, which includes modern lemurs and lorises. These small, arboreal mammals made the expanding lush, tropical angiosperm vegetation their own in the Paleocene and early Eocene. Diversification continued with the tarsiers (c. 58 million to 55 million years ago) and then the simians (c. 40 million years ago), a group that includes all monkeys and apes. As we have seen in the case of other mammals, primates found protection and opportunity in the flourishing tropical angiosperm vegetation of the Paleocene and Eocene. As with other mammals, the course of early primate evolution was seemingly characterized by teeth and body shapes that could better exploit the nutrient-rich fruits and seeds of these novel plants. And like other mammals, primates were forced to contend with shrinking megathermal forests from the late Eocene onward. The simians emerged in Asia before moving to Africa by 35 million years ago, where they became smarter, bigger, and more aggressive as they dealt with variable resources and threats in the fluctuating forest habitats of the Miocene. The primates are another clear example of how early mammalian evolution was tied up with the fates of tropical forests and the interlinking of land bridges between separating continents. However, the Miocene was only the beginning. Ongoing but shifting interactions with the forests and woodlands in and around the equator would set this group on an entirely new evolutionary course, one that would see them become arguably the most successful mammal that has ever existed. Us.[19]

Chapter 5

THE LEAFY CRADLES
OF OUR ANCESTORS

W hen highlighting the perilous state of tropical forests today, conservation agencies will often emphasize the dangers of deforestation for chimpanzees, bonobos, gorillas, and orangutans in the hope that people will sit up and take notice of the plight of our closest living relatives and their lush surroundings. Their complex social lives and ability to use tools, to show emotion, to care for their dead, and even to perform sign language make these nonhuman great apes so easy to sympathize with. From watching them play in captivity to immersing ourselves in documentaries that follow their feuding "dynasties," we can rapidly relate to these animals with whom, in the case of chimpanzees, we share 99 percent of our DNA. Through various iterations of the film *Planet of the Apes*, for example, we have grown used to seeing them as our probable planetary inheritors should we ever slip up. Somehow, however, there's a widespread feeling that these apes ultimately represent what we *were* rather than what we are, that they look and behave how we used to, and that they are somehow stuck in time, living examples of an age gone by. The same is most certainly true of their tropical forest homes, which we think of as so foreign to ourselves, environments that we abandoned as soon as we were able, early on in our evolutionary journey.[1]

It is generally accepted that sometime between 13 million and 7 million years ago, a new type of great ape began to appear in Africa: the hominins. These novel "humanlike" apes were our great-great-great-ancestors and, ever since Charles Darwin wrote about our "descent," we have associated their evolution with the leaving behind of tropical forests and the other great ape species. According to this narrative, striding out across grasslands, our predecessors started to specialize in upright walking. Moving more efficiently over long distances, they could pursue and hunt increasingly large animals out on the African savannah. With their hands now free, they could make tools and use fire to cook meat and fuel their growing brains as their canine teeth became smaller and started to resemble our own. Measurements of hominin fossil limbs, the geochemistry of hominin teeth, and discoveries of plants and animals found alongside hominins have all been used to support this so-called savannah hypothesis. From approximately 7 million years ago until the origins of our own genus, *Homo*, within the hominin line, around 3 million to 2 million years ago, the narrative has commonly been one of decreasing climbing and use of forest plants and increasing bipedal walking and hunting of grazing animals. This idea is so deeply embedded in academic thought that even the first hominins to move "out of Africa" approximately 2 million years ago are thought to have done so only when shifts in the Earth's climate encouraged the expansion of a grassland highway all the way from Africa to eastern Eurasia.[2]

Against the broad backdrop of Miocene–Pleistocene environmental change—namely, the decline of tropical forests at the expense of dramatic C_4 grassland expansion—on the face of it, this all seems to make perfect sense. Yet the latest work in paleoanthropology, archeology, and environmental science shows that tropical forests may have played a far more active and persistent role in the emergence and evolution of our early ancestors than often appreciated. The hominins certainly seem to have originated *within* tropical forests, albeit ones that were changing quickly. Our star-studded cast of ancestors were born into a world of climatic and environmental diversity, from the earliest hominins to the specialized walker "Lucy" and, later, *Homo erectus*—the hominin with

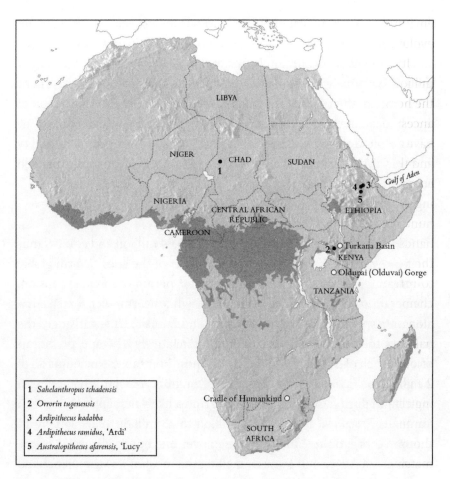

Figure 5.1. Map of major Miocene, Pliocene, and Pleistocene hominin sites in Africa discussed in the text plotted against tropical forest distributions estimated on the basis of the MODIS (Moderate Resolution Imaging Spectroradiometer) Land Cover MCD12Q1 majority land cover type 1, class 2 for 2012 (spatial resolution of 500 m). Downloaded from the US Geological Survey Earth Resources Observation System (EROS) Data Center (EDC).

the greatest geographical range prior to our own species, reaching from the United Kingdom in the west to Indonesia in the east. Although the different members of our family tree undoubtedly became increasingly at home in savannahs, continued attachment to their earlier forest origins left a trace in their hands, feet, limbs, and even the chemical makeup of their bodies. In their negotiation of the diverse and fluctuating environments around them, we can see the first traces of a behavioral flexibility that was soon to reach whole new levels as we evolved and expanded across Africa and the entire world.

DURING THE EARLY Miocene, between 23.04 million and c. 15.99 million years ago, the Earth truly was the "planet of the apes." Warm global temperatures encouraged a relatively "brief" period of forest expansion. Temperate forests inhabited Greenland, while warm, wet forest environments spread across much of Africa and Eurasia. Unsurprisingly, the primates took their chance, including the relatively new ape, or "hominoid," branch of the Oligocene simians. These hominoids, including all of the great apes as well as the "lesser apes," or gibbons, went from strength to strength during the early Miocene. From a likely first appearance with the genus *Proconsul* in Kenya 31 million to 23 million years ago, by 14 million years ago there may have been over one hundred species of ape living across Africa and Eurasia, making the most of a warm planet and flourishing angiosperm fruits, leaves, and tubers. By 10 million years ago, however, cooler middle-late Miocene temperatures and drier conditions had forced megathermal forests and their ape inhabitants into retreat. The great apes, or "hominids," suffered the most. The majority had disappeared or started to disappear outside Africa by just 2 million years later. Iconic species, such as *Sivapithecus* from Pakistan, became extinct. Orangutans were left as one of just two remaining representatives of the great apes in Asia, where they have persisted to the present day, while the ancestors of gorillas and the common ancestor of chimpanzees and bonobos began to emerge and evolve in the increasingly restricted subtropical and tropical forest belt of Africa. It was within this world of increasingly complex "mosaic" patterns of tropical forests, drier woodlands, and grasslands that the first hominins, our ancestral lineage, split off from the rest of the great apes and began to take on new, pioneering forms.[3]

Paleoanthropologists often identify fossil hominins on the basis of the similarity of their teeth and skull shape to our own, and these skeletal elements are often preserved best through the ravages of time. Menacing canines used by other great apes in threat, bluff, and social display reduced in size among hominins. Adaptations to "bipedalism," or upright, two-footed walking, are also a specific feature that might define the beginnings of our family tree and reveal something about its changing environmental context. However, poor preservation means that scientists must often make do with studying a mishmash of these features when examining fossils contesting for the title of the "earliest hominin," and where and how different hominin species moved around is frequently up for intense debate. In 1994, a team of researchers trudging through the desolate, arid environment of the Awash River basin in Ethiopia came across one of the most significant nominees for this award. Then-graduate student Dr. Yohannes Haile-Selassie of the Cleveland Museum of Natural History saw a partial hand bone poking out of the ancient silt. Subsequently, he and the rest of the team, including a seasoned veteran of hominin fossil discoveries in Africa, Professor Tim White of the University of California, Berkeley, went on to excavate the skull, teeth, pelvis, hands, and feet of a female "humanlike" skeleton dated to 4.4 million years ago. Named *Ardipithecus ramidus*, or "Ardi" for short, this is one of the most complete prehuman hominin skeletons ever found. In local Afar language, Ardi's name literally means "ground floor" and "root," and this is potentially exactly what she is. In 2001, Yohannes also went on to discover and name an even earlier ancestor of Ardi, *Ardipithecus kadabba*, in the same region, dating to 5.8 million to 5.5 million years ago. This means that her lineage reaches back further, approaching most genetic estimates of the last division of our line and that of chimpanzees approximately 7 million years ago, placing her right at the very beginning of hominin experimentation in Africa. However, as Yohannes tells me, "Just because 'Ardi' belongs to the hominin group from which our own species eventually emerged, this did not mean that she had fully committed herself to the ground. She probably spent as much time in the trees as on the ground."[4]

Ardi had a small brain, one-fifth of the size of our own and slightly smaller than that of a chimpanzee. However, the shape of her pelvis and feet means that she was definitely much better suited for bipedalism than chimpanzees. Meanwhile, Ardi's teeth and skull shape show the faintest beginnings of a trend toward those of later hominins and our own species. Yet, as Tim White argues, these features are not necessarily the most significant parts of Ardi's discovery. "It is the environment that surrounded her that truly turns existing theories of hominin evolution on their head," he says. The skeletons of animals found near Ardi's fossils show that when she was alive, tropical forest, and maybe even closed spring-fed forest, covered the now desertified region. Thanks to the rare completeness of her skeleton, we can also see that Ardi had divergent big toes and long fingers well adapted for slow climbing among the trees. In other words, and contrary to beliefs widely held by both the public and academics, "bipedalism" may have originated in tropical forests, not savannahs. More detailed analysis of the animals found around her and other fossil members of *Ardipithecus*, as well as geochemical analysis of Ardi's teeth, have shown that they may have inhabited forest patches within mixed forest-woodland-grassland settings. This would still be a far cry from popular assumptions of hominin origins out in the open.[5]

Figure 5.2. Artistic reconstruction of "Ardi" *Ardipithecus ramidus.* *Jay Matternes / Smithsonian*

Ardi is certainly not the only contender for the title of earliest hominin, and, as is usual in paleoanthropology, some have even questioned whether she is directly related to us at all. Another nomination comes in the form of *Sahelanthropus tchadensis*, a species largely represented by a series of jaw bones discovered in northern Chad by a French team led by Michel Brunet. Dating to approximately 7 million years ago, like Ardi's, its teeth and skull shape show a mixture of "hominin-like" and other "ape-like" characteristics, with some tentative suggestions of bipedal movement, in a mixed forest, woodland, and grassy lake-edge environment. A further competitor, *Orrorin tugenensis*, found by the Franco-Anglo partnership of Brigitte Senut and Martin Pickford in the Tugen Hills of Kenya, dates to 6.1 million to 5.7 million years ago. The femur, or upper leg, of *Orrorin* has a shape typical of later hominins, including our own genus, *Homo*, suggesting that it was a frequent bipedal walker. Nonetheless, once again, it retains arm bones and fingers with shapes best associated with climbing among trees, and evidence from the shape of its teeth, as well as reconstructions of the ecological habitats of other animals found nearby, suggest that it ate leaves, fruit, seeds, roots, nuts, and insects within an area dominated by drying tropical forest. Amazingly, then, although all three challengers for the crown of first hominin were found in modern deserts or arid grasslands, they confirm the ongoing importance of tropical forests during the earliest stages of our ancestors' evolution as both a safe haven from predators and a likely source of food.[6]

The fossil fragments left behind by some of the earliest proposed hominins in just a handful of locations 7 million to 5 million years ago mean that properly reconstructing our earliest evolution in Africa will always be difficult. Clearly, however, the early hominins of the late Miocene developed bipedalism at the same time as they were still climbing trees and using forest or woodland patches in the tropics. A forested origin for bipedalism is perhaps not so surprising when we look at long-term studies of modern nonhuman great ape movement. Astoundingly, the most bipedal of our living relatives are not the ground-dwelling, hunting chimpanzees but rather the orangutans, which most comfortably and frequently use bipedalism to navigate delicate branches and retrieve fruit in forest contexts. Not only that, but the fact that African nonhuman great apes only split from our own lineage around 7 million years ago, combined with the fossil record approaching this split, allows us to see that instead of being "left behind," chimpanzees have themselves evolved, changed, and survived, ending up with a preference for living in tropical forests, a specialization that has only brought them to the brink of extinction because their close relative *Homo sapiens* is destroying this forest habitat. This is a very different perspective indeed on the notions that chimpanzees are good proxies for models of early hominins and that we diverged from chimpanzees simply by walking through grasslands.[7]

ARDI WAS NOT the only, or even the first, female fossil star to rise out of Ethiopia or the Awash River basin. In 1974, another team of paleo-anthropologists, directed by Maurice Taieb, was exploring a drier part of the same valley. Two of the expedition's members, Donald Johanson and his student Tom Gray, had spent two hours scrabbling around in the increasingly dangerous heat. Just as they were about to leave and seek shelter, Johanson noticed an arm bone poking out the side of a dried-up river gully. As more bones emerged, nearly half a complete skeleton, the two men began to realize the significance of their find. In the subsequent celebrations that evening back at an exhausted but rejoicing camp, the Beatles' "Lucy in the Sky with Diamonds" inspired a more personal name for the find that, until then, was known as AL 288-1. The rest, as they say, is history. This skeleton, or "Lucy," now known as *Australopithecus afarensis*, showed that by 3.2 million years ago our ancestors were quite literally strutting around eastern Africa. Not only was Lucy remarkably complete, but careful study of the shape and dimensions of her hip, knee, and thigh bones show that, by this point in human history, hominins were now built to walk almost entirely upright along the ground of an increasingly open tropical landscape.[8]

In fact, Lucy is part of a clear, classic sequence of diverse, upright hominin bipeds from the Pliocene (5.33 million to 2.58 million years ago) to the early Pleistocene (2.58 million to 0.78 million years ago). She was preceded by *Australopithecus anamensis* in Ethiopia and also Kenya, an ancestor whose limited preserved limb bones tentatively hint at frequent bipedalism. She was succeeded by the delicate-looking *Australopithecus bahrelghazali* in Chad, *Australopithecus africanus* and *Australopithecus robustus* in South Africa, and the thick-jawed *Paranthropus boisei*, or "Nutcracker Man," in eastern Africa, all of which document the spread of specialized upright hominins across the African continent. At the end of this sequence came our own genus, *Homo*, perhaps as early as 2.8 million years ago, represented later by the energy-efficient walking of first *Homo habilis* and then *Homo erectus*—whose name directly hints at its upstanding nature. This trajectory also entailed an increasing brain size, from around 300 to 500 cubic centimeters in Lucy to 1,000 cubic centimeters in *Homo erectus*. Now comfortably walking and needing to feed larger bodies and brains, hominins also began to fashion stones into tools to procure meat and marrow around 3.3 million years ago. Traditionally, this gradual but relentless movement toward upright bipedalism, bigger brains, the need for technology, and even larger social groups has been postulated as a response to the heat, openness, and larger animals, including prey but also predators, today found in the open, arid savannah-like settings where almost all of these fossils had been discovered.[9]

As we have already seen, however, modern environments are not always a good example of what an area looked like in the past. Fossil animals found in the same geological layers as hominin fossils provided some more direct clues based on feeding preferences, and generally supported a shift to more open habitats towards the Pleistocene, though it remained hard to work out how much of a type of food a particular hominin species ate. Up stepped my own methodological field, stable isotope analysis, which relies on the fact that the abundant elements like carbon can have different atomic forms. These "isotopes" (in this case ^{13}C and ^{12}C) have different masses, causing them to react differently during important biological processes such as photosynthesis. Crucially for us, when carbon dioxide is processed by the different types of photosynthesis practiced by the C_4 grasses that dominate subtropical and tropical savannahs and the C_3 herbs, shrubs, and trees that dominate subtropical and tropical forests, the heavier isotope, ^{13}C, is more weakly or strongly discriminated against, respectively. This leads to different isotopic ratios in the plant tissues formed that are so clear as to be measurable, not just in the plants themselves but also in the well-preserved teeth of the animals that eat them, including hominins. First pioneered using hominin fossils in South Africa in the 1980s and 1990s by my former supervisor,

Professor Julia Lee-Thorp, the now significant dataset resulting from this method, covering Ardi through to *Homo*, shows a clear shift in reliance from forest or wooded habitats to savannahs through time. While this revolutionary work could not tell whether hominins ate animals feeding on C_4 plants or the C_4 plants themselves, alongside the shapes of hominin bones, fossil animals, and stone tool evidence, it apparently directly confirmed an increasing role for savannahs in our evolution between 4 million and 2 million years ago.[10]

But before we dismiss tropical forests from this part of the human journey entirely, there are more than a few hints that hominins kept connections to their ancestral homes. Where limb bones are preserved, the arms and hands of a number of hominins, including Lucy, suggest that they performed a substantial amount of climbing. Even the first well-studied member of our own genus, *Homo*, *Homo habilis*, had notably strong wrist and ankle bones that indicate some lasting connection to trees. Moreover, within a gradual trend toward more C_4 foods, the hominin stable isotope data shows significant variation. Data from *Australopithecus anamensis* indicates that it used mixed forest-woodland-grassland environments, while study of wear on its teeth reveals similarities to modern gorillas. Isotopic data from individuals of Lucy's species show that although some individuals consumed mainly savannah foods, others lived almost entirely in woodland or forest habitats. Some hominins, such as *Australopithecus africanus* and *Paranthropus boisei*, even show massive variation in dietary isotopes associated with woodland/shrubland versus grassland habitats within a given year! In fact, *Homo* also breaks the linear trend toward savannah, showing a much more mixed diet than its contemporary savannah specialist *Paranthropus*. Clearly, therefore, even though the world around them was broadly changing from one dominated by forest to one dominated by more open grassland, many hominin species and individuals were more than capable of selectively using very different environments when they were available. As we will see in Chapter 6, the hominin clade may never actually have truly completely left these habitats, right up until the origins of our species.[11]

Paleoclimatologists have also revealed complexity in the environmental context of hominin evolution when looking at broader-scale changes in the presence of grassland on the African continent from 12 million years ago. Professor Sarah Feakins at the University of Southern California has developed an innovative approach to the study of highly resistant biomolecules of plant "waxes." Studying the carbon isotope ratios of these waxes, alongside fossil pollen, blown from across eastern Africa into the ocean and preserved in well-stratified offshore marine cores from the Gulf of Aden, Sarah and her colleagues have looked at the trajectory of vegetation from across this portion of the continent, one unaffected by biases in discovered fossil location and damaging processes of land erosion. Their findings have dealt a further blow to the "savannah hypothesis." First, by combining pollen and plant wax isotopes, they found that 12 million years ago, C_3 *grasses* were already extensive across the region. As a result, the later expansion of C_4 grassland, from around 10 million years ago, actually replaced preexisting grassland ecosystems as much as forest habitats. Furthermore, "both of these processes took place *before* we see the emergence of bipedalism in the hominin fossil record," says Sarah. Second, the expansion of C_4 grasslands, when they arrived, was not a unilinear, directional one. Ocean records make clear that there were wetter and drier periods over the last 5 million years, with varying proportions of grassland ecosystems. "Ultimately, even when we 'zoom out' to this continental picture provided by the marine cores, it is clear that hominin evolution in Africa was accompanied by a complex environmental patchwork, from the first hominins ~7 Million Years Ago, through to the emergence of our genus ~3 Million Years Ago," concludes Sarah.[12]

Returning to land, using methods such as Sarah's, as well as stable isotope analysis of the fossil teeth of herbivores and carbonate nodules preserved in soils, scientists have revealed similar patchworks on diverse regional and local scales. African geography and tectonics, such as steep rift valleys, resulted in wetter conditions and the formation of lakes in certain valleys and basins, even as the eastern portion of the continent was, overall, becoming more arid. There would also have always been tree corridors along rivers and lakes among the grassland or dryland ecosystems, as well as forested mountains. This regional climatic variability also meant that C_4 ecosystems would have appeared and dominated at different points in time depending on where you were in eastern Africa. On a more site-specific basis, a reconstruction of past environments using animal fossils and isotopic analysis of ancient soils associated with *Australopithecus afarensis* fossils at the Asa Issie locality in Ethiopia suggest it was, at least in some locales, still living and dying in the same kinds of forested habitats that Ardi preferred. Mammalian fossils and ancient remains of plants also show a mixture of forest, wetlands, grasslands, and dry shrubland in association with a Pliocene hominin skull at Woranso-Mille in Ethiopia dating to approximately 3.8 million years ago. Similarly, scientists have found that tropical woodland, rather than grassland, actually surrounded some of the most important Pliocene and Pleistocene fossil and stone tool sites at the time of hominin occupation. Finally, as Sarah reminds us, "it is important to bear in mind that so-called 'savannahs' refer to a range of plant types." Savannahs today have trees and even forest patches, depending on local rainfall and geology, and this would certainly have been the case in the past. Although more open, drier environments perhaps drove the evolution of our genus, variability and a persistent use of tropical forest and woodland settings also influenced it. Adaptations to this variability definitely stood early members of *Homo* in good stead as they went on to seek new Pleistocene horizons.[13]

FROM 1.8 MILLION years ago, discoveries of fossils in the deep, dark Atapuerca Caves of Spain and the cool temperate climates of Dmanisi, Georgia, as well as stone tools in rainy Norfolk in the United Kingdom, show that some members of the striding, toolmaking, large-brained genus *Homo* had moved "out of Africa" to seek pastures new in northern and northwestern Eurasia. Beautiful pear-shaped "Acheulean" hand-axes, named after the site where they were first found in France, were in use across Africa by 1.7 million years ago, South Asia and the Levant by 1.5 million years ago, and Europe just after 1 million years ago. These remarkable pieces of stone are the breadcrumbs left behind by one of the most prolific of these early hominin adventurers, *Homo erectus*. These early hominin expansions beyond the African heartland have often been linked to grassland expansions across the Middle East, Europe, and South Asia, providing new, widespread environments for medium- and large-sized mammals that had become so important to hominin diets.

Nevertheless, one major outlier in this argument has always been the remarkable fossil finds from Southeast Asia showing that *Homo erectus* had made it to the Asian tropics by 1.5 million years ago in Java, Indonesia, implying that perhaps these new pioneers also had more forested tastes than commonly assumed.[14]

Dr. Kira Westaway of Australia's Macquarie University is one of the leading authorities on hominin migrations into the Southeast Asian tropics. A former diving instructor, she is no stranger to working in suffocating humidity, bat-filled crevices, and long, dark, seemingly endless caves in her time investigating the environments and chronologies of the remarkable variety of different hominoids now known to have roamed this part of the world during the Pleistocene. While Africa has rightly received much attention as the center of hominin evolution, after two decades of remarkable discoveries, Southeast Asia is now not far behind in terms of yielding frequent, dramatic advances in our field. Crucially, Southeast Asia provides a test case for us to determine how early members of *Homo* might have adapted to or avoided tropical forest settings. Understandable caution when applying stable isotope methodologies to rare, precious hominin fossils, an absence of relevant long-term land-based climate records when compared to Africa, and a limited understanding of how preserved animal fossils relate to often isolated hominin finds have all made this problem especially difficult to explore in the region, however. One approach has simply been to look at how hominins fit into wider patterns of changing plants and animals in Southeast Asia over the course of the Pleistocene. In particular, Kira and her team have focused their attentions on the fortunes of one remarkable Pleistocene giant, *Gigantopithecus*, the largest ape ever to have lived.

"Giganto," as Kira calls it, was found across China and perhaps also parts of Southeast Asia starting 2 million years ago. Eating tropical forest seeds, fruits, and bamboo, it could weigh a maximum of nearly three hundred kilograms (around twice the size of a modern gorilla) and proved remarkably resilient, with its last appearance in eastern Asia occurring 400,000 to 300,000 years ago. "However, something happened to this gentle colossus from around 700,000 years ago, as they were pushed back into a small part of southern China," says Kira. The range, and presumably also population size, of *Gigantopithecus* began to dwindle,

Figure 5.3. Map of known *Gigantopithecus* fossils dated to between approximately two million and three hundred thousand years ago in southern China with the reconstructed extent of tropical forest in gray. *Kira Westaway*

alongside those of other early Pleistocene animals that favored dense tropical forest habitats, including an early relative of the panda (*Ailuropoda wulingshanensis*). This seems to have been caused by changes in sea level and atmospheric carbon dioxide that resulted in the expansion of drier grasslands and woodlands across mainland and island Southeast Asia around 1 million to 0.7 million years ago in the late-early to middle Pleistocene. Assisted by land bridges across the now exposed Sunda Shelf, which connected modern islands like Borneo to continental Asia, a whole new variety of mammals, including the ancient elephant-like *Stegodon*, spread across the region. These changes eventually took their toll on "Giganto." But one ape's misfortune was another's opportunity. Scientists believe that these new grassland conveyor belts and their meaty herbivorous commuters may have taken *Homo* right into the heart of the

Asian tropics. Such a scenario would fit neatly within perceptions of an increasing hominin preference for the resources of grassland habitats.[15]

Yet, as we have already seen, Pliocene-Pleistocene hominin habitat preferences are a little more complex than such models allow. Where actual fossil plants and animals are found alongside *Homo erectus* in Southeast Asia, including pygmy hippos, turtles, deer, wild pigs, and different types of *Stegodon*, they show that it was likely living within more mixed lake-edge marsh, grassland, and woodland environments, with some tropical forest. Complex environmental mosaics hosted another star of Pleistocene Southeast Asia, the "Hobbit." Following discovery at Liang Bua Cave in 2003 by a team Kira was also part of, these diminutive hominins from the island of Flores, Indonesia, quickly became even more famous than their fictional namesakes from the Shire. Dating to 190,000 to 60,000 years old, these tiny hominins, growing to a maximum of about one meter in height, seem to descend from a small-bodied group of hominins that made it to the island 700,000 years ago and may, in turn, have been related to *Homo erectus*. The survival of this unique hominin, right up to the arrival of our own species in the region, is remarkable not only because of the potential for interaction but also because it might tell us something about their island environments. Small stature is often linked to habitation of dense tropical rainforests, and Kira and her team showed that the Hobbits' occupation of Liang Bua was most intensive during wet, more forested periods. However, this intensification likely represented these hominins' search for respite from rain. In fact, once again, fossil plants and animals left behind by the foraging and hunting of our diminutive cousins, including miniature *Stegodon*, Komodo dragons, and giant rats, indicate a preference for foraging in drier, mixed tropical woodland and grassland habitats. The change in size therefore seems to come down to the more general, well-documented process of "island dwarfism" linked to the strictly limited range of the population in an isolated location rather than to a more specific dense rainforest specialization.[16]

Although the ecological situation was certainly more complex than sweeping "savannahs," an opening of environments and an expansion of drier woodland and grassland habitats at the expense of denser rainforest clearly aided some of the earliest hominin arrivals into mainland

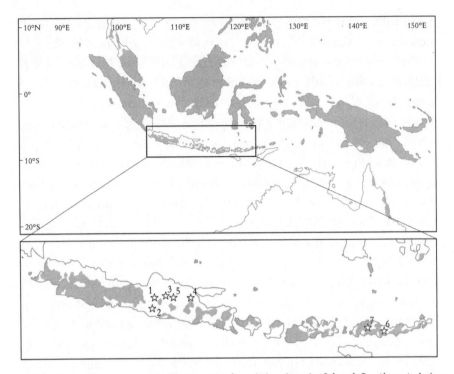

Figure 5.4. Map of major Pleistocene hominin sites in Island Southeast Asia discussed in the text plotted against tropical forest distributions estimated on the basis of the MODIS (Moderate Resolution Imaging Spectroradiometer) Land Cover MCD12Q1 majority land cover type 1, class 2 for 2012 (spatial resolution of 500 m). Downloaded from the US Geological Survey Earth Resources Observation System (EROS) Data Center (EDC). 1. Sangiran, 2. Ngandong, 3. Bapang and Trinil, 4. Mojokerto, 5. Kedung Brubus, 6. Mata Menge, 7. Liang Bua.

and island Southeast Asia. This enabled them to expand and diversify, reaching beyond the Wallace Line into island settings that remained unconnected to the mainland even during periods of low sea level, like the Philippines, Flores, and Sulawesi. We can see the true importance of these more mixed environments when we consider the eventual fates of these early hominins. Kira and colleagues have now shown that *Homo erectus* survived in the region until approximately 100,000 years ago, while the Hobbit persisted on Flores until approximately 60,000 years ago. At around 100,000 to 80,000 years ago, communities of tropical rainforest plants and animals sprang up across Southeast Asia during a period of wetter conditions. Home to tapirs, orangutans, gibbons, and sun bears, this forest is effectively what exists in humid, undisturbed portions of the region today. In 2020, my colleague Julien Louys and I compiled all of the available stable isotope data that exists for Pleistocene fossil mammals in Southeast Asia in the journal *Nature*. Studying the trends in the data, we argued that the aggressive expansion of these tropical forest ecosystems across mainland and island Southeast Asia during the late Pleistocene apparently overwhelmed our more "generalist" hominin relatives, as well as the woodland and grassland animals that had walked alongside them. Indeed, while the list of our Pleistocene hominin relatives in Southeast Asia keeps growing, with remarkable finds on a near yearly basis, only one hominin actually survived this major climatic and environmental swing back to extensive tropical canopies.[17]

SINCE THE VICTORIAN era, savannah landscapes have been the crucial backdrop for the grand story of human evolution. All the while, tropical forests remained as outdated backwaters, unattractive for increasingly upright, innovative, hunting hominins and better associated with nonhuman great apes. New research, however, has shown that tropical forests provided the vibrant cradle for the earliest great apes and, later, the earliest hominins, experimenting at a time of increasing environmental change in the late Miocene through to the Pleistocene. Hominin bipedalism itself may have evolved within and below the trees rather than out in exposed, sweltering grasslands. Certainly, drying and opening landscapes stimulated hominin diversity, specialization in bipedalism, toolmaking, and, eventually, the expansion of our earliest ancestors beyond Africa and across much of Europe, the Middle East, and Asia by the middle Pleistocene. But this was not a one-way street. Many hominins emerging in Africa kept mementos of their forested beginnings, in both their skeletons and their chemical composition, often using nutritious forest or woodland patches within complex savannahs—not the uniform, sweeping plains we are used to seeing on TV today. Similarly, when some of the earlier members of our own genus ventured into the Asian tropics, mixtures of forest, dry woodland, and moist lakeside grasslands supported their populations, large and small. These populations likely avoided monotonous grasslands just as much as they did dense, stifling tropical lowland rainforests, with trees continuing to provide a source of food and refuge for our ancestors.

In fact, climatic and environmental *variation* framed the evolution and differentiation of our hominin ancestors. Efficient bipedalism, increased brain size and social groups, and stone tool development were all key adaptations to a rapidly changing world rather than the product of monotonous existence. The Miocene-Pleistocene saw not only grassland expansion but also increasingly dramatic swings in global climate states, driven by the powerful Milankovitch cycles discussed in Chapter 2. Our ancestors adjusted to this unpredictability by adapting to the plentiful meat opportunities drying savannas had to offer. But all the time, they kept their eyes on profitable forest habitats and woodland edges—hedging their bets. This flexibility, however, had its limits. The late middle and late Pleistocene brought more "extremes" in both climates and rapidly diversifying environments. In tropical Southeast Asia, as the late Pleistocene brought with it an expansion of dense tropical rainforests, its resident hominin generalists confronted a thick, expanding green wall—one they could not ultimately overcome. Just prior to this, a new hominin emerged into this increasingly shifting "brave new world" in Africa. It went on to march out across almost all of the planet's continents, which by now were fully formed. It was, of course, our own species, *Homo sapiens.*[18]

Chapter 6

ON THE TROPICAL ORIGINS
OF OUR SPECIES

The tendrils of social media, the ubiquity of mobile phones, the cult of "celebrity" and reality TV, and our busy work and social lives all mean that we grow up in a very human-focused world. Our widespread geographical range, from the ice of Greenland to the mountains of the Himalaya, from the bustling metropolis of Beijing to the desert-fringed Dubai, leads us to simply assume that *Homo sapiens* is one of the most successful species to ever walk the planet. Not only that, but our huge global population, extensive cities, and dominating impacts on the natural world seem to give us a certain planetary "weight." Yet we are relative newcomers to life on Earth, especially when compared to the greater than 300-million-year journey of tropical forests that we have been following. The first fossil skeletons with distinctive "human" (in this book I reserve the word "human" only for our own species, *Homo sapiens*) features, including a shorter, flatter face, round skull, and large brain (over three times bigger than Lucy's), appear in Africa between 300,000 and 200,000 years ago. This fits with estimates based on "molecular clocks," akin to those used for plants and mammals in Chapters 1 and 4, respectively, calculated using modern and ancient human DNA.

Whichever way you look at it, this makes us practical newborns in an evolutionary context. Indeed, two extinct hominins we commonly think of as having eventually "failed," *Homo erectus* and *Homo neanderthalensis* (or the Neanderthals), both actually lasted longer on the Earth's surface before their demise than we have to date.[1]

It may not be too surprising, then, that paleoanthropologists and archeologists have seen *Homo sapiens* as simply the most recent point in a longer-term evolutionary trajectory of hominins and the genus *Homo* toward open, savannah contexts and away from supposedly protein-poor tropical forests. In fact, anthropologists and archeologists alike have often thought of tropical forests as "barriers" to Pleistocene human presence. Small protein-poor and difficult-to-catch prey, poisonous plants, a lack of dense carbohydrate resources, tropical diseases, and rampant humidity, not to mention poor preservation of archeological sites thanks to high water flow and acidic soils, have meant that these "green hells" have often been neglected in the story of human dispersals. Between 300,000 and 60,000 years ago, the archeological record begins to show evidence for new technologies (the bow and arrow), items made for exchange to maintain social networks, and personal ornaments and artwork, including engravings and shell-bead necklaces—all of which have been considered uniquely human capacities. Yet these innovations have been linked to the need for more efficient projectile hunting and the maintenance of useful relationships between groups living in open grassland settings prone to drying in the face of ever-harsher swings between glacial and interglacial conditions, as well as shorter-term swings in climate systems, in the late Pleistocene (Chapter 2). Alternatively, they are seen as the product of growing human populations thriving on reliable marine resources in coastal settings. As with *Homo erectus*, the eventual dispersal of *Homo sapiens* into the Middle East and Eurasia and throughout Southeast Asia has also been associated with climatic periods when more challenging environments like deserts or tropical forests retreated and rich "savannah-like" hunting grounds expanded. Either that, or it has been suggested that humans tracked rich, high-protein resources from the oceans all the way from eastern Africa, around the Indian Ocean rim, to Australia by 65,000 to 45,000 years ago and to the Americas by 30,000 to 25,000 years ago, along a series of "coastal highways." While undoubtedly new, in many ways, *Homo sapiens* has thus traditionally been considered as equally dependent on particular nonforested environments as its hominin forbears and contemporaries, which we met in Chapter 5.[2]

These narratives rely on a single, homogeneous origin and single set of either grassland or oceanic desires for humans. But one thing that has always drawn me to studying the origins of our species is that, while we are a relatively recent evolutionary arrival, we are also the only hominin that went on to occupy *all* of the planet's continents and to use *all* of the world's land-based environments. Much of this seems to have happened very quickly, during the late Pleistocene (125,000 to 11,700 years ago) and appears to call into question assumptions that humans stuck to a single, easy or familiar "route." In this chapter we will see how the latest research in archeology, paleoanthropology, genetics, and environmental science is building a more complex picture of *Homo sapiens*'s entrance onto the world stage, one that covers a number of different parts of Africa, including the tropical forests of West, Central, and eastern Africa. Not only that, but we will meet daring archeologists and paleoanthropologists who are beginning to explore tropical parts of the world that have often been ignored as unprofitable "blanks on the map." Their work is demonstrating that, as we moved beyond Africa, *Homo sapiens* repeatedly and persistently occupied some of the most extreme sets of environments the planet had to offer. Certainly our species was a continuation of a long period of hominin adaptation to variable Pleistocene environments that we met in Chapter 5. However, it is in settings such as tropical forests, rather than sweeping savannahs or uniform coastlines, that we can see how *Homo sapiens* used its new cultural, technological, and social repertoire to become the most flexible hominin yet—one that could deal with all of the climatic fluctuations, between cooling and warming and drying and wetting, that the late Pleistocene world could throw at it, and one that grew into the relentless, global force we remain today.

THE TRADITIONAL VIEW of human origins is that we evolved at one time, in one place, and in one population in Africa. In the 1980s, genetic variation in modern human mitochondria, little organelles within our cells that can only be passed down through the maternal line, was used to develop a "family tree" of human populations, linking all living people back to a single population or even just one female ancestor—"Mitochondrial Eve"—who lived in Africa between 200,000 and 140,000 years ago. Although more recent studies, using powerful modern statistics and genetic sequencers, now analyze genetic variation across the entire human genome to produce more sophisticated models, they retain the same basic principles—namely, that the variability in the DNA of modern human populations can be inserted into a simple "treelike" scheme that points back toward a single point in time and space when our species emerged and began to diversify as it dispersed across Africa and around the world. In a number of recent cases, these studies tend to zero in on KhoeSan populations, which have the most diverse human genomes seen anywhere on the planet today, suggesting that populations of *Homo sapiens* first emerged and started to diversify in the southern portion of the continent 300,000 to 150,000 years ago. In recent years, however, it has become clear that genetic models of our species' origins need to become exponentially more complex. We now know that *Homo sapiens* interbred with Neanderthals, with a second hominin group called Denisovans (named for the cave where they were found in Siberia, which itself was named after a hermit called "Denis" who once lived there), and perhaps even with a third hominin group. Not only that but they produced fertile offspring, complicating traditional determinations of "species" and leaving their mark on the genetic makeup of surviving members of our own species. Furthermore, only one Pleistocene fossil individual in Africa has produced ancient DNA, since generally hot conditions tend to break down ancient organic molecules. Justifiably, museum curators, archeologists, and paleoanthropologists are also inclined to protect limited fossils from destructive analyses, so they have approached this application cautiously. As modern DNA samples cannot reliably inform us about the geography of ancient populations, given long histories of human migrations between regions and interactions between different human populations, without ancient DNA, it is almost impossible to use modern DNA to "read back" exact past locations or timings of human origins with any level of accuracy.[3]

Paleoanthropologists and archeologists have therefore tried to get their hands on something a little more tangible than DNA to mark the date and place of human emergence: fossils. As with DNA analyses, there has been an obsession with finding the "oldest" example of something in order to crown a new region as ruler in the story of human origins. The oldest accepted fossils with distinctly "human" features have for a long time been those found, like Ardi and Lucy, in the drylands of Ethiopia, this time at the sites of Omo Kibish and Herto. Dating to 195,000 and 160,000 years old, respectively, these skeletons look morphologically "human." Animal bones found in the same place, with known ecological habits, climate records, and geological studies, suggest that these early humans were living in dry, grassland environments, apparently confirming our origins in the eastern African savannah. However, relatively complete hominin fossils, including those of our species, are always rare, scattered across the continent in different localities where chance (e.g., transport of remains into a sheltered cave system) or specific conditions (e.g., layers produced by volcanic eruptions or flooding) have shielded them from the ravages of time. As a result, some archeologists have suggested it may instead be better to try and find more commonly preserved material features, like art, symbols, and complex stone tools, to determine the "behavioral" appearance of humans. As mentioned earlier in the chapter, such evidence has tended to come from coastal settings in northern and southern Africa dating to around 100,000 to 70,000 years ago, suggesting a marine focus, although more recent work in Kenya hints at the appearance of sophisticated human exchange patterns and technology 300,000 years ago in a dry grassland setting. In both cases, research has focused on trying to find the first skeletons or novel material remains (e.g., specific stone tool technologies, projectile technologies, symbolic beads, and pigment use) left behind by our ancestors. In doing so, most attention has thus centered on the long hominin-associated savannahs or on coastlines where caves that have eroded into cliff faces provide their own perfect microclimates, sheltered from the elements, preserving traces of the past.[4]

That is, until Dr. Eleanor Scerri and her colleagues wrote a revolutionary paper in a leading scientific journal, *Trends in Ecology and Evolution*, in 2018. Although other scientists had already begun to dissent from the mainstream, this paper challenged the simple traditional view that *Homo sapiens* appeared in one place at one time. Eleanor heads the Pan-African Evolution Research Group at my own institute in Jena, and it does exactly what it says on the tin. "We believe that between 500,000 and 200,000 years ago, the key time frame for the emergence of our species, our ancestors lived in distinct, but interconnected, populations *across* Africa," says Eleanor. In other words, we did not appear at a single point in time or in a single group. Instead, millennia of partial contact between populations led to a huge raft of human diversity. The dynamic mixing of these groups was to eventually lead to our species as we know it. "To study this properly, we therefore need to look at *all* of Africa, not just a single mythical point of beginning," continues Eleanor. In fact, casting the net a little wider we can start to see a number of hominin fossils that, like those from Omo Kibish and Herto in Ethiopia, have a diverse mixture of both clear *Homo sapiens* (e.g., flat face and rounded skull) and more "archaic" (e.g., thicker brow ridge) traits, including those from Jebel Irhoud in Morocco (c. 300,000 years ago) and from Florisbad in South Africa (c. 260,000 years ago). The "full suite" of what we often consider to be distinctively *Homo sapiens* features actually only really appears to take shape between 100,000 and 40,000 years ago, though this may be a product of a lack of preserved fossils from this and the preceding time interval. Significantly, in tropical West Africa, the earliest known human fossil dates to just 16,000 to 12,000 years ago at Iwo Eleru, Nigeria, and shows just how long considerable morphological diversity was maintained between different groups of *Homo sapiens* in different parts of the continent. A reexamination of the existing genetic evidence, in light of the fact that genetic exchange probably occurred between different hominin lineages and that modern DNA gives very little insight into geographic patterning in the past, seems to support this "African multiregionalism" proposed for the evolution of *Homo sapiens* by Eleanor and her colleagues.[5]

Rather than a branching linear tree with a simple, single trunk, we therefore have fractured African roots. The same picture emerged when Eleanor and her colleagues looked at what they call a "patchwork" of cultural materials left behind by *Homo sapiens* populations across Africa in the middle and late Pleistocene. Stone tools, thanks to the materials from which they are made, are reliably preserved over deep timescales.

Not only that, but archeologists can look at their shape and how they were made to infer things about behavior and technology. Archeologists have suggested that the appearance of our own species heralded a proliferation of new forms of stone tools that could be inserted into complex projectile setups and that required more sophisticated, thought-out manufacture. These so-called Middle Stone Age toolkits have been found alongside humanlike fossils at Jebel Irhoud and Florisbad, as well as at Olorgesailie in Kenya. Importantly, the Middle Stone Age toolkit varied by region. In North Africa, the rapid expansion of grasslands under wetter climates led to growing populations of *Homo sapiens* and the widespread appearance of unique, complex, arrow-shaped (or "tanged") stone tools 120,000 to 80,000 years ago, alongside bone tools and shell beads suggestive of complex cultural behaviors. In eastern Africa, by contrast, there is huge variation and continuity in Middle Stone Age stone tools from 300,000 to 60,000 years ago. The eventual transitions from these Middle Stone Age forms to diminutive "microlith" stone tools, often linked to well-developed bow and arrow technology and hunting efficiency, between 80,000 and 40,000 years ago also occurs in highly diverse ways in different parts of Africa. Sometimes there is complete, rapid replacement; other times there is a gradual change. The appearance and disappearance of "human" productions like art and social display are equally varied regionally and often come and go in the Pleistocene archeological record right up until the Holocene. It appears, then, in contrast to the traditional story of our arrival on Earth, that there was no dramatic, consistent fossil or "material" advent of *Homo sapiens* at a particular place or time; nor was there a uniform, linear progression toward "behavioral complexity."[6]

Things get even more interesting when we look at what Eleanor and her team think drove the dynamic separation and integration of these different, evolving human populations that gave rise to all of us, wherever we live around the world. The character and chronology of the fossils, stone tools, and different material artifacts left behind by early *Homo sapiens* suggest that our species eventually appeared thanks to interaction between these diverse groups living in different parts of Africa. Eleanor and colleagues believe that the varied environmental settings of Pleistocene Africa, not just savannahs or coasts, either kept these populations separate or brought them together at different points in time. In fact, once we move away from a presumption of a uniform, single origin and environmental association, tropical forests begin to take a much more prominent role in this early, more complex evolutionary model. Indeed, as well as the fascinating Iwo Eleru fossil evidence for morphological diversity mentioned above, some of the most unique of the Middle Stone Age regional diversity comes from the modern-day tropical forests of West and Central Africa: large, heavy-duty pick-like forms, huge scrapers, and "lance-like" points have been variously associated with digging for carbohydrate-rich tropical tuber plants, like yams, as well as hunting tropical forest animals. Many of the sites where these tools, dating to between 300,000 and 40,000 years ago, have been found are today surrounded by dense tropical evergreen rainforest or drier tropical forests and woodlands. Records of ancient plants and sediments in marine cores off the coast of West Africa contain hints that tropical forests also covered these same regions in the past. More work is, however, necessary to confirm the ecologies inhabited by the populations that made these tools and the plants and animals on which they relied for survival.[7]

But while the role of dense tropical rainforests in early *Homo sapiens*'s evolution in Africa awaits definitive proof, there is growing evidence that expanding and contracting tropical forests may have shaped past human genetic and cultural diversity. Comparison of the DNA of different modern tropical forest hunter-gatherers in Central Africa with that of neighboring populations suggests that they diverged from each other at least 70,000 years ago. Not only that, but genetic differences found between different forest hunter-gatherer groups implies that they, too, diversified from each other as a result of the waxing and waning of forest habitats. At the cave site of Panga ya Saidi on the tropical coast of Kenya, where I was lucky enough to work on recovered archeological animal materials, we found evidence that a resistant mixture of moist tropical forests, woodlands, and grasslands buffered human populations against major climate change seen in the dry, grassland interior, allowing them to gradually experiment with their stone tool technology and symbolic material culture. Indeed, despite being just fifteen kilometers away from the present coastline, the site yielded no prominent evidence of the use of coastal resources for subsistence during the Pleistocene. Elsewhere, similar mixed environments promoted the movement of growing human populations in and around the African subtropics. Tropical forests were certainly not the only environment at play. Nonetheless, they were undoubtedly a key part of the complex "pan-African" skeletal, genetic, and behavioral appearance of our species.[8]

"HURRY UP, PAT, it is about to rain!" yelled my former PhD student, Dr. Oshan Wedage of the University of Jayewardenepura, as I staggered behind him through the dense, lowland, evergreen rainforest of southwestern Sri Lanka. Drenched in sweat and covered in especially cunning leeches, which drop from the treetops as they feel your vibrations, I tried to pick up my pace. Not fast enough. The heavens opened, and we were now trudging uphill through the mud, undergrowth, and fast-running streams. The rest of the animal life was too smart to be out in this weather, giving this challenging habitat an even more "barren" appearance. At that moment it was very easy to see how many leading archeologists and anthropologists, and indeed the general public, might still think that the first human populations moving out of Africa would have actively avoided tropical forests. In fact, until the beginning of the twenty-first century, the earliest evidence for definite human occupation of a tropical rainforest environment was widely considered to be limited to the very final Pleistocene (18,000 to 11,700 years ago) or early Holocene (11,700 to 8,200 years ago), when further technological innovations may have made dealing with these challenging conditions and food sources more palatable. But then, from up ahead, Oshan shouted, "We're here!" Out of the mist and water came a view of one of a cluster of cave and rock shelter sites in this part of the world that has been drastically challenging this antitropical forest bias.[9]

Sri Lanka is an island at the southern tip of South Asia, just below the Indian subcontinent and just above the equator. Sitting in the middle of the Indian Ocean, it is at a key location in models that assume a rapid "southern route" for humans from Africa all the way to Australia in the late Pleistocene. This expressway has been variously thought of as being driven by rich coastal resources, meaning that humans did not have to come inland to face more challenging environments, or by the

periodic expansion of more familiar grasslands. But since Sri Lankan archeologists, like Oshan and his earlier mentor, Dr. Siran Deraniyagala,
transformed Pleistocene archeology on the island, it has posed something of a problem for these blanket models. The site we came across at
the end of our trek was Batadomba-lena, one of three caves, alongside
the nearby Fa-Hien Lena and Kitulgala Beli-lena, that remarkably preserve the oldest human fossils anywhere in South Asia and some of the
oldest material culture that can be confidently linked with our species in
the region, dating to 45,000 years ago. As Oshan says, "Although people
have often ignored our 'Resplendent Isle,' it plays a major role in perceptions of adaptations of *Homo sapiens* as it moved beyond Africa." Instead
of on the coast or in the drier, grassland areas of Sri Lanka, the island's
earliest, clear human sites are found in inland, wet tropical rainforests.[10]

Sure, you might be thinking, but these rainforests might not have
been there in the past. It is also just an island, so maybe humans moved
around a lot, only briefly visiting the forests. It was to address these considerations that I first traveled to Sri Lanka while working on my PhD. I
instantly fell in love with the country and its people and have rarely spent
more than a year away ever since. Back in 2015, working closely with archeologists like Oshan, we applied, for the first time, the stable isotope
method I introduced in Chapter 5 to the rich record of human fossil teeth
that had emerged from Sri Lanka's rainforest caves and rockshelters.
Comparing our results to those from animal teeth found alongside the
humans, we were able to show that, unlike the African hominins, from
36,000 to 3,000 years ago, almost all of the individuals living at these sites
had isotope values that overlapped with values of animals known to be
inhabiting dense canopy rainforests, suggesting that they were fully reliant on tropical rainforest resources. This dependence continued through
periods of major climatic change, like the Last Glacial Maximum, and
even beyond the arrival of agriculture in this part of the island 3,000 years
ago. Our work proved that early expanding human populations *could* live
completely in tropical rainforests. Not just that, but they chose to, even
when grassland and coastal settings were available nearby. This specialized adaptation was also remarkably resilient, lasting over 30,000 years
and perhaps even longer. Tropical Sri Lanka was revolutionizing our understanding of how humans spread so quickly around the world. Unlike

many other hominins, they clearly did not always seek out more open, mixed settings, and they certainly did not do everything the "easy way."[11]

The next question, then, was what did these humans eat? And how did they get their food? Fortunately, unlike many tropical parts of the world, Sri Lanka had some more gifts to offer global archeology, producing long records of rich organic remains thanks to the cooler conditions inside the caves. Oshan's recent work, with a number of international partners, has studied the evidence for the technology, plants, and animals found in these archeological sites alongside our specialized ancestors in this part of the world. From the first human arrival in these rainforests 45,000 years ago, these groups were hunting small, fast rainforest animals. Amazingly, monkeys, as well as giant scampering squirrels, with characteristic "cut" and burning marks indicative of human butchery, make up around 70 percent of all mammals found at the sites from 45,000 to 3,000 years ago. Humans were also getting plenty of carbohydrates from the starchy nuts of the *Canarium* tree and maybe even from wild banana and breadfruit, while snails living in the nearby well-fed streams supplemented their diets. Oshan believes that this evidence shows that "if used the right way and with the right knowledge, tropical forests can be just as rich, if not richer, than dry grassland environments." In fact, prior to the British colonial period, some groups of the Wanniya-laeto Indigenous population of Sri Lanka still hunted wild forest game, fished, climbed to get honey, and obtained rich sugary plant resources in both the wet and dry tropical forests of the island. These were "forests of plenty," not barren "green deserts"—particularly for the highly flexible human populations that made them their home.[12]

The tools left behind by these early human groups in Sri Lanka have also provided a new environmental perspective on some key technologies and innovations associated with *Homo sapiens*. Bows and arrows and more efficient stone tools have also often been linked to African grasslands or European tundra. But in Sri Lanka, humans were recycling monkey bones to make arrow tips, while tiny "microlith" stone tools may also have been attached to projectile shafts. Similarly, clothing has often been thought of as a clever human adaptation to cold northern European or Siberian temperatures. Again, however, there is evidence of humans working animal skins, perhaps for protection against mosquitoes and leeches (!), 45,000 years ago in humid Sri Lanka. Ochre and shell beads and evidence of long-distance exchange with some human populations on the coast all show a very "human" repertoire of social connection deep in the jungles of this South Asian island. The amazing finds made in the tropical forests of Sri Lanka encourage a more diverse exploration of human dispersals during the late Pleistocene. They show that we can no longer stick to searches on coasts or in temperate and subtropical grasslands. Instead, as in Africa, we should turn our attention to little-investigated, extreme regions to reveal what it truly means to be human.[13]

MOVING AWAY FROM Sri Lanka, the last decade or so of archeological and anthropological research has begun to show the sheer diversity of tropical forests exploited and occupied by trailblazing groups of humans in late Pleistocene Southeast Asia, Near Oceania, and South America. Southern China is emerging as a potential hotspot for the earliest human fossils outside Africa. Human teeth from two sites with ages clustering around 130,000 to 100,000 years ago (although their true ages remain hotly debated) were found with animals that indicate a mixture of rainforest, mixed woodland, bamboo forest, and grassland habitats. Turning further south and east, some early human teeth on Java and Sumatra from 100,000 to 70,000 years ago have been linked to the expansion of late Pleistocene rainforests in the region, though dating these teeth and determining their relationship to animals found in the same caves is challenging at best, given that these are not archeological sites but rather collections of fossils that are the product of past flood events that washed them together, postmortem, into the locality. One of the true goldmines of Pleistocene human tropical forest adaptations is the Niah Caves on Borneo. Here, human fossils from 50,000 to 45,000 years ago are clearly linked to the hunting of wild forest boar and primates and the processing of toxic plant foods, all within a unique mixture of Southeast Asian rainforest, swamp forest, and grassland patches that may have been maintained by human hands.[14]

Unlike Sri Lanka, many early human sites in mainland and island Southeast Asia are linked to a mixture of growing evergreen rainforest and drier woodland habitats. Moving out to sea, into the Pacific, and beyond the famous Wallace Line, the situation becomes more complex. Here, groundbreaking archeologist Professor Sue O'Connor of the Australian National University has been undertaking excavations with Indonesian partners for over two decades. The islands here are generally considered impoverished in terms of animal resources. Never connected to the Southeast Asian mainland, most of these islands were home to no animals bigger than a terrier during the late Pleistocene, presenting a completely new tropical setting for humans to deal with. In fact, Sue's work on small, isolated islands like Timor and Alor, among others, has suggested that the earliest human groups practiced sophisticated seafaring adaptations. "At the site of Asitau Kuru (formerly Jerimalai) on Timor we have found one of the earliest fishhooks in the world dated to about 20,000 years ago, and in the same site we have evidence for the presence of tuna fish that usually live in open-sea environments dating back to about 45,000 years ago, suggesting that older fishhooks remain to be found," Sue states. In this case, then, it seems like a reliance on the ocean may have rapidly propelled some of the earliest human groups through this part of the Pacific to arrive in Australia sometime between 65,000 and 40,000 years ago.[15]

Together, Sue and I performed the same stable isotope analysis on human teeth from these Wallacean sites—the first time this method had been used on Pleistocene hominin teeth anywhere in Southeast Asia. The

Figure 6.1. Fishhook made from marine shell preserved at the Pleistocene site of Lene Hara on Timor, Timor-Leste. Researchers believe these same technologies would have been used by humans to capture deep sea tuna from the earliest occupation of this part of the world. *Sue O'Connor*

results initially confirmed what she had expected. The earliest human tooth we had available, dating to 45,000 to 39,000 years ago, showed an isotope value that we would expect from someone eating fish and mollusks from the sea. However, from here things got more complex. During the dry conditions of the Last Glacial Maximum explored in Chapter 2, tropical rainforests retreated in the face of more open, dry forest and grasslands. A return to wetter conditions in the terminal Pleistocene and Holocene saw them rapidly expand. At this point growing human populations seem to have shifted their attention inland. From around 20,000 years ago, the stable isotope data shows that humans were increasingly reliant on these supposedly "impoverished" forest environments. What were they eating? Giant rats appear to be the answer! At around five kilograms, these rodents weighed about the same as an average house cat, and while to us they might not seem too tasty, they would have provided valuable protein. It is also highly likely that poor preservation conditions mean we are missing the large amounts of soft fleshy fruits and starch-rich tubers (i.e., the underground storage organs like we are used to devouring from sweet potatoes) from plants that humans must have used as they sought nutrition beyond the sea. Once again, *Homo sapiens*, albeit slightly belatedly, found a way to adapt to even the most challenging and isolated of tropical forest environments.[16]

The pattern continues into the frosty mountains of New Guinea. Here, over 2,000 meters above sea level, humans encountered montane tropical forests that can receive snow in winter. Leading experts on the region, like Chris Gosden, have said that during periods like the Last Glacial Maximum, they would have been particularly "cold, difficult, and unpromising" as temperatures fell by as much as 5°C.[17] Nevertheless, from 50,000 to 45,000 years ago, a number of sites, particularly within the Ivane Valley of Papua New Guinea, show that early groups of humans occupied the region. Glenn Summerhayes and his team have faced death-defying flights in small planes to try and find the sites, deep in the heart of New Guinea, that allow them to discover how these Pleistocene populations survived under difficult conditions. Using a microscope to explore the soils at these archeological sites, they have found fossil starch grains left behind by carbohydrate-rich yams. Charred, protein-rich *Pandanus* nuts, apparently cooked by humans on a fire, have also been found at the sites, while the characteristic stone axes found in this part of the New Guinea Highlands have even been associated with the deliberate modification of forests by humans. Alongside the likely pursuit of marsupials in more open settings, humans refined their strategies to survive in these lofty environments.[18]

Tentative linkage of increasing frequency of charcoal to human arrival and burning of the landscape, as well as archeological evidence for the occupation of cave sites, suggests that humans had reached the dry and wet sclerophyll (a type of vegetation with hard, densely packed leaves along stems that famously includes eucalypt trees) subtropical and tropical forests of northeastern Australia by around 45,000 to 35,000 years ago, though intensive occupation of the wet rainforests of this region seems to have started around 8,000 years ago. After "the Antipodes," the tropics of Central and South America were the last to be reached by dispersing humans during the Pleistocene. Arriving in the Americas 30,000 to 20,000 years ago, human populations seemingly rapidly dispersed across these regions by around 18,000 to 14,000 years ago, meeting the vast array of environments this part of the world has to offer. In Central America, terminal Pleistocene communities apparently focused on lowland, seasonal tropical forests and dry, cool highland forests. At the famous site of Monte Verde in Chile, humans made use of the diverse dense, temperate rainforest and open, dry forests of Patagonia. By around 13,000 years ago, *Homo sapiens* had also made it to the rainy, humid evergreen rainforests of the Amazon Basin. Finally, humans reached the supposedly inhospitable montane forests of the high-altitude Andes by 11,500 years ago, adjusting to the Peruvian montane forests and to the low oxygen of heights above 4,500 meters above sea level that has left its mark on the genetic makeup of populations there today. Tropical forests, then, in the Americas and around the Pleistocene world, provide a crystal-clear window into the early human taste for diversity and adversity, rather than monotonous, simple routes of environmental adaptations.[19]

TROPICAL FORESTS WERE, of course, not the only show in town. Archeologists working in some of the most extreme environments around the world, including deserts, high-altitude habitats, and those beyond the Arctic circle, have all enriched our understanding of the immense threshold of human flexibility during the late Pleistocene. In the Sahara, the Arabian Peninsula, and northern India, human entrances into these modern deserts seem to have been linked to changes toward wetter climates about 100,000 to 70,000 years ago. However, in the Kalahari and Namib Deserts of southern Africa and the Central Deserts of Australia, Pleistocene humans seem to have also prospered under dry conditions with limited water. By 80,000 years ago, humans had adapted to the cold, patchy environments of the African "Mountain Kingdom" of Lesotho as well as the highlands of Ethiopia, and by 40,000 to 30,000 years ago, they had even reached the icy Tibetan plateau. Bones with cutmarks found at the 72nd parallel north even tentatively show the advance of humans into the Arctic Circle by 45,000 years ago. Of course, *Homo sapiens* used savannahs and coastal settings as well. They would have been foolish not to. But all of the above evidence, like the diverse tropical forest examples we have seen, show that our species was not a one-trick pony and was, almost literally, *everywhere* by the close of the Pleistocene.[20]

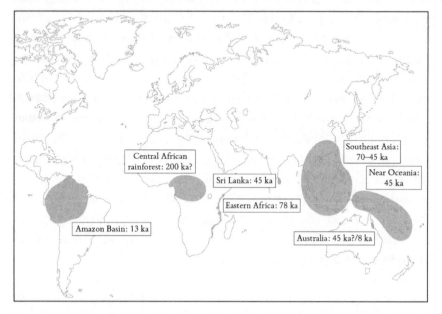

Figure 6.2.   Map showing the dates of earliest occupation of different extreme environments by *Homo sapiens* during the late Pleistocene.

With this in mind, the new technologies, new social capacities, new ornaments, new cultural materials, and new exchange systems and ways of passing down knowledge that are often considered uniquely "human" are best seen as a wide-ranging catalog for dealing with diversity rather than a one-size-fits-all tool for a single habitat. The advantage is obvious. Different, growing communities, all specializing in very different environments, meant that the fluctuating climates of the late Pleistocene would be highly unlikely to affect *Homo sapiens* at the species level. Populations of our human ancestors almost certainly disappeared locally (or were "extirpated"). In the Levant, for example, early evidence for human arrival from Africa between 200,000 to 100,000 years ago is followed by a marked absence until around 43,000 years ago. However, seemingly unlike the Neanderthals, the Denisovans, the "Hobbits," and *Homo erectus*, the *overall* human population generally kept growing, with communities in certain environments picking up the slack when other environments turned hostile. Using personal ornaments and social networks to communicate between populations inhabiting very different environments provided a crucial social fallback in times of stress. In this way we survived and also continued to expand. With greater populations and greater numbers in different parts of the planet, we began to interact more intensively with the natural world, finding new, more manipulative ways of feeding ourselves and buffering the unpredictable Pleistocene world we had been born into. By the fall of the Pleistocene curtain, we had adapted to all of the land-based environments the world has to offer. What we did next would change them forever.[21]

Chapter 7

FARMED FORESTS

F ollowing their arrival across a number of the world's environments during the late Pleistocene, humans began to make increasingly intensive interventions into the natural world in order to feed their growing populations and complex social networks. Using fire to shape the types of habitats in their surroundings, moving particularly nutritious plants, and carrying animals beyond their wild ranges as handy protein packages, human populations began to reconfigure the distribution of particular species and even entire ecosystems. During the terminal Pleistocene and Holocene transition (c. 12,000 to 8,000 years ago), and perhaps also before, these interventions became even more intrusive. Human selection for docile, social animals or carbohydrate-rich, clustered plants eventually led to morphological and genetic changes in certain species that became formally "domesticated" as part of what has been called the "origins of agriculture." Western perceptions of what this "agriculture" should look like—namely, expansive fields of cereals like wheat or barley and dense, ranging herds of cattle, sheep, and goats—have meant that much archeological, archeobotanical (the study of plants preserved in archeological sites), and zooarcheological (the study of animals preserved in archeological sites) research into its origins has centered on the dry river valleys of the

"Fertile Crescent" in the Middle East, where our familiar domesticates emerged before spreading across Europe during the Neolithic (c. 9,000 to 4,000 years ago).[1]

Tropical forests have, by contrast, frequently fallen by the wayside of such discussions. How could the dense green tangle and poor acidic soils of these environments provide our daily bread, our "usual" at the pub, or our regular pints of milk or blocks of cheese—let alone preserve traces of plants and animals we might use to retrace the footsteps of the first farmers? Surely these forests have to be avoided or removed for such activities to be successful? But here is the problem. We are so stuck in our Euro-American worldviews of what "agriculture" should be—a settled economy, fully reliant on domesticated plants and animals, extensive land clearance, and monotonous field systems—that we miss other, equally ingenious, and equally impactful roads to "food production." If we instead focus on the farming *practices* that lead to domestication, we open ourselves up to a whole new series of possibilities. Taking this approach, we can look at cultivation as human behaviors that promote the survival, reproduction, and growth of certain plants and at herding as human behaviors that shape the mobility, population, and anatomy of certain animals to provide a wider lens for viewing how our species began to insert itself into the natural world. Instead of relying on a narrow, confining definition of "agriculture," by looking at the experience and process of farming or "food production," we can gain a truly global perspective on one of the most pivotal periods in human history—one that, of course, can include some of the most biodiverse environments on the planet: tropical forests.[2]

Far from being unproductive sideshows, tropical forests display some of the earliest human manipulations of plants and animals in the name of food security as early as 45,000 years ago. These diverse habitats include montane or seasonally dry forests with penetrating light and fertile soils, where burning can be easily practiced, as well as wet rainforests. And they potentially hosted some of the earliest examples of deliberate cultivation across the terminal Pleistocene–Holocene boundary, as humans changed landscapes and planted trees with sugar-rich fruits and tubers in places they would not normally grow. We will see how archeobotanists and zooarcheologists have shown that some of the most used domesticates around the world today, produced by these cultivating and herding behaviors, actually have their origins in tropical forests. In fact, a significant portion of your weekly shopping lists comes from tropical forests without you even realizing. Eggs, cornflakes, tortillas, sugar lumps, chocolate bars, ham and pineapple pizza, pepper, cigarettes, and the food spreads or cosmetics that contain palm oil are just some of the diverse products that owe their existence to the "jungle." In some cases, tropical farming practices did take the form of "agriculture," with novel field systems leading to prehistoric deforestation and major ecological changes that provide a stark warning to us today. But in others, domesticated resources were slotted into existing, sustainable hunting, gathering, and fishing economies. Whereas we once considered tropical forests to be environments where "farming" was doomed to failure, we now know that over the last 10,000 years, human societies have modified tropical forest plants, animals, and entire landscapes the world over to provide many of the supermarket items we all take for granted today.[3]

TROPICAL FORESTS HAVE a reputation for having been some of the most "pristine" environments prior to industrialization. However, they actually have one of the longest records, if not the longest, of human modification. As early as 45,000 years ago, with the arrival of our species in the tropics and subtropics of Southeast Asia, Near Oceania, and Australia, humans were burning tropical forest ecosystems to promote grassland patches and open up forest floors. This enabled them to pick and choose a cornucopia of resources from a variety of open areas, drier forests, and denser rainforest. Some groups were even managing plants by moving them. Starchy plants, like yams, were potentially carried across the Wallace Line as early as 40,000 years ago. Meanwhile the range of plants like the sago palm, various yams, and taros across mainland and island Southeast Asia, and perhaps even Australia, by the start of the Holocene certainly has the ring of human handiwork about it. Even tropical animals were not left in peace by our early ancestors. In Near Oceania, by 20,000 years ago, seafaring people carried furry bandicoots and cuscuses (not to be mistaken for couscous!) as snacks—moving them beyond their natural homes and onto more isolated islands, such as those of the Bismarck Archipelago, where they were to thrive. Why, then, have archeologists and the public alike so often ignored tropical forests as possible crucial locations where *Homo sapiens* performed early food-production experiments? It certainly hasn't helped that many productive tropical plants are soft and fleshy, not prone to preservation over long periods, while bones often face the harsh acidity of the soil upon burial. Unlike more temperate or arid areas, where charred seeds or ample animal remains allow the observation of gradual changes from wild to domesticated features, similar detective work in the tropics has therefore certainly been at something of a disadvantage.[4]

Nevertheless, archeologists are stubborn and used to working with what they have. Archeobotanists and geoarcheologists, like Professor Tim Denham of Australian National University, have turned to their microscopes to try and identify uniquely shaped starch grains or characteristic microscopic silica structures (phytoliths) left behind by particular plants in tropical soils. While the work is difficult and prone to contamination from modern foods eaten in and around the laboratory or excavation sites, these painstaking methods can revolutionize our understanding of food production in more challenging environmental contexts. In fact, using these approaches, Tim and his colleagues think that they have found a key, early heartland of independent plant cultivation and domestication in tropical forests. This heartland not only challenges the time line of global food production but also causes us to face our own biases and assumptions as to what "agriculture" looks like—located, as it is, in the middle of a wet, sticky swamp. As Tim says, "At Kuk Swamp, in the Wahgi Valley of highland Papua New Guinea, archeological excavations from the 1970s and 1990s have allowed us to reveal one of the earliest farming landscapes found anywhere in the tropics." Marking out and excavating two hundred trenches, archeologists have uncovered a series of "phases," from between 10,000 and 4,000 years ago, that document the increasing lengths humans went to in order to modify the landscape to make it better for the cultivation of some of the world's modern tropical favorites: taro, banana, and even sugarcane. In fact, this evidence makes Kuk Swamp one of the earliest sites where farming practices by *Homo sapiens* have been observed—not in temperate or dry river valleys or grasslands but in highland, humid, tropical forest wetlands around 1,500 meters above sea level.[5]

Phase 1 of Kuk Swamp, dated to approximately 10,000 years ago, has been famous since archeologist Jack Golson identified a series of drainage ditches. Back in the 1970s and 1980s, Jack put the cat among the archeological pigeons by suggesting that these ditches were a deliberate attempt by humans to drain the local wetland and help the growth of carbohydrate-rich plants like taro and banana. This early evidence has proven controversial, with a lack of direct verification of plant cultivation, suggestions that the ditches may, in fact, be natural, and an absence of similar landscape modification anywhere else in New Guinea raising some reservations. Nonetheless, banana phytoliths were found in this phase, in a more open environment than they would usually appear, hinting that some human gardening was in progress. Not just that, but a number of environmental records from the terminal Pleistocene and early Holocene have shown an increased incidence of human burning 10,000 years ago and the opening of forest across montane New Guinea, including directly around Kuk Swamp, conveniently providing the perfect environmental storm of forest patches for bananas, open forest floors for taros and yams, and grasslands for sugarcane. Some permanent settlements have also been identified in the region, suggesting growing populations and humans investing more in particular parts of the landscape, perhaps linked to the initial cultivation of these few key, managed crops, though Tim remains skeptical about these. The jury also remains out on this earliest phase of food production at Kuk Swamp. If proven, however, it would be some of the first deliberate food production and cultivation found anywhere in the world.[6]

Modern genetic evidence has certainly shown that an independent domestication of taro, the greater yam, and bananas occurred in Near Oceania before expanding out around the world. Today, the Eumusa banana found in the region is one of the most used domesticated bananas globally. Tim and his team of "microbotanists" (named for the diminutive size of the plant parts they are working with) have provided clear evidence by Phase 2 (c. 7,000 to 6,400 years ago) and Phase 3 (c. 4,400 to 4,000 years ago) of the presence of starch grains and phytoliths of taros, bananas, yams, and precursors of sugarcane, planted in direct association with clear, planned, and extensive drains and ditches dug by the local human population, during a time when people also started transitioning to a more sedentary lifestyle. The well-thought-out maintenance of plants that liked forests and plants that liked more open grasslands would have provided a set of important food stores for growing human populations facing fluctuating climates in the cool highlands at the terminal Pleistocene–Holocene transition. It will always be challenging to try and identify exactly when these plants became "domesticated" based on these tiny traces, but archeologists around the world increasingly accept that the early Holocene plant-management strategies at Kuk Swamp show a step change in human influence on plant life cycles and growing conditions—a new relationship with some of the plants many of us regularly have in our kitchens or put in our tea today. Kuk Swamp's importance to world archeology was, in 2008, confirmed by the granting of UNESCO World Heritage Site status for the entirety of the site's remarkable 116-hectare span, making it a true giant of prehistoric farming.[7]

Figure 7.1. Present-day mound gardening of sweet potato in New Guinea.
Robin Hide, Tim Denham

Kuk Swamp is significant not just because it calls into question all of our assumptions about the timing and location of "agricultural" origins but because it also challenges our assumption about what the earliest farming societies looked like. The cultivation of bananas, tubers, and sugarcane at Kuk was not the rapid product of a single "inventor" or "inventors." Instead, it was the end point of a long process begun by the first humans in Near Oceania who burned forests to maintain profitable mixtures of resources. Certainly, this reached new levels at Kuk, but it did not happen overnight and had a long, tropical trajectory. Furthermore, the development of cultivation at Kuk did not lead to sweeping changes. Not everyone instantly took to cultivation, and only later in the Holocene did it become a permanent fixture of New Guinea economies. With a team of researchers, I have worked at the site of Kiowa, in the same environments and at the same altitude as Kuk Swamp, to show that human populations just over one hundred kilometers away continued to hunt possums, cuscuses, and tree kangaroos between 12,000 and 500 years ago, even as groups at Kuk were developing and intensifying "food production." Evidently, then, not everyone saw the increasing tending of specific plants as the revolutionary, world-shattering idea we imagine it to be today. Looking on at the new experiments being undertaken at Kuk, some people, in one of the next valleys over, were more than happy to continue to shape forest environments more broadly through burning, to gather nuts, and to hunt forest prey in a manner that, as we saw in Chapter 6, had enabled them to gain their first footholds in this tropical region. We often assume that farming was an inevitability, something humans grappled to attain once discovered. Here, in tropical New Guinea, however, it was just one more way of managing landscapes that had reliably provided for humans for thousands of years.[8]

FROM THE EARLY Holocene (11,700 to 8,200 years ago) onward, a number of different societies began to perform farming practices in tropical forests. These interventions were to result in some of the planet's key dietary "staples." Today, maize, or "corn," is the most produced grain in the world, and it has long provided the basis for many of the increasingly dense and complex societies that appeared across the Americas during the Holocene. Although it is often portrayed as an ideal arid-adapted crop, one that might sustain us to the brink of drought-filled doomsday scenarios, microbotanists, like Professor Dolores Piperno of the Smithsonian Tropical Research Institute and Smithsonian National Museum of Natural History, have shown that its origins actually lie in the warm, wet seasonal tropical forest habitats of Mexico. Traditionally, maize domestication has been dated to 6,000 years ago based on charred cobs found in ideal preservation conditions in the semiarid highlands of Oaxaca alongside squash, gourds, and beans. "However, these cobs almost certainly do not represent the first cultivation of maize, and similar remains are simply not preserved in the wetter areas its wild ancestor is known to inhabit," Dolores tells me. Instead, applying the same microbotanical approaches that were performed at Kuk Swamp, Dolores and her team discovered earlier evidence for maize cultivation 9,000 years ago in the tropical, humid Balsas River valley, in agreement with recent modern genetic evidence for the timing and geographical range of maize domestication. Maize is not, however, the only important modern domesticate the tropical forests of the Americas had to offer.[9]

The stifling conditions of the wet tropical rainforests of the Amazon Basin were once thought to have made any type of farming activity absolutely impossible prior to colonial and industrial clearance. Indeed, past human impacts on the Amazon rainforest have generally been thought to have been minimal. Nevertheless, starch, phytolith, and genetic research has now shown that the significant root crop manioc, or cassava, was first cultivated and domesticated there, in a process that began as early as 10,000 years ago. Falling just shy of the potato, a domesticate that itself emerged in the montane tropical environments of the Andes, manioc is today the second-most important root crop globally, feeding growing populations across Latin America, Africa, and Asia. The importance of the Amazon for prehistoric food production does not stop there, however, as it was also the site of the domestication of pineapples, avocadoes, and peanuts during the middle to late Holocene. Chili peppers were also most likely domesticated just south of the Amazon Basin in drier tropical forest and woodland habitats. Meanwhile, cacao was probably domesticated, or at least semidomesticated, in the Amazon Basin (although the Central American tropics have also been suggested as the geographical origins for this key ingredient of the chocolate that many of us cannot do without today).[10]

Sweet potato, an increasingly popular food in trendy restaurants across Europe and North America, was also cultivated in the American tropics, somewhere between Colombia and Venezuela in the south and Mexico in the north. Although poor preservation of yet another squishy but rich tuber has made tracking down its origins challenging, starch grains of sweet potato, from residues of tools interpreted as having been used for plant processing, have now been dated to 7,700 years ago in the montane tropical forests of Colombia, and firm data exists for domesticated sweet potato 4,500 years ago in Peru. By 1,000 to 500 years ago, this root crop was also being cultivated not just across Latin America but throughout the Pacific among the precolonial societies of the Cook Islands, Rapa Nui (Easter Island), and Aotearoa (New Zealand). It is also worth noting that prehistoric societies across the tropics of the Americas also manipulated many wild, semidomesticated, or domesticated trees. The modern genetics and distribution of the hulking Amazon nut (Brazil nut) tree, for example, has been shown to map closely onto ancient archeological settlements, suggesting that humans carried and planted protein-rich Amazon nuts in a form of sophisticated agroforestry. Peach palm was also domesticated somewhere in the Amazon Basin. Finally, beyond foods, per se, the Neotropics have also produced some more "recreational" crops. The Madeira River valley in the southwestern Amazon Basin was the origin of one of the major sources of modern crime across Latin America and, indeed, globally: coca (the constituent of cocaine). It also witnessed the domestication of the tobacco plant that fills many a cigarette or pipe around the world today.[11]

The Llanos de Moxos area in the Bolivian Amazon perhaps best encapsulates the complete shift in understandings of the tropical rainforests of the Amazon Basin as centers of early plant cultivation. Today this sweeping region, characterized by a highly seasonal climate, sees significant portions of its "savannah-like" landscape flooded for much of the year. However, all is not as it seems. When a team of scientists led by Dr. Umberto Lombardo from the University of Zurich cast their eyes over high-resolution satellite data for the area, as well as computer-based models of how the terrain differed over space, they identified some unusual alterations to the surface of the land. Together they pinpointed over 6,600 sites that they call "forest islands"—raised, forested patches that, when visited by archeologists, also revealed clear traces of human presence and enrichment of soils with their waste. These human-made mounds would have provided areas safe from the annual flooding—areas put to good use by past human populations. Publishing in the renowned journal *Nature*, Umberto and his colleagues demonstrated that the soils of these "forest islands" revealed a long record of human manipulation of plants. Heart-shaped manioc phytoliths dated to 10,350 years ago, while spherical squash phytoliths dated to 10,250 years ago. Whether these plants were fully domesticated by this time or whether this was just the beginning of a long process to produce crops that act as staples for populations of the Amazon Basin today remains unclear. However, as study coauthor Professor José Iriarte of the University of Exeter argues, "This evidence, alongside the tropical forests of south-western Mexico, undoubtedly means that this region should be considered, alongside the heavyweights of China and the Middle East, in global discussions of plant cultivation and domestication following the last glacial period."[12]

Turning our attention to the tropics of the "Old World," South and Southeast Asia have also produced a number of domesticated plants and animals found in many of our kitchens today. In the humid state of Odisha in India, a mixture of environments, from sweeping riverbanks to tropical forests, seemingly provided the setting for humans to locally domesticate Indian varieties of rice, millet, gourds, and types of cucumber. Okra, a plant familiar to anyone who has eaten a number of native Indian dishes, was also domesticated in this part of India. Delicious, succulent mangoes originated as a tree crop in the tropics of Southeast and South Asia. Meanwhile, black pepper, which flavors a vast variety of cuisines and dishes around the world, from morning eggs to spicy "curry" dishes, owes its domestication to cultivators inhabiting the tropical coastlines of Kerala in southern India. Cinnamon, another crucial spice for budding and professional chefs alike, appears to have been cultivated slightly further south, in Sri Lanka. The first mention of this spice 5,000 years ago indicates that local merchants exported it to ancient China and, later, to the classical Mediterranean world. It even makes an appearance in the Bible. Finally, although the exact origins of domesticated

Figure 7.2. Forest islands at Llanos de Moxos, Bolivia. *Umberto Lombardo*

citrus precursors of the lemon remain hazy, and citrus fruits are notoriously difficult to keep track of archeologically, it seems most likely that they were domesticated somewhere in Southeast Asia, within their wild range of growth. Although, in all of the above cases, dating the exact arrival of a "domesticated" form of these plants is immensely challenging, relying on rough, modern genetic "clocks" and fragmentary remains left in tropical soils, we can posit that much of our food and drink would taste very different without these tropical bounties shaped by prehistoric human hands.[13]

Tropical forests may also seem unlikely homes for domesticated animals, but they played a hand in the appearance of two of the animals most relied upon by farmers around the planet today. Humans use water buffaloes more than any other domesticated animal, with the large populations of Asia relying on these hulking beasts for milk, yoghurt, and cheese. The two "types" of these slow but strong beasts, the "river" and "swamp" buffaloes, today wallow in waterlogged settings flanked by tropical and subtropical forests, consuming freshwater plants. Although their origins remain obscure and they have received little attention compared to Euro-American cattle, deliberate human herding of these bovids seems to have started around 5,000 years ago in India and 4,000 years ago in Southeast or East Asia. The chicken represents perhaps a still more secure tropical forest domesticate. Humans have morphed this bird into one of the most common animals on the planet today, with a population of over 25 billion. While you would not know it to look at them, they are descended from the red "jungle" fowls that have long stalked the tropical forests of Asia. A recent genetic study, based on the genomes of 863 chickens and a sampling of all species of wild jungle fowl, shows that the domestic chicken emerged from a subspecies of jungle fowl that lives in the tropics of southern China, Thailand, and Myanmar. After their domestication, these poultry were moved throughout Southeast and South Asia, further interbreeding with other jungle fowl species. While its history is complex, the chicken, which forms the basis of many a fast-food meal, sandwich, or Sunday lunch today, undoubtedly originated in the tropics.[14]

ALL OF THE above examples refer to the domestication of plants and animals *within* tropical forests, as hunting-and-gathering communities living in these environments gradually increased their involvement in the natural world to become food producers. In many cases, as at Kuk Swamp, this meant that domesticated plants and animals supplemented existing consumption of wild plants and animals as part of diverse economies that operated on the basis of clear local knowledge of tropical forest life cycles and tolerances. In the Amazon Basin, for example, prehistoric farming practices represent a sophisticated early form of agroforestry. "Gardens" of manioc and, when it arrived, maize were cultivated at the same time as managed, but intact, forests were used for fruits, nuts, and wild animals. Wild Muscovy ducks and freshwater turtles may even have in some way been "herded" as part of these varied approaches to landscape management. Although, as we will see later, this increasingly intensive food production shaped the soils, geography, forest structure, and biodiversity of the Amazon Basin, it seems to have had a relatively limited impact on overall past tropical forest cover. Similarly, diverse and sustainable strategies of food production have also been documented in early and middle Holocene New Guinea and island Southeast Asia. But what happened in prehistory when farming, particularly in the form of field-based agriculture, was introduced *into* tropical forests from the outside? Did it lead to massive deforestation, landscape degradation, and cultural shifts, as it does today, as people tried to force their way of life onto these environments? Or did past societies manage to adapt external practices to local situations?[15]

A classic example of a more "agricultural" introduction into tropical forests is the spread of the type of rice most common around the world today. Genetic and archeobotanical research suggests that rice emerged in the Yangtze River basin in southern China around 9,000 years ago from a wild, wetland grass. From here, the *Oryza sativa japonica* rice sub-species diversified and moved both north, to the edge of its lowermost temperature tolerance, and south, toward the tropical forests of South and Southeast Asia. By about 4,000 years ago *Oryza sativa japonica* had reached India, adding to the indigenous forms of rice (*Oryza sativa indica*) already cultivated on the Ganges floodplain since 5,000 years ago, as well as mainland and island Southeast Asia. Here, rice, although deriving from a wild subtropical grass that requires significant rainfall, also must grow in the open, seemingly necessitating forest clearance if it is to be cultivated. This is particularly the case as more intensive "wetland" farming techniques were applied to rice from around 6,000 years ago in China in order to harness its energy potential, with terracing and rice paddies requiring significant alterations to the landscape and year-round waterlogging. Foxtail millet was also domesticated in China and became rice's partner in crime as it moved southward as part of a package. As this crop generally prefers drier, open conditions, it would have placed further demands on cultivated land, particularly in the tropics.[16]

Unsurprisingly, the expansion of these crops into the tropical forests of Southeast Asia has generally been linked to large-scale forest clearance, resulting in the dramatic landscapes, such as the UNESCO-protected Ifugao rice terraces of the Philippines, that tourists flock to see today. Nevertheless, pinpointing their arrival has been challenging given that miniscule rice and millet grains are rarely preserved in archeological sites, and the diligent sieving required is only a fairly recent archeological practice. The earliest archeobotanical evidence for rice in mainland Southeast Asia dates to 4,000 to 3,500 years ago in the coastal portion of Thailand, with rice and millet reaching further inland by 3,000 years ago. In island Southeast Asia, the first evidence for rice dates to roughly the same time in Borneo, although some of this evidence is based on impressions left by pieces of rice in pottery. In areas with drier, seasonal tropical forests, there is evidence of burning to clear the way for these new, productive crops. That said, destruction was not guaranteed. In the lowland evergreen rainforests of mainland and island Southeast Asia, rice and millet farming was often combined with the ongoing cultivation of yam and taro tubers, as well as starchy sago palms, in smaller, more open forest patches. Alternatively, farming was ignored altogether by forager communities more interested in harnessing the local productive resources than investing energy to clear forests that had long provided them with shelter and food.[17]

Figure 7.3.   Image of the UNESCO-protected Banaue rice terraces in Ifugao on the island of Luzon in the Philippines. These terraces likely have ancient precursors in the region that may have been used for the growth of taro prior to rice (see also Chapter 10).   *Getty Images*

The deep forests of Central Africa are another important setting where we can explore what happened when incoming agricultural strategies adapted to local tropical conditions. The arrival of farming in this part of the world has often been linked to the so-called Bantu expansion, a migration of populations from West Africa that all speak languages within the same Bantu family. These people are thought to have brought the crop pearl millet with them as well as iron technology, sweeping southward through the African continent, all the way to South Africa, during the late Holocene (4,200 years ago to the present). Pearl millet is an arid-adapted grass that appears to have been domesticated in the river valleys on the southern edge of an expanding Sahara (just after its last "green" phase) around 4,000 years ago. It then expanded south, moving with Bantu-speaking populations and reaching northern Ghana by 3,500 years ago, the rainforests of Cameroon by 2,500 years ago, and the Democratic Republic of the Congo in the last 2,000 years. Due to assumed vulnerabilities of this crop to warm and wet conditions, scientists have associated its appearance with the dramatic late Holocene "rainforest crisis" in Africa around 2,500 years ago, when lowland rainforest seems to have rapidly retreated in favor of more open settings. Some scientists believe this change was a climatic phenomenon; others think it was actually caused by iron-wielding, Bantu-speaking farmers who indiscriminately cleared the landscape to promote the production of their favorite food source.[18]

Growing evidence suggests, however, that disastrous forest clearance need not have accompanied the arrival of pearl millet in the Central African tropics. Scientists undertaking experimental farming today have shown that this crop can be effectively grown in wet, humid plots next to rainforest. Bantu-speaking communities and pearl millet also continued to move and occupy different tributaries of the Congo River long after the end of the supposed crisis. Interestingly, the first pearl millet in Ghana is joined by plants like oil palm and yams, which had their origins in African tropical forests, suggesting a more mixed strategy. As we will see in Chapters 13 and 14, oil palm is now part of a global economy that fuels products as diverse as nut-based chocolate spreads and face creams. Local contributions also seem evident in the use of groundnuts, castor beans, and gourds, which were also domesticated in the tropics of Africa and continued to provide important food sources from these environments. In the absence of evidence for pre-Bantu communities, at least in the western Democratic Republic of the Congo, these combinations of resources seem to show rapid farming ingenuity and local adaptation on the part of new human arrivals. In a recent study I was involved in, working closely with a prolific team at the University of Cologne who had undertaken decades of fieldwork in the Congo Basin, we were lucky enough to study Bantu-speaking populations moving into the far, forested reaches of the region. Here, the earliest arrival of pearl millet occurred as part of a diverse strategy that included heavy reliance on freshwater fish and the use of wild and domesticated forest plants. Similar "mosaic" strategies of food production continue to be sustainable in this part of the Congo Basin, showing that there is no necessary, single path to farming in tropical forests, even for incoming populations.[19]

THE DILIGENT WORK of archeobotanists (especially microbotanists), zooarcheologists, plant geneticists, and archeologists has demonstrated that tropical forests are crucial sites for scientists studying the earliest prehistoric farming practices on a global scale. It is in tropical forests that humans apparently first began to interfere with habitat structures, plant growth, and animal distributions. It is also in tropical forests that some of the earliest examples of farming practices occurred. Next time, before you put sugar in your coffee, before you peel a banana, before you apply moisturizer to your skin, or before you tuck into some sweet potato fries, remember just how many of our modern tastes and styles are fed by human activities in these environments that occurred throughout the Holocene. The "Fertile Crescent," the dry highlands of Mexico, the rivers of North America, the great valleys of China, and the Saharan border of Africa have all proven key centers of domestication and agricultural origins, particularly in terms of animals and crops farmed in the way we think farming should happen from the view out of our Euro-American bubbles. However, tropical forests also witnessed some of the most significant and effective examples of food production as prehistoric societies harnessed the huge diversity of species these environments had to offer. This is particularly the case when we move away from the popular myth that tropical forests are solely characterized by dense, light- and soil-poor rainforests and look at the diversity of more seasonal forest environments that hosted some of the earliest examples of tropical food production discussed above. The changes to the landscape might not have been obvious to us, but they were certainly real.

In fact, considering the issues of land-use sustainability we face around the world in the twenty-first century, tropical forests give us a remarkable lens for viewing some of the most ecologically savvy of farmers there ever were. Locally domesticating plants and animals, early tropical farmers managed these forms of food production within forests that were by and large left standing and their wildlife left abundant, although their structure and species distributions may have been altered. Even some past farmers who moved into tropical forests with domesticated plants and animals from elsewhere, which often required more open, drier conditions, developed sustainable, mixed approaches that combined a wealth of local, wild tropical resources with these useful new species. These examples provide food for thought when we think of some of the disasters that governments, companies, and even our own choices wreak on these environments today in the name of "development," "infrastructure," or "productivity." The past shows that tropical forests can be used and adapted, even within frameworks of farming, without being removed forever. However, some prehistoric societies certainly did remove the forests, literally clearing the way for types of "agriculture" that would have felt more familiar to us. This could have major consequences in the form of destabilized landscapes and soil erosion, collapses in biodiversity, and increased vulnerability to extreme weather events. This was particularly the case when new desires for "food production" washed up on the shores of more isolated, tropical islands.

Chapter 8

ISLAND PARADISES LOST?

We are used to seeing tropical islands, like the forests they are home to, as untouched "deserts" or blissful paradises. Whether you are a rum-sodden Jack Sparrow, an inventive Robinson Crusoe, a delivery man with a volleyball, or a troubled Jack Shephard trying to lead a ragtag group of "lost" lottery winners, conmen, artists, and rock stars, tropical islands represent an inescapable nightmare of being stranded in maddening isolation. Reality shows like *Survivor* often use tropical islands as the setting for testing the survival skills, social interactions, and morals of twenty-first-century societies. For many of us, they are the dream, but perhaps impractical, holiday: clean blue oceans, fine white sand, and quiet palm trees—places where we could finally "get away from it all." For some sickeningly wealthy people who can afford to actually buy these landmasses, this is already a heavenly reality. Either way, it is probably a good thing that we widely view these tropical islands as arenas generally inaccessible to humans—especially as we tend to also stereotype them as being biologically and ecologically fragile. They are often small, so their soils and forests can change rapidly across much of the island surface in the face of external threats. While they are home to high levels of plant and animal biodiversity, unique species are often present in low numbers. Plants often have limited capacities to disperse their seeds widely. Meanwhile, isolated island animals frequently have few defenses against predators and competitors. Close bonds between species also mean that a cascade of extinctions can occur should one begin to dwindle. Together, this all means that island ecologies have been seen as particularly vulnerable to the invasion of new, competing species or predators, including *Homo sapiens*.[1]

These perceptions of islands have also pervaded academic thought. In archeology and anthropology, island ecosystems have often been contrasted with their mainland counterparts as uncompromisingly sensitive to past human actions. Think of, for example, the biggest celebrity of tropical island extinctions, the dodo of Mauritius, which disappeared in the seventeenth century AD shortly after the first permanent European settlements were constructed on this Indian Ocean island. On the one hand, seafaring hunters have been seen as decimating any island animal life encountering our species for the first time. Without protective strategies, and with no hope of escape or reinforcements from distant shores, mammals, birds, amphibians, and reptiles could very quickly be erased from history. On the other hand, islands have been framed as inherently "resource poor," with human societies only able to succeed by bringing new, domesticated plants and animals along for the ride. These same food producers faced further, if different, issues of sustainability, however. Limited island soils were prone to erosion and the loss of nutrients. Smaller forests, if not given time to regrow, could also rapidly disappear, leaving a barren outlook for settlers reliant on construction materials and firewood. Sedentary farming societies also often, intentionally or unintentionally, brought along hitchhikers, like dogs, pigs, and rodents, that wreaked their own havoc on local plants and animals. Ultimately, in both cases, tropical islands have been seen as particularly prone to forces of human "destruction." The collapse of island environments, as well as their prehistoric human societies, must have been inevitable. After all, the increasingly expanding infrastructure, settlement, and profit-driven agricultural strategies of our own societies are unleashing irreversible changes on these habitats today.[2]

Nevertheless, if we want to properly study how past humans negotiated the colonization of uninhabited tropical islands, and if we want to determine the degree to which their increasingly intensive interaction with the natural world had impacts on island ecosystems, compared to human-driven changes seen in the twenty-first century, we need hard data. We also need to strip away our own value judgments of which activities and changes are inherently "good" and "bad." Particularly, given what we have seen in Chapter 7, we need to take a step back from assumptions that all forms of "food production" look like the monotonous, tree-clearing agricultural fields and pastures we see today. Instead, we can investigate islands as places where people experimented with the introduction of new plant and animal species, alongside the foraging, fishing, hunting, and managing of the available natural resources. Focusing on some of the most classic of our island paradise fantasies, across the Caribbean, the Pacific, and the coastlines of Africa, we will now delve into the blossoming subfield of island archeology. We will see how archeobotany, zooarcheology, and modern and ancient genetic analyses are providing insights into how humans constructed new island "niches" for themselves, transporting landscapes, as well as domesticates, from mainland settings. We will accompany scientists working in some of the most remote regions on Earth to look at how environmental records of landscape change, from lakes, swamps, and archeological sites, can show the changing scale and intensity of the subsequent impacts of different practices on local or "endemic" island plants and animals, as well as the soils and forests these populations came to depend upon for cultivation and raw materials. Certainly, difficult situations arose, sometimes in settings many hundreds or even thousands of kilometers away from the next human population. Yet adaptable oversight and innovative combinations of local resource use and domesticated plants and animals could prove remarkably resilient in what have often been considered the most delicate of tropical environments. All it needed was some good old-fashioned on-the-ground understanding of local ecologies—something that Euro-American societies are often disturbingly lacking in today.[3]

WE START OUR island circuit in the Caribbean, a series of seven hundred islands well known for their tropical vegetation, spotless beaches, diverse Indigenous cultures, and film-famous buccaneers. Surrounded by the Caribbean Sea and North Atlantic Ocean, just east of North, Central, and South America, this region has been occupied by a variety of hunting and farming communities over the course of the last 8,000 years, providing ideal "laboratories" for observing how different human economies and forms of social organization can impact tropical islands. Professor Scott Fitzpatrick of the University of Oregon is well and truly an island archeologist and has had the "arduous" task of excavating in the Caribbean for much of his academic career. The envy of most of us, Scott has made his name exploring how oceangoing human societies colonized, adapted to, and impacted islands across a region many would pay handsomely to visit just once. "Beyond being a wonderful location to work, the Caribbean is ideally placed to test whether the combination of humans plus islands inescapably equals ecological catastrophe," he says. This issue is particularly pressing for communities and governments across this part of the world given that deforestation, threats to unique plants and animals, and the degradation of coral reefs all run rampant in the twenty-first century. In fact, entire marine and land-based ecosystems are expected to disappear in many parts of the Caribbean within the next twenty years if action is not taken to curb pollution, global warming, overfishing, and tourism. Are these modern phenomena? Or can we perhaps learn from ancient precedents?[4]

The first human inhabitants of the Caribbean arrived via two different entry routes. The so-called Lithic group potentially came from Central America, arriving in the Greater Antilles (the northern chain, which includes Cuba and Hispaniola) by c. 6,000 to 5,000 years ago. The other, "Archaic" group moved from South America, arriving first in Trinidad 8,000 years ago before moving into parts of the Greater Antilles, the northern Lesser Antilles, and at least two islands in the southern Lesser Antilles, Barbados and Curaçao, between c. 6,000 and 3,000 years ago, and eventually expanding throughout the Caribbean. These groups were mixed hunter-gatherer-fishers, making tools from stone and, in the case of Archaic groups, relying heavily on coastal resources, as revealed by large mounds of human-accumulated shells (or "shell middens") located on the coast, which likely sustained increasingly settled populations. While we are unsure of their population sizes, these Archaic populations certainly left their mark. An extensive survey of environmental cores obtained from wetlands and lakes across the southern and eastern Caribbean indicates that these groups were active horticulturalists that cleared forests, perhaps using fire from their earliest arrival in order to manage environmental variation, though this remains disputed. Certainly in Trinidad, a present-day island that would have been connected to the South American mainland during the early Holocene (11,700 to 8,200 years ago), the opening up of patches of forest aided the growth of South American domesticates, such as maize, sweet potato, and chili peppers, which have now been found at Archaic sites from as early as 7,700 to 6,000 years ago.[5]

These land-altering Archaic populations have been implicated in the extinction of native fauna in a region that has witnessed more Holocene (11,700 years ago to the present) mammal extinctions than any other. For example, the deliberate human burning of forest in Antigua may have stimulated the disappearance of three endemic small bat species from this island. Direct hunting of slow-reproducing large mammals, like native primates and the giant ground sloth, is also thought to have led to the extinction of species that had survived in the Caribbean well into the Holocene. Their disappearance, following Archaic arrival, certainly piques the interest of archeological investigators. Nevertheless, this remains debated, and there are few sites with clear evidence of hunting by these earliest settlers, with a combination of early Holocene climatic variability, human-induced habitat changes, and perhaps limited hunting more likely combining to bring these animals down in a more gradual manner. In fact, these early Caribbean populations were remarkably in tune with their new island environments. As well as domesticates, they moved well-suited native wild species between the different islands, including wild avocado, wild fig, and rodents. They cleared forests, not just to make way for domesticates, but also to promote nutritionally useful native taxa such as wild plantain, yams, the starchy roots of maranguey, arrowroot, and the seeds of certain palm trees. On Puerto Rico they even seem to have modified soils to promote fertility by 3,000 years ago.[6]

The later "Ceramic" period of human expansion across the Caribbean seems to have brought an intensification of certain types of activities along with it. Beginning 2,500 years ago in Puerto Rico and the Lesser Antilles, prior to spreading out across the rest of the Greater Antilles by 1,500 years

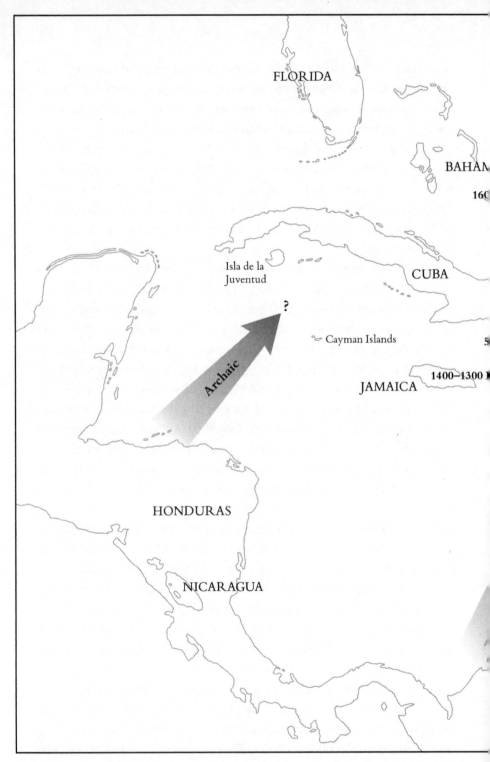

Figure 8.1. Map of the Caribbean showing the main waves of human colonization of differen islands, with dates (the arrows show routes of dispersal that would have required seafarin The first human inhabitants of the Caribbean ("Archaic") are thought to have arrived fro South America, making it to Trinidad 8,000 years ago before reaching the Greater Antille

ATLANTIC OCEAN

Turks & Caicos

Great Inagua

GREATER ANTILLES

HISPANIOLA

British Virgin Islands

US Virgin Islands

Vieques

Ceramic

Anguilla

St Martin

Leeward Is.

?

?

PUERTO RICO

St Croix

5000–2500 BP

St Kitts & Nevis

Barbuda

Antigua

Guadeloupe

Caribbean Sea

Montserrat

Dominica

2500 BP

Martinique

LESSER ANTILLES

St Lucia

Aruba

Bonaire

Curaçao

Los Roques

Margarita

St Vincent & Grenadines

1700 BP

Carriacou

Grenada

Barbados

5000 BP

Tobago

Archaic

Ceramic

Ceramic

8000 BP

Trinidad

COLOMBIA

VENEZUELA

the northern Lesser Antilles, and Barbados ~6,000 to 3,000 years ago. A later "Ceramic" period of human expansion from South America, which reached Puerto Rico and covered the entire Lesser Antilles, occurred around 2,500 years ago. Here, BP stands for "calibrated years Before Present" (the standard format for radiocarbon timescales). *Scott Fitzpatrick* ago, this wave of Early Ceramic populations, with an apparent genetic link to human populations in the Amazon Basin, involved the ancestors of the Taíno Indigenous group of the Caribbean. As the name suggests, these populations introduced pottery to the region, although others have suggested it had already arrived with preceding Archaic groups. They also introduced dogs and guinea pigs across its islands and moved the agouti (a sort of long-legged guinea pig), peccary (a piglike mammal with hooves), opossums, and armadillos out into the Lesser Antilles. It was also during this wave of migration, archeobotanists can show, that South American crops truly took hold across the Caribbean, and maize, beans, manioc, peanuts, sweet potato, tobacco, chili peppers, and guava can be found across the islands, fueling the establishment of settlements that were increasingly distributed across island interiors as well as the shoreline, unlike their more coastal-focused Archaic predecessors. The extensive farming, clearance, and horticulture practiced by these Ceramic groups had visible impacts on Caribbean ecosystems. By the time Europeans arrived, Indigenous populations were cultivating nearly one hundred species of plants; they had cleared forest in many island areas, transformed hillsides with terraces, and formed mounds for growing manioc.[7]

Unsurprisingly, this more intensive occupation and settlement would have significant impacts on local island ecosystems. In Puerto Rico and the Virgin Islands, exhaustion of local crab resources stimulated a switch to consumption of mollusks, which were in turn overharvested in Jamaica from 1,000 years ago. On the island of Puerto Rico, zooarcheologists have shown a reduction of fish body size and a decline in the length of reef food chains at Ceramic sites over time, which indicates overfishing. Evidence for the consumption of native amphibians, reptiles, seabirds, and rodents, often in unsustainable fashion, is also clear at a number of sites across the Lesser Antilles. These issues are visible in many parts of the Caribbean today, particularly in relation to the overuse of reefs and forest clearance for agriculture. Nonetheless, the degree to which Ceramic populations led to sweeping and permanent change remains unclear, especially as estimating their population sizes is difficult. Many small animals being used by these groups, such as now threatened tiny "hutias" and other mouse-like rodents, survived until after the arrival of European colonizers in the Caribbean. On the islands of Anguilla and Nevis, precolonial populations were sustainably, albeit intensively, harvesting marine mollusks. We will never know how the Ceramic populations would have continued to fare on their island homes. Indeed, it is only with the spread of disease and atrocities perpetrated by European colonizers, which resulted in mass mortality among the Indigenous populations, that we begin to see a clear and complete disruption of long-term human-environment interactions and the onset of profit-driven extraction of resources and labor that we will turn to in Chapter 10.[8]

IN CONTRAST TO the multiple waves of human cultures with different economies moving into the Caribbean, the human colonization of the Pacific Islands by people speaking languages within the Austronesian group has often been considered a more sweeping, classic example of a migrating human population with a uniform, introduced farming economy. Ancient DNA of human skeletons from the isolated Near Oceanian and Polynesian islands of Vanuatu and Tonga has shown that their earliest human occupants, buried with so-called Lapita pottery, descended from populations in East Asia. Moving from Taiwan, human populations, with this characteristic "dentate" style of pottery, reached the Philippines by approximately 4,000 years ago. True Lapita pottery, tempered from sand and shell, then appears on the Bismarck Archipelago by approximately 3,350 years ago before moving through the Solomon Islands, Vanuatu, and New Caledonia, reaching Fiji, Samoa, and Tonga by approximately 2,700 years ago. From that point there was something of a pause, with around 2,000 years passing before Polynesian descendants of these early populations went on to colonize Rapa Nui (Easter Island), Hawai'i, and Aotearoa (New Zealand). These expert seafarers navigated thousands of kilometers of open water, using sophisticated outrigger canoes like those made globally famous in recent times by Disney's *Moana*. In many cases, they were the first human feet ashore, and what they brought with them could have major consequences for ecosystems, particularly on islands where there was no easy trip home and adventurous exploring groups could quickly become cut off.[9]

Despite its common association with the expansion of the Austronesian language groups across island Southeast Asia, there is no direct

evidence for domesticated rice at any early Lapita site in the Pacific Islands. Instead, one of the main early domestic arrivals on these island
outposts is an animal. Modern and ancient DNA evidence shows that
domesticated pigs made the journey with Lapita-carrying populations
out across the Pacific. When we see these animals rummaging around
human settlements or farms, we do not often consider that they can actually have widespread environment-altering impacts. Today, feral pig
populations on Pacific Islands reduce rare, native tropical plant biodiversity as a result of their indiscriminate foraging, and they encourage
invasive species to take root by reducing local competition and dispersing seeds through their digestive tracts or by carrying them on their skin.
The foraging and trampling of pigs can also increase soil erosion, and
scientists have invoked pigs, which are immensely important to many
Pacific Island cultures and economies today, as a cause of increasing
landscape instability on many islands following the arrival of humans.
Although without rice, these colonizing populations also carried horti-
cultural "transported landscapes," using fire to open forests and promote
the growth of other introduced plant domesticates such as taro, yam, and
banana. Deliberate deforestation for houses, canoes, pigpens, and other
structures led to further landscape changes. On Tonga, Vanuatu, and the
famous Rapa Nui with its haunting statues, tropical forest tree species
declined in abundance following human colonization, further exacer-
bating soil erosion as root systems were dragged away. Zooarcheologists
have also observed heavy human pressure placed on reef resources by
these nautical maestros, particularly in later periods as settled populations expanded.[10]

The spread of Lapita pottery across the Pacific Islands also brought
with it one of the most prominent examples of an island-altering ecological stowaway. The Pacific rat is commensal to humans, meaning
that it can rapidly become entwined with human settlements, relying
on human food and traveling closely with human groups where it is
able. Archeological and genetic evidence shows that during the expansion of Austronesian-language-speaking populations, the Pacific rat was
transported onto virtually every single island in the Pacific. The results
were often game changing for island ecosystems. These rats were large,
fast-breeding, omnivorous, and numerous, used to feeding off human

scraps. Alongside direct hunting by humans, rats posed a major threat to land and sea birds given their particular fetish for the eggs of endemic birds. Unable to reproduce in safety and unused to such menace, as many as 2,000 species of bird went extinct following human occupation on tropical Pacific Islands. These feasting rodents may also have driven the extinction of native terrestrial snails and insects. Rats, gnawing on seeds and nuts, have also been implicated in declines of local plant communities, including the reductions in forest cover witnessed on Rapa Nui. As this latter example has often been used to show, intended and unintended human introductions to Pacific Island ecosystems could together lead to the rapid transformation of island forests into landscapes with no trees, reduced diversity of land-based wildlife, slipping soils that posed obstacles to food production, and declining fish and shellfish. On islands separated from other landfall by as much as 2,000 kilometers, human population collapse could become all but guaranteed.[11]

Although this dramatic "doomsday" scenario sells books, it also glosses over a number of instances in which arriving farmers and horticulturalists tuned their strategies to their new island homes. Dr. Monica Tromp of the University of Otago has applied the latest scientific methodologies to human remains from the island of Vanuatu to revolutionize our understanding of the diets of some of the earliest Lapita producers. As we have seen, human colonization of much of the Pacific is often characterized as involving a complete overhaul of island ecosystems. "However, by studying microscopic remains of plants trapped in the gunk between ancient peoples' teeth we gain another perspective on the region's earliest settlers," says Monica. Instead of domesticated plants, Monica found that the vast number of preserved plant particles in this ancient tartar came from wild rainforest trees. This is direct evidence that tropical forests not only supplemented but often also dominated the diets of the first settlers of Vanuatu. While cultivated crops were also introduced, Lapita-carrying seafarers knew the value of tending to the existing wild resources. This story is not unique to Vanuatu, and the microscopic remains of rainforest trees have also been documented at Lapita sites elsewhere in Near Oceania as well as Micronesia. Indeed, many signals of "clearance" and burning may also indicate the promotion of local useful plants, as much as incoming domesticated crops and animals, as populations shaped local forests to their liking. Elsewhere, the use of wild animals such as fruit bats, turtles, cuscuses, rodents, lizards, frogs, snakes, and crocodiles, some (albeit relatively few) of which survive to this day, has also been documented. Meanwhile, zooarchaeological evidence for the successful, diverse use of marine resources, from different ocean habitats, has provided a more sustainable, island-specific insight into fishing practices.[12]

In fact, it should not be too surprising, given the sheer environmental diversity that spans the Pacific Ocean, but Lapita-carrying (and later Polynesian) settlers were clearly flexible as they felt out their new, yet also familiar, surroundings. They managed local and introduced tree crops, foraged for plant resources and hunted local animals, and introduced new domesticates into tended gardens. The fact that these populations inhabited this vast oceanic space, from the Bismarck Archipelago to western Polynesia, in just a few hundred years shows that they were able to ingeniously and rapidly adapt to a variety of island settings. Moreover, the paucity of remains of introduced domesticates like pigs, as well as chickens and dogs, in early Lapita sites may show the difficulties (or choices made) when establishing stable domesticated communities across this diverse space. In Near Oceania, in and around New Guinea, for example, Lapita settlements are located on the coast or on small islands, tuning in on marine resource use, perhaps as a way of avoiding the existing occupants of this region, who had been using inland resources since the Pleistocene (c. 45,000 years ago). Further east, on remoter islands, Lapita sites are located some distance inland, and it is increasingly evident that they managed and used forest environments at the same time as investing in opened spaces for cultivation and pig penning. The type of island could also make a difference. Volcanic islands, like those of Hawai'i, could provide productive waters for fishing and, where large enough, nutrient-rich soils for cultivation. Indeed, archeological research on this tropical island chain has shown the ingenuity of past Hawaiian field systems that adapted to grow different cultivars and use different strategies on "leeward" (facing away from prevailing winds and often drier) and "windward" (facing toward prevailing winds and often wetter) sides of ridges. Complex systems of ancient aquaculture, in the form of stone-constructed pools, were also developed in Hawai'i to manage marine resources. Meanwhile, tiny coral atolls, like Pitcairn Island, had poor soils but vast reefs for fishing. Raised limestone, *makatea* islands, like those of the Solomon Islands, Fiji, and Tonga, had high levels of inland biodiversity, which, while potentially threatened by populations of humans and their animal companions, could also represent a significant resource if harnessed in the right way.[13]

Though shaped by their environments, these expanding human populations could also make their own decisions when the going got tough. Tikopia, an island of just 5 km^2, was settled by people with Lapita pottery approximately 2,900 years ago. Over 2,000 years of forest clearing for the growth of yams and taro, the management of pigs and dogs, and the use of marine resources led to a drastic reduction in local biodiversity and forest cover and the increasing intensity of soil erosion on what was previously a fertile island. Around three hundred years ago, the local population decided enough was enough. They removed pigs and dogs from the island, managed forests instead of cutting them down, and eased pressure on vulnerable fish species. They also managed their population size. The result was relatively sustainable existence up to European colonization. Similarly, as noted above, the classic tale of Rapa Nui is usually one of human population explosion, complete deforestation, warfare, famine, and collapse by AD 1680, with stone faces of *moai* statues left behind on a now largely depopulated island as monuments to the overexploitation of resources. However, detailed archeobotanical analysis and renewed dating efforts have shown that, although humans and rats certainly did impact the forest between AD 1200 and 1650, this was not the whole story. From this point, the local population, which likely never grew to more than a few thousand, began using grass and ferns for fuel, stopped the creation of their iconic monuments, shifted to joint management of wild plants and domesticates, and persisted alongside some remnant tropical trees. The innovativeness of these populations can be seen in their use of mulch from volcanic rocks to supplement poor soils and their formation of underground *manava* gardens to protect their crops from wind and sea spray. In fact, it looks like it was actually the arrival of European colonizers on the island in AD 1722, with guns, diseases, and slavery, that led to the almost complete collapse of theretofore undoubtedly adaptable Indigenous Polynesian populations.[14]

THE EXPANSION OF Austronesian-language-speaking communities across the Pacific leads us to a surprising but neat segue into the final stop on our island tour, Africa's coastal islands. Madagascar, one of the last large landmasses to be occupied by humans, sits around four hundred kilometers east of Mozambique. Remarkably, its modern Malagasy population has, among its ancestors, some of the same Austronesian speakers traveling through Southeast Asia who gave rise to the first settlers of Polynesia. In perhaps the single-greatest feat of seafaring in human history, beginning around 1,000 years ago, human populations from Indonesia, engaged in wider Indian Ocean trade networks, brought their language, crops, and genes all the way to the African continent. But we are getting ahead of ourselves. The timing of the first human settlement of Madagascar is, remarkably, not settled, and current suggestions range all the way from 10,500 to 1,200 years ago. The most recent detailed study, based on radiocarbon dates from a number of archeological sites, as well as clear evidence of butchered animal bones, suggests that humans were present by at least 2,000 years ago. Whatever the case, these first settlers are famous for a very different reason than the aforementioned Austronesian-language-speaking arrivals. Carrying stone tools from the African mainland and altering forests through burning, these groups have been linked to the disappearance of a variety of Madagascar's enigmatic "megafauna," including giant lemurs, humongous elephant birds, and native hippos. By 2,000 years ago, long prior to any intensive record of landscape burning, crop cultivation, and clear dense settlements, human activities—either direct hunting or indirect landscape changes—had possibly begun to have major implications for large animals across the spiny thickets, dry deciduous forests, and montane forests of Madagascar. By five hundred years ago, all wild animals larger than ten kilograms had disappeared from the island, which, back in the 1980s, served as a classic example of the "blitzkrieg" impacts of human hunting on megafaunal populations encountering humans for the first time.[15]

As Professor Kristina Douglass of the Pennsylvania State University, a seasoned explorer of Madagascar's forests, explains, however, "The story is, once again, much more complicated than that." Archeological sites have reported finds of giant lemurs and hippo bones with cutmarks from between 2,300 and 1,100 years ago in different parts of the island.

Figure 8.2. Artistic reconstruction of the now-extinct elephant birds of Madagascar. *Velizar Simeonovski*

Meanwhile, Kristina and her team have shown that prehistoric elephant bird eggs bear clear signs of human exploitation. Ultimately, though, these direct traces are few and far between. Peaks in deliberate human burning may have occurred at various times in the different biomes of the island between 1,700 and 1,100 years ago as populations expanded into different parts of the interior, though these changes seem to suggest a patchwork of human advance rather than a relentless frontier. Furthermore, even if the later dates of human occupation (c. 2,000 years ago) are taken, then human populations and megafauna coexisted without any obvious signs of decline for at least 1,000 years—longer if the earlier dates are accepted. Many of the large megafauna likely actually persisted on the island until at least 1,000 years ago. Not only that, but when these larger animals did begin to disappear, they did so in different ways and at different speeds across the hugely contrasting forest environments of Madagascar. As a result, it seems highly unlikely that human hunting alone eliminated Madagascar's megafauna, with more complex, multi-factor explanations, including recurring climatic variability and periods of drought, more probable.[16]

A more recent "subsistence shift hypothesis" proposes that a change from a hunting and foraging lifestyle toward food production, notably pastoralism, imposed more sweeping, island-wide pressures on these Malagasy megafauna, particularly as populations expanded and Madagascar became part of enlarging Indian Ocean networks from around 1,000 years ago. The first more permanent "residential" settlements appear from 1,300 years ago on the northeastern coast. In the next half millennium, villages moved to the interior as well as the shorelines, spanning from the eastern evergreen rainforests to the arid spiny bush of the southwest. Over approximately the same period, cattle and later goats arrived, and we see an increase in forest clearance and grassland presence across Madagascar as these key herd animals were left to overgraze and stop tree regrowth, as they do in many parts of Africa today. From seven hundred years ago, rice, the greater yam, and coconuts brought from Southeast Asia were grown in wetlands. Later, sorghum and cowpea from eastern Africa were used in more arid regions. These farming activities, alongside growing populations that began to produce elaborate settlements and hilltop fortifications, would certainly have increased the strain on native animals, perhaps even acting as the final tipping point for the disappearance of many endemic species by five hundred years ago against a backdrop of increased aridity. Nevertheless, as we have seen, different species declined at different rates and in different areas, with many on their way out prior to this point.[17]

In fact, the earliest food producers of Madagascar were remarkably resourceful. They combined the cultivation of incoming Southeast Asian and eastern African domesticates to different extents with the use of wild plant resources, depending on their environmental contexts and social situations. The same is seen across the other eastern African islands caught within a wider, flourishing Indian Ocean trading network. On near coastal islands, like Zanzibar and Pemba, Asian crops were typically rare relative to African millets, sorghum, and cowpea, with evidence of persistent hunting and fishing. In Madagascar itself, villages adapted to local contexts. Some practiced hillslope and wetland cultivation, while others preferred the herding of cattle and goats. Coastal communities focused on fishing, which still drives Malagasy economies and supplements farming diets today. Overgrazing and agricultural landscape transformations continue to increase soil erosion and deforestation and have helped make the remaining native lemurs some of the most threatened primates in the world. Yet, prior to five hundred years ago, human populations remained relatively small. So too did their herds of goat and cattle. The large-scale agricultural landscape modifications challenging sustainability on the island today only truly began following European arrival as colonizers and growing Indigenous polities grappled to control the increasingly contested tropical landscape.[18]

Turning to the opposite corner of the African continent, although the Canary Islands are just north of the Tropic of Cancer, their moist, marine climate supports extensive tropical and subtropical vegetation, and they provide another useful example for exploring prehistoric human impacts on island environments. The Canary Islands are a volcanic group of islands in the Atlantic Ocean just off the coast of northwestern Africa and include a number of favorite European holiday destinations. Their earliest human inhabitants, collectively known as the "Guanches" by later Spanish colonizers, arrived approximately 2,500 years ago. We now know that these first settlers, although previously characterized as stone-tool wielding foragers, had a rich toolkit of pottery, varied lithics, bone, leather, shell, wood, and other fibrous plants. Genetic, linguistic, and material comparisons indicate that these first inhabitants had a clear affinity to North African Berber populations and brought goats, sheep, pigs, and the trinity of dogs, cats, and mice with them to all of the Canary Islands. They also cultivated novel crops like barley, wheat, lentils, beans, figs, and peas carried from the mainland. As seen in Madagascar, and indeed in many of the islands we have touched upon so far, these farmers used fire to clear and open the landscapes for their introduced domesticates. This resulted in drastic reductions of forest cover and in "lost forests" in parts of islands or even across whole islands. Certain native tree species, like the laurels and the strawberry tree, disappeared on islands like Fuerteventura. Growing sheep and goat populations, on small islands, halted forest resurgence and led to soil erosion, while introduced mice and cats sent native rodents, large and small, into decline. Direct hunting also seems to have resulted in the demise of a large native lizard and a flightless quail.[19]

Scientists have suggested that, unlike on some of the other examples we have touched upon, a general lack of useful native plant resources and limited numbers of native large animals forced human arrivals to intensively pursue food production on the Canary Islands, with the corresponding consequences. Yet, again, the story is more complicated. A clear distinction can be seen between small, flat islands like Fuerteventura and Lanzarote, whose dry forest vegetation was more vulnerable to human clearance and all but disappeared, and the wetter of the small islands, like La Gomera and La Palma, and the larger islands of Tenerife and Gran Canaria, where forests remained standing upon European arrival. In the case of La Gomera, a sediment core collected from a former lake bed demonstrated that there was no clear signal of deforestation following the arrival of humans and that previous climate changes had led to greater forest disturbance. On La Palma, prehistoric humans changed their choice of tree species for fuel, recognizing the growing problem near their sites and venturing higher up into the mountains to ease the pressure. On Tenerife and Gran Canaria, the remaining forest regions might be seen as even more impressive given that these latter two islands had developed larger populations and a clear social hierarchy by the time European travelers coasted by their shores. Overall, the arrival of farming practices on the Canary Islands had stark consequences for local animals, plants, and landscape stability. The issues of deforestation and overgrazing still plague these islands today. Nevertheless, prior to European arrival, these impacts were dispersed and varied by island and location. As we will see in Chapters 10 and 11, Spanish colonization, ranching, and sugar manufacturing, driven by imperial demand, initiated truly sweeping changes across the archipelago, including perhaps local climatic shifts, that stay with its inhabitants to this day.[20]

TROPICAL ISLANDS HAVE often been thought of as ideal, isolated "laboratories" in which to study the impact of prehistoric human arrival on previously "pristine" ecosystems. While by no means a complete tour of all tropical islands, our trip around the Caribbean, the Pacific Islands, and the shores of Africa has shown us that hunting and gathering and, particularly, forms of farming practice can have major, lasting impacts

on the native plants, animals, and landscapes of these often remote settings. Over the course of the Holocene, tropical islands have experienced
a massive loss of endemic birds, amphibians, reptiles, and mammals,
driven by human hunting, habitat change, and the invasive species humans brought along with them, either accidentally or deliberately. Rapid
soil erosion and a loss of forest cover occurred as populations grew and
farmers sought openings to plant more productive crops or societies
sought new landmarks to differentiate status and act as cultural symbols.
Entire ecosystems and environments could change, or even disappear,
as new, economically useful plants were carried or promoted and others
fell away. Beyond the land, overfishing and overuse of shellfish led to a
dramatic reduction in their size and also to the shortening of marine
food chains. These laboratories show us that prehistoric human impacts
could leave a massive lasting impression on plants, animals, and entire
island landscapes. Consequential changes to soils and available wild resources could also leave the settlers themselves facing an ultimatum of
population collapse or island abandonment.

Nonetheless, the growing field of island archeology has also shown
that these outcomes need not be inevitable and can indeed be averted
through flexible decision making even in the most challenging of scenarios. Many of the societies we have met, including those that brought
farming "packages" with them, adapted their lifestyles to their new situations or adapted the new situations to fit their lifestyles in what has
become known as human "niche construction." Some complemented
their domesticate diets with local tropical forest plants and animals or
fish from the reefs or deep blue seas off their island coasts. Some managed the local landscapes to ensure a continuity of wild, native species
alongside more tentative introductions. Where changes in ecosystem
dynamics became obvious, groups could clearly also decide to rid themselves of domesticates and reduce deforestation, setting themselves
on a new course toward sustainability. As a matter of fact, the starkest warning from these island examples is not that island environments
were inherently vulnerable in the past and remain inherently vulnerable
now. It is rather that in the past, populations were able to change their
land use and their economies as they saw the landscape change around
them. They noticed soils sliding, prey dwindling, and forests receding

and could make decisions. Today, the forces of poverty, climate change, political priorities, and large industries wash up on island shores, staying the hands of the local inhabitants. As each example discussed above hints, the disruptive arrival of European empires and mercantile interests invariably brought these pressures of global consumer desires along with them, something we will explore in more detail in Chapters 10 and 11. However, before turning to the beginnings of a "globalization" that defines our current relationship with the tropics, it is worth first exploring the massive Indigenous urban societies that developed *within* tropical forests.[21]

Chapter 9

CITIES IN THE "JUNGLE"

Visions of "lost cities" in the jungle have plagued Western imaginations since Europeans first visited the tropics of Asia, Africa, and the Americas. From the *Lost City of Z* to El Dorado, a thirst for finding ancient civilizations and their treasures in perilous tropical forest settings has driven innumerable ill-fated expeditions. This obsession has diffused into Western societies' popular ideas of tropical forest cities, with overgrown ruins acting as the backdrop for fear, discovery, and life-threatening challenges across a number of popular video games (the *Uncharted* series), horror films (*The Ruins*), and novels (*The Jungle Book*). Throughout all of these depictions runs the pervasive idea that all ancient cities and states in tropical forests were doomed to fail, that the most resilient occupants of tropical forests are small villages of poison-dart-blowing hunter-gatherers, and that vicious vines and towering trees or, in the case of *The Jungle Book*, a boisterous army of monkeys will inevitably claw any significant form of monumental human achievement back into the suffocating green from whence it came. This idea has not been helped by best-selling books or apocalyptic films that focus on the blockbuster "collapse" of particularly enigmatic societies, such as the Classic Maya. Ultimately, the decaying stone walls, the empty grand structures, and the deserted

153

streets of these tropical urban leftovers act as a tragic warning that our own way of life, communities, and economies are not as infinitely secure as we would often like to assume.[1]

The situation has not been all that dissimilar in academic considerations of the potential of tropical forests to sustain ancient urbanism. On the one hand, intensive agriculture seen as necessary to fuel the growth of cities and powerful social elites has simply been considered impossible on the wet, acidic, nutrient-poor soils of tropical forests. On the other hand, where the obvious rubble of cities can simply not be denied, in the drier tropics of North and Central America, South Asia, and Southeast Asia, catastrophic ecological insolvency has been thought to have been inevitable. Deforestation to build massive buildings and make way for growing populations, an expansion of agriculture across marginal soils, and natural disasters such as mudslides, flooding, and drought, like today, must have made experimentation with tropical cities a big challenge at best and a fool's gambit at worst. Overhauling these stereotypes has been difficult as the large, multiyear field explorations usually undertaken on the sites of ancient cities—for example, in Mesopotamia or Egypt—face unique tropical trials. Dense vegetation, mosquito-borne disease, poisonous plants and animals, and torrential weather have made finding, reaching, and excavating past urban centers in tropical forests arduous. Where organic materials, instead of stone, might have served as a construction material, the task becomes even more taxing. As a result, research into past tropical urbanism, its forms, economies, and extents, has somewhat trudged along behind similar research in the semiarid and arid zones of Mesopotamia and Egypt and the sweeping river valleys of East Asia.[2]

Yet, as we saw in Chapters 7 and 8, many tropical forest societies found immensely successful avenues to food production in even the most challenging of circumstances. While these might not look like the "agricultural" fields we think of as supporting cities and states today, they could sustain impressively large populations and social structures. We now turn to some of the most famous examples of supposed tropical disaster, the Classic and Postclassic Maya, the Khmer Empire of Cambodia, and the cities of northern Sri Lanka, to see how the last two decades of intrepid archeological exploration, applying the latest science from both the land and air, have stripped away canopies to provide new, more

favorable assessments. We will find that not only did precolonial tropical cities flourish, but they were actually some of the most extensive
urban landscapes anywhere in the preindustrial world—far outstripping
ancient Rome, Constantinople/Istanbul, and the ancient cities of China.
We will find that even the supposedly most "pristine" forest region, the
Amazon Basin, was actually home to millions of people extended across
urban-like settlement networks. Ancient tropical cities could be remarkably resilient, sometimes surviving many centuries longer than colonial
and industrial period urban networks in similar environments. Although
they could face immense obstacles and often had to overhaul their social
structures and settlement locations to beat changing climates and their
own eventual overuse of the surrounding landscape, they also developed
completely new forms of what a city could, and perhaps should, be. Extensive, interspersed with nature, and combining food production with
social and political function, these ancient layouts are now catching the
eyes of twenty-first-century urban planners trying to come to grips with
tropical forests as sites of some of the fastest-growing human populations around the world today.

AS WITH "AGRICULTURE," we tend to view the concept of a "city"
through a blinkered Western lens. Based on our own experiences, we
think of cities as compact, densely populated areas, home to administrative and political elites, full of bustling trade and manufacturing, and
fed by vast agricultural fields and animal herds often located at some distance from the city boundaries. For over fifty years, since the archeologist
Gordon Childe coined the "agricultural" and "urban" revolutions as two
major stepping-stones in human history, the same has been broadly true
of archeology. Vast Near Eastern "tell" mound sites, formed from humongous accumulations of mudbricks and refuse and representing the
earliest cities dating to around 6,000 years ago, as well as the cities of the
classical world and ancient China, have acted as the standard for tracking
the origins and expansion of a settlement pattern that, to a large extent,
still shapes many of our lives today. This "model" of social complexity
unsurprisingly seems somewhat out of place in tropical forests, where
sweeping fields of uniform crops, grazing animals, and dense, built-up

Figure 9.1. The ancient Classic Maya ceremonial center of Tikal, Guatemala.
Michael Godek / Getty Images

settlements can lead to drastic deforestation, subsequent soil erosion, and eventually starvation and social disintegration. As a result, where such seemingly "compact" cities have been identified in the tropics—for example, in the case of the Classic Maya of southeastern Mexico, Guatemala, Belize, and western Honduras and El Salvador—scientists have tended to assume the worst: that these cities and their farming focus on certain key crops were too much for their tropical forest landscapes, leading to degradation, an overthrow of rulers, and ultimate abandonment.[3]

Maya urban forms began to appear around 800 BC in the so-called Preclassic period. Cities, monumental stone architecture, and writing all gradually emerged at certain key political centers led by kings and fed by the North and Central American staple crops of maize, beans, and squash. Things truly took off during the Classic period, especially in a region known as the southern lowlands (including northern Guatemala, Belize, and southeastern Mexico). Between AD 250 and 900, growing populations, more cities, more monuments, and more inscriptions appeared. The leaders of the famous cities of Tikal and Calakmul, populated by as many as 120,000 people, engaged in warfare, extended political influence, and harnessed long-distance, high-value prestige resources in ways that would make many of our more devious administrations proud today. Although many of these urban nodes were located on soils particularly suited to productive maize agriculture, there was one problem. Annual rainfall variation in the region could be as high as 2,000 millimeters (for comparison, the average annual rainfall recorded at Greenwich Observatory London between 1981 and 2010 was 621 mm), and the geology often made the capture and storage of precious water through the dry months challenging. As a consequence, major droughts witnessed in lake records from the region have been argued to have brought the system to its knees in the Terminal Classic period, between AD 800 and 900, in the southern lowlands. No response was possible given that the large centers and their political classes had already overreached, cutting down trees to fuel their monuments and planting their corn on poor soils as the more fertile regions "filled up." Precariously sustaining their populations across the heavily altered landscape, they could do nothing. People lost faith in elites, construction stopped, famine ensued, and the Classic population dispersed itself across the landscape to eke out a living in a drier world. Passing by these former centers a few hundred years later, Spanish explorers such as the infamous Hernán Cortés did not even deem what was left significant enough for comment, and the ruined cities awaited eventual "rediscovery" by intrepid American archeologists of the nineteenth century.[4]

So often goes the classic story of the Classic Maya. Amazingly, however, as early as seventy years ago, surveys of Classic Maya cities by the archeologists William Coe and Gordon Wiley had already questioned such an interpretation of their urban organization and landscape placement. Far from being compact, we now know, even the best-known of Maya centers, like Copán and Tikal, practiced what has elsewhere been called "agrarian-based, low-density urbanism."[5] This effectively means that the population was relatively dispersed (one to ten persons per hectare) rather than dense. Instead of fields being outside and politics inside, fields were located throughout the urban infrastructure and among residences. And rather than being small focal points, cities sprawled over 100 km². To take Tikal as an example, recent survey data has illustrated a network of moats, dwellings, reservoirs, and pyramid clusters that

Figure 9.2   The central temples of the pre-Angkorian urban complex of Sambor Prei Kuk, Cambodia. *Visualization by Damian Evans from Lidar elevation data acquired by the Cambodian Archaeological Lidar Initiative in 2015*

extend out from a single hill for up to 200 km² into the surrounding landscape. Innovative aerial survey methodologies have now made similar findings elsewhere across the Maya world, from Copán to Caracol. In almost all instances, instead of isolated urban buds, scientists have found vast landscapes of small and large centers connected by dispersed agrarian landscapes, residential areas, causeways, and complex, interlinking systems of dams, reservoirs, sinkholes, channels, and swamps that supported growing populations through even the driest of seasons. As leading "Mayanist" Professor Lisa Lucero of the University of Illinois puts it, "Having been there for over a millennium already, the Classic Maya knew the importance of water and of fertile agricultural soils, the latter dispersed in variously sized pockets, mirrored by a dispersed agricultural settlement. This low-density approach to cities was a logical, innovative solution."[6]

The Classic Maya also had far more diverse and sophisticated economies than has often been appreciated. Archeobotanists have shown that alongside the key crops, the planting of avocados, pineapples, sunflowers, tomatoes, and manioc added to a more dispersed settlement and lifestyle than conveyed by popular imaginings of endless rows of maize. Analysis of modern forests growing around Classic Maya centers also shows that the occupants of these urban networks actively managed tropical forest plants to promote economically useful species. They did not stop at plants. The Classic Maya also penned, fed, and fattened wild turkeys and deer for use as key protein sources. Overall, instead of rolling agricultural fields of single crops, scientists now present diverse "forest gardens" as the sustaining source of these cities. Based on ethnographic study of and testaments from Maya communities today, this type of cultivation, called *milpa* (or *kol* in the local Yukatek language), involves the use of multiple crops and the movement of fields, allowing different parts of the forest to grow back and patches of soil to rest and restock before planting begins in a locality again. We also know that instead of indiscriminately planting in soils of all types, the Classic Maya actually followed rich veins of particularly productive soils known as "mollisols," giving their field systems a winding appearance that snaked along rivers and up slopes. They even added special plants, like water lilies, to reservoirs. These plants are incredibly sensitive to water quality, only growing under clean conditions, and allowed people to monitor the buildup of stagnant water and thus guard against disease. "Ultimately, the wider Classic Maya landscape can be imagined as a patchwork of seasonal wetlands and well-drained forests used for palms, dyes, fruits, animals and construction materials, and open areas filled with diverse fields (*milpas*)," states Lisa.[7]

The difficulty of maintaining large populations and complicated social and political structures in a highly seasonal tropical landscape, where water was often in demand, did eventually take its toll in many parts of the so-called Maya heartlands. Despite well-accepted catastrophic scenarios, it is immensely unlikely that any single, dramatic drought brought about the end of a given city. Nonetheless, detailed research by climate scientists, working on growing records of past precipitation from lake sediments as well as the gradual buildup of stalagmites in nearby caves, has shown an increasing seasonality in rainfall amounts, an increasing number of droughts, and a gradual downward trend in precipitation. The extent of pre-Columbian deforestation is not agreed upon. However, pollen records from sediment cores from throughout the southern lowlands suggest varying degrees of deforestation and forest management at each of the hundreds of Maya centers, which, in some cases, may have exacerbated climatic drying. In the southern lowlands, where surface water was hard to find at the best of times, many cities, including Tikal, saw failing agricultural returns, growing malnourishment, and stress. Since kings claimed such close ties to the gods, in the face of droughts and failed crops, their source of political power was called into question, especially as intercity violence grew more frequent. People refused to work on monuments, seeing that they made no difference. Ceremonial centers were abandoned in the southern Maya lowlands, the ruins left for scientists arriving centuries later. Looking at the sequences of certain sites, this may certainly seem like a rapid, uncompromising disaster. But was it really? Given such long-term Classic Maya knowledge of their ecosystems, their well-tuned economies, and their sophisticated water management, things that twenty-first-century urban systems may often lack, is a sweeping, rapid "collapse" truly likely?[8]

In fact, Maya centers actually flourished into the Postclassic period (AD 900–1520) (Chichén Itzá, Uxmal), some right up to Spanish arrival, simply locating themselves near sinkholes fed by groundwater or coasts in the northern lowlands, near the relatively few lakes and rivers in the southern lowlands, and later in the highlands of Guatemala (e.g., Q'umarkaj [Utatlán]). Even in the drying southern lowlands, we can see that the wealthy and powerful suffered—but what about everyone else? In many areas, it actually seems like the independent farmers, who were the key links in the chain of vast Classic Maya urban systems, persisted, albeit with much-reduced population sizes. In the region of El Pilar, surrounding the ceremonial center of Tikal, "forest gardens" were managed by farming communities. Remarkably, this management and the presence of large numbers of farmers continued through the rise and fall of Tikal itself. This diverse *milpa* agriculture also persists among the Indigenous Maya communities that still occupy many parts of the region today. These groups still practice traditional manufacturing and landscape management. Urban archeologists, like our societies in general, tend to focus on the "rich" remains of elite structures, which can make a big media splash and leave a big mark. But when your tropical city is built on sprawling networks of independent farmers and craftspeople, you can often miss the remarkable resilience inherent in the foundations of the system beyond the more obvious stars of the show.[9]

THE GREATER ANGKOR region of Cambodia is another prominent, tourist-attracting set of ruins emerging out of a "jungle." Gap-yearing students, jetsetters, elderly tour groups, and interested locals all flock toward the prominent, monkey-covered temple of Angkor Wat, which became the religious center of the Khmer Empire in the twelfth century AD. Few of them, however, realize that this monumental shrine is actually just the tip of the Angkorian iceberg. Urbanism began to emerge in this part of the world from the first millennium BC. Initially this was based on walled and moated towns of five to twenty km², with a focus on the management of water to feed their growing rice fields through the often severe regional dry season, which boomed into the first millennium AD. From the eighth century AD, the region of Angkor took this to new levels. Beginning as two separate centers, Yasodharapura, the new capital of the Khmer Empire, the dominant state of mainland Southeast Asia at the time, had formed by the ninth century AD. The capital, characterized by massive earthen-built reservoirs known as *barays* and a series of walled administrative palaces and Buddhist and Hindu temples, thrived for half a millennium until the fourteenth century AD. Previously, like tourists do today, many archeologists had focused on the seemingly important, compact "ceremonial" centers, like Angkor Thom and Angkor Wat, built by the various rulers of the empire.[10]

But then two things happened. First, the French archeologist Christophe Pottier and his local collaborators undertook decades of on-the-ground survey work, even in the face of security issues with holdouts from the Khmer Rouge rebellion in the 1990s. In doing so, they were able to plot the presence of vast numbers of construction features, large and small, across an entire Greater Angkor region. Second, a new method arrived: light detection and ranging (LiDAR). One of the leading experts applying this methodology in tropical archeology today is Dr. Damian Evans of the École française d'Extrême-Orient, Paris. Dense vegetation has often hindered the identification of ancient buildings in tropical forests, particularly in Southeast Asia, where dwellings were often constructed

Figure 9.3.   Aerial overview of the Angkor Wat temple complex.   *Damian Evans*

from wood or bamboo, leaving behind just subtle changes in soil color or elevation. However, as Damian describes, "LiDAR allows us to virtually 'strip away' vegetation. Using a laser scanner attached to an aircraft, we carpet the terrain below us with pulses of lasers, collecting billions of points. Some bounce back from the trees but others slip through, allowing us to build a model of what lies beneath." What they have found at Angkor is simply mind-boggling: an urban residential area of over 1,000 km^2 has emerged alongside a modified landscape of 3,000 km^2. This makes Greater Angkor the most extensive preindustrial settlement complex anywhere on the face of the Earth, and larger even than cities such as Paris today. It also revolutionizes our understanding of how this ancient megacity operated.[11]

As in the case of the Classic Maya, where LiDAR has also exposed vast, settled urban landscapes, rather than being solely composed of compact ceremonial centers, Angkor was another sprawling example of agrarian-based, low-density urbanism. At ceremonial centers like Angkor Thom, LiDAR filled in the apparently spacious courtyards to reveal a densely packed religious support team living in wooden pile houses within the stone walls. Beyond focal points, numerous sprawling domestic mounds, smaller shrines, and rice fields extend from right up against the boundaries of sites like Angkor Wat, out across the lowlands, and into the hills. As in tropical North and Central America, the urban populace used the forests, exploiting fruits in managed orchards, wild palms, and vegetables, ensuring some forest cover persisted. Rivers and ponds provided fish, while zooarcheologists have shown that pigs, cattle, and chickens snuffled, brayed, and clucked through the busy streets. This expansive human and animal population could only be supported among the seasonally dry forests that line the Siem Reap River delta by the human creation of a vast water-drainage system. In addition to buildings, the earlier surveying and later LiDAR work together show a dendritic network of water transport and storage, from giant reservoirs and temple moats to small, locally maintained channels. Ingenious design and application of sandy clay to parts of the probing tendrils of this vast construct allowed it to deal with both droughts and floods, and the Angkorian leviathan persisted through many a dry season and political crisis until the fourteenth century AD.[12]

Unsurprisingly, this vast urban metropolis placed a strain on the tropical landscape. Human-induced deforestation and soil erosion have left a clear signal in paleoenvironmental records from the region. However, the agrarian urban spread dispersed these impacts across the catchment so that it did not intensively impact any particular area. Angkor's time did, however, eventually come to an end. During the late fourteenth century AD, increasingly extreme climatic swings between drought and deluge ruptured parts of the water network and took their toll on farming yields. Palaces and temples were abandoned, and the urban landscape broke up as farmers saw no reason to stay within a dysfunctional state-guided system of competing elites. But, again, the result was more interesting than complete "collapse." The rulers moved to new, more compact cities in the Phnom Penh area, where the current capital of Cambodia still stands to this day. Meanwhile, the diligent farmers populated more numerous small towns along the banks of the Mekong River and Tonlé Sap lake, preferring more stable water courses for their future endeavors. As with the Classic Maya then, the governing system certainly failed, particularly in the wake of climatic challenges facing a region prone to seasonal variation in rainfall. The eventual abandonment and upheaval seen at Greater

Angkor is nothing to sniff at. However, the elite thought up a new strategy and moved to new sites of power. Meanwhile, the farmers, who had
fueled the sprawl from its beginning, continued on a landscape that had
provided for them for so long, albeit focusing their efforts on more profitable areas for cultivation and herding. Once we cast our eyes away from
the overgrown walls of temples, then, we can begin to see how Greater
Angkor represented a vast, highly resilient approach to urban living in
the tropics.[13]

A comparable state of affairs can also be found in one of the greatest ancient urban tourist attractions in South Asia, Anuradhapura of the
UNESCO "Golden Triangle" in Sri Lanka. Construction of this capital
of the Sinhalese king Pandukabhaya seems to have begun around 500
BC. Monumental Buddhist temples ("stupas"), including the largest brick
structure anywhere in the world, monasteries, palaces, and state-directed
reservoirs to support intensive rice agriculture characterize this center
from the third century AD. The water "tanks" are particularly impressive, with the largest, Nuwarawewa, built in the first century AD, reaching
as large as 9 km^2. By the ninth and tenth centuries AD, Anuradhapura
dominated the political landscape of the island. The three great monastic
centers of the capital could have sustained 30,000 religious worshippers
at any given time, documented by a number of immense preserved stone
vats that could have each fed rice to 1,000 people every mealtime. Com-
parably, only 1,000 worshippers were invited into the Hagia Sophia for
prayers when it was converted into an active mosque in 2020. However,
once again, it has taken dedicated survey work to see past Anuradhapura's stone and brick behemoths to the 500 km^2 sprawl that extended
out into the wider countryside, housing an estimated 250,000 people at
its peak. That's an area over four times larger than the modern city of
Liverpool. This spread was fed by timber and stone canals that linked
the central tanks to water storage constructions in the hills nearly one
hundred kilometers away. While these larger layouts were state directed,
as at Greater Angkor, small field systems, reservoirs, and alterations to
rivers were undertaken by local farming groups that occupied the surrounding landscape, independent of the larger state building works and
political power shifts. These farmers also developed gardens and moved

their plots of land on a frequent basis to maintain intact parts of dry tropical forest in order to diversify their diets and landscape use beyond rice.[14]

Political and climatic fluctuations eventually spelled the end of ancient Anuradhapura as a ceremonial and political center. The Sinhalese capital moved first to the nearby Polonnaruwa in the twelfth century AD, before rulers abandoned the drier north altogether by the thirteenth century AD and moved south to the wetter tropical forests of Dambadeniya and eventually Kandy. This may have been stimulated by fluctuations in the Indian Ocean monsoon system, which brought intense drought to the reservoir-dependent cities of the north, a region that faces marked, arid dry seasons in the summer months to this day. Nevertheless, although the state system faded locally, through relocation it persisted in the region as a whole. Furthermore, the more independent farmers once again ploughed their own course beyond the end of the political structure. In the later twelfth century AD, after the rulers had already left, there is evidence that some dogged farmers reoccupied different parts of Anuradhapura and dusted down the monuments. Moreover, although British colonial bureaucrats who arrived in the north in the nineteenth century found ruin, they also found continued small-scale rice farming and water management across the Anuradhapura region, which, incidentally, still persists as a city, albeit fueled by tourism. In fact, having been occupied for around 1,700 years as an urban center, ancient Anuradhapura stood far longer as a Sri Lankan capital than the current city of Colombo, which, while occupied by multicultural traders previously, was only truly established as a major urban center following Portuguese, Dutch, and British colonization from the sixteenth century AD onward. Given that this period is also far longer than most of our European cities have stood the test of time, the Sri Lankan example should once again make us reflect on our common assumptions that tropical forests make bad homes for urbanites.[15]

THE LOWLAND EVERGREEN and semievergreen rainforests of the Amazon Basin are another challenge altogether for urban societies compared to the seasonally arid dry tropical forests of Central and North

America, South Asia, and Southeast Asia. In fact, the heat, sticky humidity, and acidic, often-flooded soils led some archeologists and anthropologists to assume that both cities and settled agriculture were environmentally impossible. Most surviving Indigenous communities in these forests were, after all, living in small villages with no clear evidence of social hierarchy. Meanwhile, today, expanding infrastructure, development, plantation agriculture, and ranching undoubtedly do untold harm to these environments. Nevertheless, Chapter 7 has already shown us that more innovative, ecologically sensitive paths to food production were underway in the Amazon for much of the Holocene. In fact, renewed survey by a number of dedicated teams of archeologists, anthropologists, environmental scientists, and Indigenous communities, as well as tragic deforestation stripping away protecting forest, has revealed substantial "garden city" landscapes of earthworks, clustered structures, and "road-like" paths across the Amazon Basin. In and around the Xingu River, these settlements reached their peak between AD 1250 and 1650, just prior to European arrival. Intriguingly, they show a similar pattern to the cases of agrarian-based, low-density urbanism we have already visited. In each case, clear, larger towns at central points were surrounded by monumental wooden walls and ditches and connected to a number of satellite villages by pathways cleared through the forest. Instead of resulting in mass deforestation, these settlements were separated by intact forest used for the management of fruit-bearing trees, ponds for farming freshwater turtles and fish, and more open fields for manioc and maize.[16]

Many years of research have also documented a similar series of urban-like settlements at Marajó Island, right at the mouth of the Amazon River. During the first millennium AD, mound sites were produced by the sustained occupation of large domestic residences, burial areas, elaborate pottery, and waste disposal. Increasing in size and density until the start of the fourteenth century AD, the population in this region at its peak may have been as large as 100,000. Whatever the true size, as in the "garden cities" of the Xingu, archeobotany, zooarcheology, and stable isotope analysis of human remains have shown that the occupants of Marajó Island, as well as the nearby Maracá region, fed themselves through a mixed system of foraging of diverse plants, hunting of wild animals, and fishing or penning of freshwater resources in and around intact rainforest, with some limited cultivation of manioc and maize in more open patches perhaps occurring in addition. The now famous settlements of Santarém, this time at the mouth of the Tapajos River, present a similar scenario. Here, a subsistence combination of manioc and other root crops, alongside tree fruits, fish, and the occasional, perhaps ritual, use of maize, once again drove the success of an interlinking network of denser, occupied settlements and more dispersed satellite villages. Interestingly, in this case, each satellite may have had its own specialism, with some showing a particular focus on the capture of rainforest animals. Ultimately, together these Amazonian forms of agrarian-based, low-density urbanism may have enabled populations of between 8 million and 20 million to have existed across Amazonia by the time of European arrival. Given that the estimated population of all of Europe in AD 1492 was between 70 million and 88 million, this represents a considerable number of human lives sustained within what must undoubtedly be seen as substantial urban types of settlement, albeit very different from our usual ideas of cities.[17]

Global discussions of urbanism have perhaps neglected these Amazonian examples because they lack the more obvious stone architecture that many more classic examples display. Yet they had their own "monuments." We met the early cultivators of the Llanos de Moxos and their "forest islands" in Bolivia in Chapter 7. However, from the start of the eleventh century AD, their interventions in the landscape to protect their key crops from flooding became even larger in scale as extensive mounds, earthworks, and ditch systems emerged. More ceremonial constructs by so-called earthwork builders have now been found across northeastern Bolivia, in the uplands of the Ecuadorean Amazon, in Brazil's Acré State in southwestern Amazonia, and across the eastern Amazon. Perhaps the most resilient "monuments" of Amazonian urban life, however, are the vast changes made to the soils to support increasingly large populations. As Professor Eduardo Neves, a leading Amazonian archeologist based at the University of São Paulo, puts it, "The *terra pretas* (now known as Amazonian Dark Earth) soils are perhaps the late Holocene Amazon's greatest legacy." These clear layers of dark soils, full of charred bones and ceramics left over from occupation, seem to be deliberate attempts to maintain soil fertility through controlled forest burning and the mixing in of fertile waste products. They were likely created by food producers within these wider urban systems, who kept moving their fields around in a so-called slash-and-burn system to try and maintain forest cover and also allow soils to rejuvenate locally every so often. Amazonian Dark Earth soils are almost always found in the growing settlement sites of the first and second millennia AD in the Amazon. "Their enormous extent, now known to span across the eastern and western Amazon, stands as testament to the spread of populations, and their 'garden cities,' across this region," continues Eduardo.[18]

Beyond their being yet another example of more dispersed and integrated types of urbanism than we are used to, these Amazonian settlements are interesting for another reason. Despite showing many of the hallmarks of urban networks (craft production divided into different, dependent satellites, significant population size, and physical and ecological monuments), they almost completely lack evidence of a ruling class. The sites of Marajó and along the Xingu River have no distinctive fancy "palace" residences, no obvious division of "wealth" in terms of personal property, and no hoarding of elaborate grave goods by certain individuals. "Nobles" and directed "warbands" noted by European chroniclers imply some kind of centralized control, though the biases of those observing could also have impacted these accounts. Indeed, overall, surviving images and feasting contexts seem to be based more on shamanic rituals and interaction with forest environments than a display of overt political power. A similar state of affairs has been documented in subtropical Mali at the site of Jenné-jeno, where large numbers of people, complex satellite settlement networks, and walled mounds appeared by the ninth century AD, seemingly without a clear ruling class. Likewise, archeological records document dispersed but connected settlements in the Middle Senegal Valley throughout the first millennium AD. Ethnography in West Africa, as well as in Bali in Southeast Asia, has shown that complex works and large numbers of people may have been coordinated by a "heterarchy." In these systems, power is spread more horizontally, rather than vertically, with decisions made by representatives of different crafts, ritualistic specialists, or families. Such a structure may have existed in the archeological examples of Amazonia and West Africa and may even have helped the sustenance and maintenance of widely spread urban populations. Either way, the more obvious divine rulers of the Classic Maya, the Khmer Empire, or Anuradhapura are currently nowhere to be seen.[19]

The degree to which the "garden cities" and their other urban-like cousins impacted Amazonian environments is currently unclear. Vast earthworks, forest burning to produce human-modified soils and maintain open patches for crops, and the selection of certain economically useful trees must have had some wide-reaching effects, particularly given the potential number of people undertaking these activities. Nonetheless, as of yet, environmental records taken across the Amazon Basin have provided no unified message. Some scientists have argued that, in Bolivia, "earthwork builders" simply used already open grassland landscapes produced by drier climates just before 2,000 years ago. Meanwhile, further south, environmental archives seem to show no evidence of permanent clearance in connection to ceremonial landscape modifications. Others, however, have debated these findings, and the case very much remains open. In general, most records of Amazonian burning during the time frame of these "garden cities" seem to fit with a shifting form of farming (known as "swidden"). The temporary clearance of certain patches for a few years of cultivation, prior to movement to a new area, supported dynamic, mosaic landscapes rather than complete clearance, and no evidence suggests severe soil erosion or environmentally induced collapse. In fact, many of these urban forms had been in place for 1,000 years when European colonists arrived. As we will see in Chapter 10, it was then that everything changed. Ultimately, all that remained were the small villages of Indigenous peoples met by anthropologists in the twentieth century, who assumed them to be timeless and pristine like their forest homes. All traces of their previous urban glories were erased from history until the last three decades of archeological research and scientific methodologies, alongside better consultation of these same Indigenous communities and recognition of their stewardship and knowledge, raised them from the Earth, challenging widely held misconceptions.[20]

AGRARIAN-BASED, LOW-DENSITY URBANISM does not characterize all ancient cities in the tropics. Compared to more "compact" city forms, which persisted from 6,000 years ago in the Middle East through to medieval Europe and also emerged in the wake of the fall of some of these monstrous urban expanses in the tropics, it is certainly rarer and, on balance, more prone to disappear from a region. However, agrarian-based, low-density urbanism provides an important avenue for looking at past cities in tropical forests as both obviously possible and incredibly inventive. The diverse use of wild plants and animals in managed forest patches, fishing in rich freshwater settings, and the mobile use of open areas for crops provided sustainable pathways to food production sufficient to fuel what were some of the largest cities in the preindustrial world. Dispersing the impacts of growing populations across satellites tempered the loss of unique tropical biodiversity and soil quality in the face of growing human pressures. They could certainly face challenges, particularly in seasonally dry forests where human-induced deforestation and climate change could tip the scale between precarious viability and failure. Nevertheless, even in these instances, urban centers and ruling classes emerged elsewhere on the landscape. Meanwhile, ecologically experienced food producers persisted, often in large numbers, on the still-productive, though challenging, landscapes that had fed them for so long, even as monuments crumbled around them. Not only that, but while many of these types of cities ultimately fell to ruin, their hardiness shows in their often long spans of existence—longer than five hundred years in the case of Greater Angkor and some Classic Maya urban centers and nearly two millennia for Anuradhapura in Sri Lanka. This is far longer than the majority of industrialized tropical cities, as well as many modern European and northern North American cities, providing a very different perspective on our assumptions about their inevitable "collapse." Indeed, the timeless strength of agrarian-based, low-density urbanism remains an attractive model for present-day urban planners looking for "green cities" that balance urgent conservation and environmental needs, political and cultural infrastructure, and growing urban populations across the twenty-first-century tropics, something we will return to in Chapter 13.[21]

There are many, many more examples of cities ruled over by pre-colonial states and empires that emerged across the tropics in North, Central, and South America (e.g., those of the Triple Alliance of the Aztec Empire and the Inka), mainland and island Southeast Asia (e.g., Bagan of the Pagan Kingdom, Borobudur of the Sailendra Dynasty in Java), the Pacific (e.g., on Tongatapu at the center of the Tu'i Tonga Empire), and West and Central Africa (e.g., the centers of the Kingdom of Benin and Oyo Empire of western Nigeria, the Ashanti Empire of Ghana, and the Kongo Kingdom of the Democratic Republic of the Congo) during the late Holocene, some of which we will meet in the next two chapters. A significant number of these, like the examples of the Amazon and the Postclassic Maya, were still flourishing at the time of European contact. European visitors often even actively admired them. So why, then, do we now tend to think of tropical forests as so hostile to large, food-producing human populations? Why do all of our popular assumptions fit ruins and small, isolated Indigenous bands of foragers to these environments rather than lively, humming streets, residential blocks, and monumental constructions that could stand the test of time? The answer probably has something to do with what happened next. Arriving on ships across the horizon, Europeans did not just bring their notebooks with them to the tropics. They brought new diseases, new crops, new animals, new ways of using and seeing the natural world, and a political, religious, and social agenda that sought to "progress" from anything that had gone before. In the traumatic clash of worlds that followed lie the origins of our globalized, but unequal, world. It is here that our modern economic, political, and climatic reliance on the tropics, no matter where on the planet we live, began.[22]

Chapter 10

EUROPE AND THE TROPICS
IN THE "AGE OF EXPLORATION"

W e were not discovered," read the statement released by a coalition of Indigenous peoples in Mexico's western state of Michoacán on October 12, 2020—the anniversary of the day when Christopher Columbus (or Christóbal Cólon, as he was actually known) first landed in the Americas.[1] "Columbus Day" has been a federal holiday in the United States since 1971 and forms part of a wider public consciousness in Europe and northern North America that has tended to hold up Columbus, as well as other individuals, such as Ferdinand Magellan, Vasco de Gama, and Walter Raleigh, as fearless pioneers of an "Age of Exploration" that spanned from the fifteenth to the seventeenth centuries. Particularly within the tropics, they are often credited with "finding" mineral wealth, exotic plants and animals, and new, productive lands that were ripe for the taking—boosting the global economic and political standing of expanding European empires as they did so. Away from individuals, particular royal lineages similarly frequently take the spotlight as the active shapers of "new worlds," inventing new ways to travel the oceans and wielding unparalleled weapons, as they sought to extract new tropical resources and labor in a bid to gain religious and political prestige back home on the competitive stage of Europe. Certainly, when Pope Alexander VI divided up the lands of the Americas and Asia between the Catholic Monarchs of Castile and Aragon and the Portuguese in 1493 as part of the Treaty of Tordesillas, he had little thought for how the people already occupying these lands might react. This European focus can lead us to assume that existing Indigenous inhabitants were passive or insignificant throughout this process—which, as we have already seen in Chapters 8 and 9, could not have been further from the truth.[2]

In the fifteenth century, Europe was actually something of a backwater. In the thirteenth and fourteenth centuries it had benefited from a renewed flow of spices and commerce along the Silk Roads that united Europe, the Middle East, eastern Africa, and Asia following the expansion of the Mongol Empire and its successors and the relative peace and stability that ensued (the so-called *Pax Mongolica*). However, by the fifteenth century, the expansion of the Ottoman Empire based in Constantinople had blocked direct European contact with Central and East Asia and these crucial lines of commerce. Meanwhile, in the tropics, powerful empires had emerged. In the Americas, Tenochtitlan (now Mexico City) was the center of the Triple Alliance of the Aztec Empire. In Asia, Vijayanagara was the capital of the Hindu Vijayanagara Empire in southern India. In Africa, the city of Gao was the leading city of the powerful Islamic Songhai Empire in Mali. This is not to mention the significant numbers of other empires, kingdoms, and polities that also occupied the tropical portions of the Americas, Africa, and Asia, jostling with these larger powers for prestige, wealth, and survival, as well as the innovative, resilient island societies we met in Chapter 8. Even in the Amazon Basin, Chapter 9 has shown us, well-populated "garden cities" sprawled across lowland evergreen rainforests, and the latest estimates put the human population of the precolonial Americas at 60.5 million, just shy of Europe's estimated total at the time. As the last two decades of historical and archeological research have revealed, tropical landscapes at the time of Columbus's voyages were certainly not empty. Instead, they were very much full of active rulers, merchants, and food producers who took part in vast exchange systems that already spanned across the Americas, across Africa, and across Southeast Asia prior to European arrival. Perhaps unsurprisingly, many Latin American nations therefore question the validity of celebrating misguided, and often murderous, European individuals instead of the Indigenous achievements that preceded them.[3]

So what changed? How did powerful tropical states and individuals, in stark contrast to their European counterparts, end up as the "people without history"?[4] Contact between Europeans and the diverse societies inhabiting the tropics generated shockwaves that have echoed through the landscape of global history ever since. We will now zoom in on tropical forests as the crucial theater for the emergence of a new, interconnected world. Following the latest archeological, ecological, and historical thinking, we will see how the interaction between Europe and the tropics brought disease, warfare, forced relocation, and demands for labor and land that catastrophically impacted Indigenous populations and their traditional management activities. Not only that, but through both deliberate and unintentional "Columbian exchanges," plants, animals, people, and beliefs began to regularly move between cities as far apart as Madrid, Mexico City, and Manila, reconfiguring entire tropical ecosystems. The outcomes for landscapes and societies were shaped not only by the political, economic, and religious situation back in Europe but also by politics within and between Indigenous states, by local merchants seeking to tap into new global flows of goods, and by Indigenous food producers and hunter-gatherers incorporating or resisting new crops and animals. However, by the end, the varied, and often violent, processes of colonialism and the overall European control of global flows of wealth across oceans led to a masking of the previous human achievements in the tropics. Ultimately, this collision of worlds left us with the pervasive Euro-American assumption that tropical forests could only be successfully occupied by small-scale, mobile, hunting-and-gathering communities—"green hells" to be profitably converted rather than productive, manageable "forests of plenty."[5]

IN THE FIFTEENTH and sixteenth centuries AD, the expansion of the Spanish (in the form of the combined Catholic Monarchy of Castile and Aragon) and Portuguese empires began the joining of two "worlds" that had been separated since the division of the Pangaea supercontinent

176

Figure 10.1. Map of the Iberian empires and regions mentioned in Chapter 10.

Tropic of Cancer

Macau

Spanish
East
Indies

Goa

Portuguese
Ceylon

Malacca

Ternate

INDIAN OCEAN

Tropic of Capricorn

Extent of Spanish Empire
Extent of Portuguese Empire
Boundary of Brazil

Europe (Iberia) and the tropics in the Age of Exploration

Canary Islands

a) Spanish arrive on the Canary Islands in 1402

b) Diseases rip through Indigenous populations in 50 years following first arrival

c) Gruelling campaigns against Indigenous Guanche populations across the different islands, see final subjugation of Tenerife in 1496

Neotropics

a) Columbus lands and establishes settlement on Hispaniola 1492

b) Columbus returns to find settlement sacked in 1493

c) First permanent settlement on the mainland in Venezuela in 1501 (Cumaná)

d) Cattle, sheep and goats start to arrive in the Caribbean and Central America (1501–50)

e) Portuguese reach coast of Brazil in 1502

f) 150 years following Iberian arrival witnesses various waves of disease destroy Indigenous populations (a little later in Brazilian Amazon)

g) Initiation of sugarcane planting in fields in the Caribbean within 25 years of landing

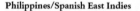

Philippines/Spanish East Indies

a) Spanish attempt at colonizing the Philippines begins in 1565

b) First Spanish settlement at Cebu in 1565

c) Spanish conflict with 'Moro' populations including Borneo Sultanate, Maguindanao Sultanate and Sulu Sultanate. Ongoing resistance to Spanish rule between 1565 and 1898

d) Spanish take Manila (present capital (1571)) and establish Spanish East Indies, which includes bits of Taiwan, Guam, the Marianas islands and Palau

Africa

a) Mid-15th century, caravel invented and Portuguese reach West Africa

b) Expansion of sugarcane cultivation on Madeira in 1455

c) São Tomé and Príncipe found in 1469 and sugar plantations begin there

d) From late 15th century, expansion of Portuguese trading with West and Central African states

e) Portuguese traders settle at sites like Elmina under the watch of local elites

f) Maize reaches Africa around 1500

Figure 10.2. Timeline of events discussed in Chapter 10.

h) Spanish with Indigenous allies defeat Triple Alliance and claim sovereignty over Mexico – 1519–1521 (Aztec)

i) Spanish and Indigenous allies capture Inka leader Atawallpa in 1532

j) Potosí mine (Bolivia) opens in 1546

k) Final defeat of the Inka Empire following gruelling war in 1572

l) Portuguese establish captaincies and sugarcane plantations are expanded rapidly in areas such as Pernambuco

m) Gold discovered in Minas Gerais, Brazil, in 1693, causing a rapid influx of population until 1739

n) Portuguese settle region that is now Amazonas state and formally incorporate into empire in 1750

e) Manila 'galleon trade' begins in 1571

f) Introduction of sweet potatoes, horses, etc. between 1630 and 1800 AD. Sweet potatoes taken up widely by Indigenous populations

g) Initiation of wet rice agriculture on the Ifugao terraces between 1630 and 1800 AD, with staunch resistance to Spanish rule in the Highlands

c

f, g

Asia

a) Vasco da Gama rounds the Cape of Good Hope between 1497 and 1498

b) Portuguese reach south-western India in 1498

c) Portuguese establish forts on Indian and Sri Lankan coastline and capture Goa amid a series of conflicts with local rulers 1503–10

d) Portuguese capture of Malacca in 1511 disrupts thriving spice trade between states in India, the Middle East and China

e) Portuguese conflict with China until China allows Portuguese traders to settle, enabling a new flow of trade between Europe, Goa and China 1545–57

f) 17th century, Portugal fades from prominence in the Indian Ocean due to defeats to local rulers and Dutch assaults

| 1700 | 1750 | 1800 | 1850 | 1900 |

Year

approximately 150–65 million years ago. Of these empires, the Spanish quickly became the largest. Following the gradual occupation of the Canary Islands (1402–1496), and Columbus's landfall at Hispaniola in the Americas in 1492, Spain expanded its pantropical grip across the Caribbean, North and Central America, and South America and, toward the end of the sixteenth century, parts of Southeast Asia. So how did they do it? Gunpowder, horses, steel, and shipbuilding innovations certainly helped. So too did the ruthless enslavement of local populations that began almost as soon as Columbus stepped ashore. The maneuvering of local political rivals, often using the Spanish to settle their own scores, was also key. Yet one of the most terrible products of the so-called Columbian exchange, reconfiguring power dynamics both regionally and globally, was the fact that Europeans did not just bring themselves to the tropics. As we now know from extensive archival research of records left by colonizers and Indigenous writers, as well as the recent groundbreaking ability of archeogeneticists to extract DNA not just from ancient humans but also from the diseases that debilitated them, they also brought along the microbial causes of deadly measles, smallpox, typhoid, influenza, and bubonic plague. Where populations had never before been exposed to them and had little to no immune response, the result could be catastrophic. Reconstructing exact population changes, pre- and postcolonial arrival, using census data, archeological sites, and other forms of historical records is notoriously contentious, and fiery debates relating to estimates of precolonial populations in the Americas have raged for the last fifty years. Nonetheless, demographic and earth systems modeler Dr. Alexander Koch of Hong Kong University tells me that he and his team's latest compilation of published estimates from colonial census figures, archeological data (e.g., numbers of buildings, settlement size), and predictions based on assumptions as to how many people a given environment could hold "suggest a staggering ~90 percent of the Indigenous population of the Neotropics was wiped out within 150 years of Columbus's arrival by repeated waves of these diseases." That is nearly 55 million people.[6]

Two prominent, but very different, tropical examples highlight this reality. In 1492, Tenochtitlan (now, for the most part, buried under the present-day capital of Mexico, Mexico City) was one of the most densely populated cities on Earth. Tenochtitlan, the home of the Mexica ethnic group, lying just south of the Tropic of Cancer, sat at the center of the so-called Aztec Empire, based on the Triple Alliance of three city-states, which extended out across the Valley of Mexico and into the subtropical and tropical lowlands and highlands beyond. The complex drainage system of this often-flooded area, specialized raised fields known as *chinampas* for growing a diversity of crops, including maize, chili peppers, and beans, and canals for transport as well as fish formed the basis of a vast 14 km² anthropogenically constructed "floating" urban landscape. Royal courts, the gold, silver, jewels, and exotic feathers of wealthy elites arriving through long-distance exchange networks, detailed inscriptions and texts, vast markets, and giant temples built for human sacrifices both impressed and appalled the Spanish arriving at the capital in 1519. Numbering just 630, the Spanish, headed by Hernán Cortés, defeated this giant through political and biological opportunism. First, after initial encounters and skirmishes, the Tlaxcala, the traditional enemies of the Triple Alliance, entered into an alliance with the Spanish. This was made easier by the fact that an enslaved Indigenous woman from the Gulf Coast of Mexico, La Malinche, gifted to the Spanish by a Maya state, acted as an interpreter, advisor, and intermediary for Cortés. Backed by military support from their local allies, the Spanish advanced onto Tenochtitlan, where they somehow managed to abduct King Motechuzoma I from right under the noses of his subjects. Following retaliation and ejection from the city, the Spanish returned, with reinforcements, and another ally, this time from the nearby city of Texcoco, to lay siege to the capital. Nonetheless, really, given the ongoing staunch Mexica resistance they faced, if it had not been for a smallpox pandemic tearing through Tenochtitlan prior to the crucial battle for the city, something documented in the colonial archival records of the time, they might never have succeeded. Further pandemics, alongside forced labor of enslaved Indigenous peoples in mines and brutal public executions, reduced and pacified the population of the Valley of Mexico, making the subsequent maintenance of control more feasible than it might otherwise have been.[7]

A similar end awaited the Inka Empire among the montane tropical forests and grasslands of the High Andes. By area, in 1492 the Inka

had the largest empire on Earth, spanning from lowland evergreen rainforest in the Amazon Basin to the deserts of Peru. A humongous 6,000
km^2 (that's an area greater than 1 million soccer pitches!) of agricultural
terraces identified by archeological landscape survey have been linked
to precolonial Andean populations, perhaps even giving the mountain
range its name (from *los Andenes* for steps or terraces) (though see
A. Borsdorf and C. Stadel's *The Andes: A Geographical Portrait* for discussion of alternative origins). Many of these likely supported an Inkan
agricultural system based on crops of potatoes, quinoa, and maize and
domesticated llamas, alpacas, and guinea pigs. A sophisticated archeological and historical record of pictographs, winding roads, well-placed
warehouses, and knotted-string counting systems that acted as a means
of conveying tax obligations, census records, calendars, and military orders enabled the Inka to transport goods produced in different corners
of the empire to where they were needed. All of this was ruled out of the
mountainous city of Qosqo (now Cusco), with its glittering rulers, monumental plazas, giant stone blocks covered in plates of gold, shrines, and
tombs. In 1532, an even smaller Spanish party, 168 men led by Francisco
Pizarro, moved on the Inka territory, having already seen its riches in
1526. While horses and cannon certainly helped, Pizarro, like Cortés,
benefited from the local political situation. The Inka Empire had just
emerged from a grueling civil war. Pizarro made the most of the fact that
the winning ruler, Atawallpa, left his mountainous citadel and met him at
Kashamarka (Cajamarca), where Pizarro captured, blackmailed, and then
killed him, throwing his subjects into disarray. Pizarro progressed onto
Qosqo. Local factions within the empire saw an opportunity for advancement and supported the Spanish advance. Nevertheless, as Pizarro's own
journals attest, he never felt that he was in a strong political position.
Once again, this improbable conquest (it took until 1572 to complete)
was spurred on by rampant smallpox, typhus, and influenza epidemics,
vividly recorded in text and images by Inka and Spanish chroniclers alike,
which had already killed Atawallpa's father, as well as 200,000 subjects,
and continued to catalyze civil unrest and eventual capitulation to the
Spanish. Even then, though, the conquest of what was to become the
"Viceroyalty of Peru" took decades, with the mountainous and forested
terrain frequently invalidating Spanish military advantage.[8]

The Canary Islands, in many ways, provided the subtropical stepping-stones for the Spanish on their path to the Americas. And here, once again, the spread of diseases was crucial to conquest. Control of the Canary Islands significantly reduced the voyage length for the Spanish to reach the Americas (in much the same way the Azores islands facilitated long-distance Portuguese voyages). As Friar Espinosa, a sixteenth-century Spanish historian and chronicler, wrote, "If it had not been for the pestilence it [the conquest of the Canary Islands] would have taken much longer."[9] Even then, fierce resistance by well-organized societies, particularly on the islands of Fuerteventura and La Palma, meant it was a long, hard, and often fortuitous slog to victory for the Spanish. Similarly, the first island footholds in the Americas within the Caribbean were only taken following tides of disease that ripped through resident populations. Just fifty years after Columbus's arrival on Hispaniola, the local Taíno population, estimated at between 100,000 and 1 million people, which had initially put up forceful resistance to Spanish settlers, had all but completely vanished thanks to the spread of disease, warfare, and enslavement. Slightly later in time, the rampant documented advance of smallpox and measles throughout Brazil from the sixteenth century onward, along with enslavement of Indigenous groups, murder, and forced relocation to towns that exacerbated microbial transmission, reduced a once staunch resistance against Portuguese expansion into the evergreen rainforests of the Amazon Basin that numerous European expeditions had shown themselves completely unequipped to deal with. In the absence of these microbiological stowaways, it is hard to see how some of the most magnificent cities and largest, most powerful states that existed anywhere on Earth at that time could so rapidly succumb to invasion. Clearly, in many of the cases mentioned, Indigenous political rivals often exploited European arrival for their own advancement. Nonetheless, cruel European corralling of Indigenous populations as enslaved labor and malnutrition resulting from warfare and the breakdown of existing political and social structures, as we know from more recent pandemic experiences, undoubtedly would have paved the way for perhaps the worst epidemiological disaster ever recorded.[10]

These undoubtedly horrific impacts are now well attested, and these diseases have become a major part of discussions of the process of early

European colonization. However, this does not mean that the exact same patterns of Columbian exchanges played out everywhere in the tropics, and Europeans and Indigenous populations both negotiated these new connections in a variety of different ways. For example, in the case of Spanish arrival in the Philippines in 1565, Miguel López de Legazpi and his navigator, Andrés Ochoa de Urdaneta y Cerain, settled Cebu and had conquered Maynila (now Manila) by 1571. While European diseases may still have had an impact, the existence of greater local immunity, dispersed Indigenous settlement patterns, and different political structures led to distinctive patterns of Spanish presence and colonization. Slow, morbid warfare against the existing sultanates of Maguindanao, Lanao, and Sulu failed to yield Spanish economic or political control in these parts of the archipelago. Indeed, these fights arguably never ended, given ongoing battles today between the Philippine government and Islamic extremists on the island of Mindanao. The Spanish committed many atrocities in the Philippines. Even during the first five years of settlement on the island of Luzon, soldiers sacked villages, seized food, and enslaved inhabitants. Similarly, in the late seventeenth century, Indigenous converts to Christianity disappeared from founded missions in the Cagayan Valley as they sought to avoid documented ill treatment from a garrison of soldiers at Itugud. Once again, Indigenous resistance is evident. Archeological excavation and scientific dating methods have shown that the majestic UNESCO-protected Ifugao rice terraces we briefly met in Chapter 7, while used for taro in precolonial times, were expanded and repurposed by large Ifugao populations that moved into the highlands to escape and resist Spanish rule in the lowlands. Early Portuguese interactions with West and Central Africa show a similar pattern. Here, local kingdoms such as those of Benin and Kongo actually seem to have grown in response to new trading opportunities in the fifteenth century, materially displayed in the form of remarkable ivory objects deliberately designed for export to Europe and European pipes found in dwellings at the site of Savi in Benin. The Portuguese participated in intermarriage and local politics and constructed their buildings adjacent to existing royal enclosures in traditional architectural forms, only ever really maintaining a firm, settled foothold on the coast and largely relying on African associates to extract resources and enslaved labor from locations further inland. In this case, as we will see in Chapter 11, the coordinated expansion of the transatlantic slave trade, particularly between the seventeenth and nineteenth centuries, would take the truest toll on local populations across the African tropics and encourage, alongside the discovery of quinine as a potent antimalarial drug from the nineteenth century, later European conquests into the interior.[11]

The "Great Dying" of the Americas, as it has become known, turned the global demographic and political status quo on its head. Once thriving populations collapsed, with between approximately 80 and 95 percent of Indigenous populations lost in different parts of the Neotropics, based on a number of different estimates. Today, the sheer scale of this destruction is hard for us to comprehend. The decline of Indigenous populations, from the Caribbean to Central America, from the Andes to the Amazon, led to European writers later documenting failing or small populations in tropical forests that still dictate our stereotypes of these environments today. The initial arrival of the diseases may have been unintentional. However, the historically documented abuses that occurred at the hands of European invaders, which exacerbated and followed this disaster, were not. For example, the Quechua nobleman Felipe Guaman Poma de Ayala (or Huamán Poma) wrote a famous treatise in Spanish titled *El primer nueva corónica y buen gobierno* (or *The First New Chronicle and Good Government*) that chronicled and denounced the ill treatment of Indigenous populations within the former realm of the Inka in Peru. The ongoing marginalization of Indigenous groups up and down the American tropics stems from these initial, disastrous, genocidal encounters. Archeological and paleoecological evidence for the active introduction of new plants, animals, and land tenure had similar impacts on traditional use of tropical environments. Much of the Americas, as well as the Canary Islands, saw widespread abandonment of well-honed strategies for food production in subtropical and tropical forests as European agricultural species and ideas were imposed on landscapes and Indigenous populations dwindled. Nonetheless, groups that did survive across the American tropics, as well as populations in Asia and Africa, often actively resisted relocation, dictated trade, and gave or did not give the local ecological knowledge that would enable European colonizers to settle different regions. The arrival of Europeans,

as well as the diseases, plants, animals, and worldviews they brought
with them, had major, sometimes irreparable, impacts on societies and
landscapes in the tropics. Yet, given the often slow and limited arrival of
Spanish or Portuguese settlers, Indigenous populations, as well as new
cultures and groups formed in the melting pot of an interacting world,
played a major role in how these things were transposed onto varied
tropical canvases.[12]

THE CREATOR OF the term "Columbian exchange," historian Alfred
Crosby, in 1972 revolutionized historical research into European colonialism by focusing on both the cultural *and* biological consequences
of the meeting of two worlds. Crosby not only was one of the first to
highlight the potential role of transported European diseases in shaping
a "new world" but also discussed how plants and animals from both sides
of the Atlantic affected landscapes and cuisines the world over. In the
context of the tropics, the arrival of the Spanish, who took a particular
interest in implanting their own forms of "agriculture" and "pastoralism,"
could have major cultural and environmental results. Wheat, as well as
grape vines, were particularly important to them, given their association
with Christian sacraments. Historians have discussed how many Spanish settlers, including Hernán Cortés, attempted to plant these Middle
Eastern and Mediterranean crops, which ultimately failed, particularly
in humid, lowland forest areas. Several faced severe malnutrition as a
result, and historical records document how, despite resistance, they
were forced to eat the local staple crops of maize, manioc, peanuts, or
pineapples like their Indigenous neighbors. Indeed, a number of these
Neotropical crops still form the basis of cuisines in many North, Central,
and South American nations today. Other crops introduced by Europeans were more successful. By the sixteenth century, observers in Peru
recorded how lettuce, cabbage, radishes, peas, onions, and turnips were
all being cultivated in the Andean highlands. Meanwhile, the Spanish
and Portuguese also brought ecologically well-suited crops from their
other tropical outposts in Africa and Asia. As a consequence, bananas,
mangoes, coconuts, and oranges are all successfully cultivated up and
down the American tropics today and, particularly in the case of oranges

in Florida, have even often become popularly associated with this part of the world. In this period of new oceanic connections, the tropics not only witnessed arrivals from outside their latitudinal boundaries but also a reshuffling of the plants grown and animals raised in different corners of their equatorial realm.[13]

More dramatic environmental consequences occurred as a result of the introduction of domesticated animals, including sheep, goats, cattle, donkeys, and horses. Ranching of cattle and sheep, as well as the keeping of pigs, became perhaps the single greatest economic practice across the Neotropics, from Mexico, across the Caribbean, and south into the Amazon. In Mexico, the introduction and dramatic expansion of sheep, consuming all vegetation in their path, has been shown, through paleoecological analysis of lake records and historical reconstructions of herding, to have led to a drying out of the landscape, reduction of forest cover, increased soil erosion, and the introduction of invasive species through their dung. Similarly, a pollen record from a dried-up river meander bed in the Dominican Republic shows how the introduction of cattle by the Spanish resulted in a decline in tropical forest density on Hispaniola. Cattle were, and still are, a main driving force behind the clearance of the Amazon Basin from the sixteenth century onward. Even closer to home on the Canary Islands, where sheep and goats were already present, Spanish introduction of new animals, like donkeys, reduced the landscape to such an extent that, in 1491, it is recorded that the Spanish decided to ride out on donkey hunts on Fuerteventura to keep the land profitable! The rabbits introduced on Porto Santo likewise consumed the Spaniards' crops, causing them to abandon the island. Nevertheless, in Mexico, zooarcheological analysis of animal remains from colonial-era farming sites has shown how Indigenous communities could effectively combine these new agropastoral species within traditional *milpa* forms of food production. Similarly, in the Caribbean, analysis of linguistics and historically documented Indigenous cosmologies show how domesticated animals brought by Europeans were incorporated into long-held traditional schemes of adopting and taming a variety of animals, including parrots and manatees. Sadly, all too often Indigenous population decline and maltreatment dampened the influence of these more ecologically canny approaches.[14]

The Columbian exchange did not just operate in one direction, and many crops domesticated in the Americas also made it back to the Old World. We might often associate capsicum or chili peppers with spicy Indian food, Chinese Sichuan cuisine, or Korean Kimchi cabbage. However, as we saw in Chapter 7, these were first domesticated in the Neotropics. Historical records show that they were brought across by the Spanish to Europe and Africa, becoming the basis of Hungary's national spice "paprika," before being carried eastward by the Portuguese in the early sixteenth century to be deeply embedded in the cuisines of South and Southeast Asia. Meanwhile, tomatoes, shown by botanists to have been domesticated in South and Central America in the first millennium AD, are first mentioned in European texts in 1544. Cultivation of this fruit in Italy in the sixteenth century paved the way for the creation of some of the most globally distinctive sets of cuisines on the European continent. The Portuguese also likely introduced maize into Africa in the sixteenth century, while manioc was similarly brought to the African continent as part of this process. Cacao, cultivated by farmers in Central America at European arrival, became a taste of kings and queens across much of Europe and a widely distributed drink in the new globalized capital of Madrid, where it remains a famous tourist attraction as a drink in combination with *churros* to this day. You might associate vanilla with Madagascar; however, historical archives and botanical research show that it too comes from the Neotropics, arriving in Africa via Europe through first Spanish- and then French-mediated hands. The onset of global addictions to cocaine and tobacco, as we saw in Chapter 7, can also ultimately be linked via the Columbian exchange to the Neotropics, although the latter had already been passed up to North America prior to the arrival of the French and British, who developed a liking for it. Zooarcheological and genetic analyses also demonstrate that guinea pigs and turkeys, common pets or Christmas dinners across western Europe and northern North America today, arrived from the American tropics as a consequence of the new global networks that emerged between the fifteenth and seventeenth centuries.[15]

Much discussion of the Columbian exchange often focuses on the Atlantic. However, by colonizing the Philippine Archipelago, the Spanish formed a regular global exchange system that united the Americas,

Europe, and Asia. A "galleon trade" was created between Manila and Acapulco in Mexico (then known as New Spain), which, from 1573 saw two fully loaded ships take products from Asia to the Americas and back again in most years. While wheat was eventually successfully cultivated in the Philippines at Spanish direction later in the seventeenth century, less direct European control over agrarian activities than in much of the Americas led to prominent, locally directed incorporation of novel plants and animals, both by Indigenous populations and the more independent religious orders, into subsistence activities. Sweet potatoes arrived from North and Central America and quickly became widely used in local swidden fields, having similar properties to natively cultivated yams and taros. Their journey from the Neotropics is highlighted by the fact that sixteenth-century accounts show that Indigenous communities in the Philippines called them *camotes*, the Nahuatl (and Uto-Aztecan language) word for the crop. Other arrivals included tomatoes, papaya, pineapples, agave, squash, and tobacco, as well as potatoes, whose global journey we will read more about in the next chapter. Zooarcheological analysis of sites in the Philippines shows that goats were already present in the Philippines in the pre-Spanish period. However, sheep and horses were likely introduced by the Spanish. Meanwhile cattle keeping was more limited here, at least prior to the eighteenth and particularly the nineteenth centuries. As Filipino zooarcheologist Dr. Noel Amano of the Max Planck Institute for the Science of Human History puts it, "Long-term Indigenous use of water buffalos for traction and manure, that still continues in the rural areas of the archipelago today, meant that the new arrivals from Europe provided far fewer local benefits away from a few wealthy, Spanish landowners who could keep thousands of them in profitable herds."[16]

As the example of the Philippines demonstrates, the spread of crops and animals from their regions of domestication into novel ecologies was neither sweeping, uniformly European-dictated, nor necessarily desirable. Archeobotanist Professor Amanda Logan at Northwestern University has further highlighted this point in the context of the spread of maize into tropical West Africa. The arrival of maize has frequently been seen as a "grace" for a continent often framed as doomed to scarcity. Its rapid growth and high yield provided the opportunity to produce a large surplus and support local populations, and monoculture fields of maize continue to be promoted as a means of alleviating food insecurity across Africa in the twenty-first century. Nevertheless, working with the Banda community of West Ghana, Amanda has elegantly compiled ethnographic interviews, historical information, and archeological and archeobotanical data to challenge these narratives. First, shifting cultivation of indigenous crops of pearl millet and sorghum, alongside tubers, legumes (e.g., cowpea), and the collection of wild shea nuts, was more than capable of supporting African communities in the region at the time of European arrival. Maize was virtually ignored until later centuries, suggesting that people had plenty to eat. Second, maize can prove incredibly costly for soil fertility, is vulnerable to pests, and can struggle under dry conditions that periodically impact the seasonally dry forest areas of western Ghana, as they did at the time of European arrival. Finally, maize became important more quickly along the coast. Here, African farmers profited from growing it in large quantities to meet increased demand from Europeans, who required it to feed themselves, as well as the economic demands of new, capitalist business initiatives centered on human souls. As Amanda puts it, "An obsession with Africa as a continent lacking in food security, and in need of surplus from introduced Eurasian and American crops, goes completely against millennia of African farming innovations that effectively managed environmental perturbations and that processes of colonialism have prejudiced over the last five centuries."[17]

Indeed, the introduction of new crops and animals has, as we have seen in the Canary Islands and the Americas, often left legacies on landscapes that are not entirely positive. In the case of Africa, a focus on monoculture crops such as maize can leave farmers more vulnerable to climatic change. In China, the arrival of maize and sweet potato, with their very different climatic and soil tolerances when compared to Old World crops, offered the opportunity to expand agriculture into new earthy horizons. Sweet potato arrived from the Philippines in the late sixteenth century and provided a crucial food source for local farmers reeling from climatic-induced devastation of their rice fields. Sweet potato and maize, which arrived via Portuguese-held Macao, also supported emigrating Chinese farmers moving as far southwest as Sichuan and as far northwest as the Gobi Desert. By the eighteenth century agricultural capacity and populations in these previously sparsely utilized regions skyrocketed. Paleoecological records and historical observations document the increasing deforestation and soil erosion that resulted. The rich nutrients and carbohydrates provided by these new crops certainly, in part, contributed to China becoming the most populous country on the face of the planet. However, geomorphological study of ancient soil quality, in combination with archival references to soil instability, shows that these new crops replaced tree cover. In doing so, they contributed to the washing away of unprotected soils, the loss of nutrients, an increase in major flooding, and, ultimately, malnutrition of overextended rural populations that the country is still grappling with in many areas today. The Columbian exchange certainly operated in various directions between the now linked worlds, both in terms of cultural influences and potential environmental and nutritional repercussions.[18]

Although the Columbian exchange is often popularly described as a product of nations or empires consciously "transporting" and using different crops or animals (and I am also guilty of that in the above!), the examples of West Africa, the Philippines, and China show that, ultimately, it entailed a network built upon the backs of a variety of merchants, food producers, travelers, writers, and government officials from all continents and cultures. Seeds and animals simply were not enough; their appropriate management, especially in the tropics, required different traditions of cultivation and herding that came in the form of people moving. To take one revealing example, between the sixteenth and nineteenth centuries, approximately 12 million to 13 million Africans were forcibly removed from their homes and transported across the Atlantic as part of the transatlantic slave trade. We will see in the next chapter how this represented the onset of increasingly globalized, capitalist, and racist exploitation of landscapes and labor in the tropics. However, it is also worth noting that the Africans who were forcibly shipped, and whose descendants now make up significant proportions of many nations in the Americas, brought with them yams, black-eyed peas, watermelons, and plantains, as well as knowledge of how to farm them, working small "plots" at the edges of the plantations on which they were forced to work. Not only that, but as many of these individuals had already been farming in fields with maize and manioc back in West Africa from the sixteenth century, they also brought a cultural tradition of combining these different foods into new, sustainable systems and eclectic cuisines. Rice cultivation in North America has been suggested to have arrived in a similar manner from West Africa, rather than Asia. These are far from the only examples, but they clearly show that often-neglected, marginalized groups were also key players in the formation of new post-Columbian ecosystems and economies.[19]

EUROPEAN COLONIZERS BROUGHT along their own belief systems and concepts of land use as well as new biological entities. Intense, direct control over the natural world, concepts of property, ideas for how a settlement should be laid out, and the priorities placed on maximizing output from a given portion of land represented a new, frequently traumatic experience for tropical landscapes and their occupants in what has become known as "ecological imperialism." We can find one of the more obvious examples of this new ideology of exploitation in the form of Iberian desires for Neotropical silver and gold. Seeing the rich use of ores by Indigenous populations, the Spanish set about fervent survey and extraction. Beginning in the mid-sixteenth century, after just one hundred years, Spain alone had extracted three times Europe's previous silver wealth. The scale of these operations shows in the size of mining cities that emerged, such as the high-altitude city of Potosí, Bolivia, which by 1660 was home to 100,000 people—more than Madrid and Rome at the time. Forced movement of enslaved individuals and the flocking of Europeans, among others seeking to make their fortune, led to rampant deforestation to fuel both mining and construction. Similarly, when rich gold deposits were found in the current state of Minas Gerais in Brazil in the late seventeenth and early eighteenth centuries, the region's dry tropical forests were destroyed to make way for mines, pastures, crop fields, and villages to support the thousands of people seeking quick enrichment. In the case of silver, more insidious environmental impacts also came about as a result of the use of mercury to more efficiently extract this precious metal from the sixteenth century onward. At mining locations such as Huancavelica in Peru, the release of mercury into soils, water systems, and the atmosphere, not to mention the lungs of forced laborers, had lethal, lasting consequences visible in the chemical contents of the soils of the exploited regions to this day, with ongoing impacts on the health of contemporary inhabitants. Some of these effects were documented in pre-Columbian periods in sediment cores. However, the invading Iberian powers undoubtedly ramped them up to whole new levels to support expanding colonial infrastructure, warfare, and their economic and political designs back in Europe.[20]

Colonial Iberians also brought their ideas of where and how to live to their new tropical colonies. In many cases, the Spanish built their administrative centers on top of the Indigenous cities they so admired. However, they often neglected the traditional knowledge that had dictated landscape use prior to their arrival. Hernán Cortés famously refused to abandon Tenochtitlan despite the fact that, without Indigenous knowledge of water and field management, the colonial residents were vulnerable to stagnant water and seasonal flooding. Similarly, in the Philippines, the Spanish built many towns and churches in river valleys, then later abandoned them due to flooding. In other cases, the Spanish, and their Portuguese counterparts, made their own towns and cities, often laying them out around central plazas, with key, wide streets and specified sectors for craft activities, just as they had back home. Environmental historian Shawn Miller has noted that, amazingly, nearly 50 percent of people in Spain's American colonies lived in cities by 1600, something not achieved in England until the mid-nineteenth century. Although this figure is likely problematic, given that many Indigenous people deliberately fled into the hinterlands and rural areas to avoid census recording, it does show the emphasis that the Spanish placed on attaining their urban ideals. Across the wider landscape, analysis of historical records shows that the Spanish also tried to organize the populations in their new tropical realms for maximum political and economic control. This included the system of *reducciones* (or *aldeias* in the case of Portuguese realms) from the sixteenth century, which forcibly relocated many Indigenous people into settlements. This enabled easier taxation, census counting, enculturation, and recruitment of Indigenous populations through *encomiendas*, royal grants of effectively enslaved labor to chosen Spanish elites. In the case of the Philippines, this involved centralized *cabacera* towns that were frequently built around a church and town hall, with the former holding immense power over the labor and disciplining of local Indigenous populations. These systems often also benefited Indigenous *alcaldes* (mayors) and *gobernadorcillos* (low-level governors) who sought to gain influence. Still visible in the building plans and architecture of key cities such as Vigan and Taal today, these novel forms of settlement organization could all impact the landscape, demanding wood for construction, more intensive cultivation of fields for tribute and for larger, urban populations, and greater intensity of domesticated animals in a given locality. European ideologies as to how people should live were thus transplanted from the other side of the world into the tropics.[21]

While the effects of mining and new types of settlement and organization could leave deep scars upon local tropical landscapes, more pervasive changes occurred as a result of the imposition of plantation agriculture. The Spanish realized that many economically seductive crops could be more productively grown in their new tropical lands, away from native pests. Prime among these examples is sugar, which originated in

Figure 10.3. The streets and buildings of Vigan City in the Philippines are today a World Heritage site and show the way the Spanish sought to impose their ideals of urbanism in this part of the tropics. *Jason Langley / Alamy*

the tropics of New Guinea. As early as 1493, on Columbus's second voyage to the Americas, the Spanish brought sugar for cultivation in the Caribbean, and by the sixteenth century, it was also being grown across the Canary Islands, Mexico, and Central America. The Portuguese, meanwhile, had introduced sugar to the Atlantic islands of Madeira and São Tomé and, by 1516, the shores of Brazil, planting it directly within former Indigenous field systems as a morbid symbol of what was to come. While not always grown in a strict "monoculture" at this early stage, sugar sucked nutrients from soils and led to vast deforestation for both space and the fuel for its refinement. By 1550, Madeira's forests were in tatters, and sugar production had to be abandoned; historian Jason Moore has estimated a cumulative deforestation total of around 155 km^2 in just over a century. By the 1600s, the forests of the Canary Islands and Brazil's Atlantic coast faced similar pressures. The same type of overexploitation can also be seen in the Spanish focus on "ranches" of livestock. Here, herds of hundreds of thousands of animals were used for profit rather than subsistence. These models of land ownership, land use, and demand-driven operation were new to these landscapes and still haunt them to this day. This is not to mention European merchants' new market-driven patterns of exploitation of wild animals such as marine mammals off tropical coasts, civet cats hunted for their skins and scent glands, and parrots captured for their feathers.[22]

These changes in perception of the natural world, ownership of the land, and intensity of land use to feed wider, global tastes and demands undoubtedly had massive pantropical implications. However, once again, as the historians Amélia Polóna and Jorge Pacheco have highlighted, it is important to balance the global picture with the local result. As we have seen, environments, existing Indigenous populations and traditions, and different forms of European rule could shape the ultimate outcomes of colonial Iberian tropical endeavors. Ecologies, soils, and histories of use molded the nature and impacts of sugar production in different tropical island and mainland contexts. In contrast to the Americas, gold mining in Africa followed Indigenous patterns of exploitation until the nineteenth century, as the Portuguese, mainly located on the coast, were unable to interfere directly in the mines. Similarly, different patterns of *reducción* and *aldeias*, seen in historical and archeological records of colonial-period settlements, shaped demography and the spread of disease to variable extents between the Caribbean, Mexico, the Philippines, religious missions in Brazil, and traditional settlements in Africa. Finally, Indigenous populations, instead of being "uncontacted" and "isolated," could be highly active in participating in and shaping these new economic systems. More traditional approaches to food production often fed the workers and enslaved people laboring in new agricultural field systems; meanwhile historical texts state how hunter-gatherers in India and Brazil brought honey and jaguar skins, respectively, to the worldwide market. Not only that, but a number of European invaders intermarried with Indigenous populations. Although this frequently entailed its own form of power dynamics, often involving European men marrying the daughters of local elites, it did eventually result in the forming of new colonial identities and cultural mixes or, in the case of some Europeans, the abandonment of previous ways of life and the joining of new families in tropical forest environments.[23]

The Philippines provides a particularly elegant example of some of these more local colonial experiences. In the present-day capital of Manila, although the Spanish erected a stone-walled town to develop a clear plaza and compact urban system that was more familiar to them, they were repeatedly distressed by the appearance of a sprawling "second" Manila outside its walls. This so-called Extramuros, or "city beyond the walls," began developing as early as 1583 and included a district, known as the Parián, inhabited by Chinese merchants and their families eager to access silver from Acapulco. It eventually dwarfed the official city, and the Spanish tried several times to remove and destroy it. Similarly, as Filipino archeologist Dr. Grace Barretto-Tesoro of the University of the Philippines puts it, "The enculturation of Indigenous populations occurred at a different pace and to different extents across the archipelago and was often dictated by local populations." For example, Indigenous populations on Mindanao, known as the *Lumad*, took advantage of Spanish presence to counteract the powerful Muslim groups of the southern Philippines. Back on the terraces of Ifugao, growing, politically powerful populations actively traded with the Spanish while also raiding and resisting them. The precolonial environmental and cultural situation of the archipelago also fashioned Spanish interactions with the environment. As environmental historian Greg Bankoff has elegantly noted, endemic termites, or "white ants," meant that only thirty to forty species of tree could be profitably felled for forts, shipbuilding, and construction by Indigenous populations and Spaniards alike. Thus, although certain key species became threatened, archival reconstruction of forest cover through time shows that blocks of overall forest remained until nineteenth-century commercialization. These local colonial ecological and cultural experiences occurred across the new Iberian realms.[24]

THE DECAPITATIONS AND removals of statues of Christopher Columbus up and down North America in 2020 show the degree of controversy that his legacy continues to engender. More and more Indigenous activists and institutions are calling for the replacement of Columbus reverence with a celebration of precolonial Indigenous societies and an acknowledgment that the actions of European "explorers" and their consequences for tropical cultures, populations, and landscapes are by no means cause for fanfare. The rampant advance of European diseases, alongside active enslavement, murder, abuse, and invasion, meant that the previously significant population of the Americas did not reach its former levels until the nineteenth century. The introduction of new plants and animals often permanently reconfigured ecosystems, leaving lasting legacies and conservation challenges to this day (as we will see in Chapter 13). Yet the agency of Indigenous and other marginalized groups, tropical diseases, and difficult geography could still dictate or even prevent Iberian expansion into much of the Amazon Basin, the Caribbean, North and Central America, West and Central Africa, South Asia, and Southeast Asia between the fifteenth and seventeenth centuries. In many cases, local elites, some of whom were subordinate to existing tropical empires, grappled to make use of incoming European soldiers to further their own political advantage. Precolonial exchange systems spanning continents, as well as trades in enslaved individuals, already existed in many parts of the precolonial tropics, meaning that Europeans often tapped into established economies to apply their new ideas of towns, mining, and profit-making agricultural and pastoral extraction. However, Europeans ultimately possessed the keys to this new globalized world. Following the decimation of Indigenous populations across the Americas, with a monopoly on naval power and technologies, Europeans alone routinely and regularly plied the world's oceans. Intending to maintain global trade and dominance, they were left in a position to exploit the resources, labor, and landscapes of the tropics on an unprecedented scale.[25]

Although it is seldom discussed, between the fifteenth and seventeenth centuries, to some extent, Iberian royal monopolies and administrative flaws could actually limit the transformation wrought on tropical landscapes by colonial enterprises. In Brazil, the Caribbean, and Mexico, kings and queens reserved forests and certain tree species for state interests in building ships or fortresses. The king of Portugal controlled sixteenth-century whaling activities off Brazilian shores. Indigenous enslaved labor on new Spanish plantations was also initially regulated by royally decreed *encomiendas* granted to chosen nobles. Meanwhile, in the Philippines, while the church was highly active in the settlement of the archipelago, the Spanish monarchy did everything it could to control the "galleon trade," which sent silver and gold from the Americas to the shores of the Philippines and China and saw silk and porcelain return in the opposite direction. Indeed, you might have noticed that the Iberian empires themselves have also had something of a monopoly on this very chapter. This is because the expanding Spanish and Portuguese empires, between the fifteenth and seventeenth centuries, with some limited exceptions, primarily initiated and sought to control novel economic and cultural exchanges into and within the tropics. These systems would have been unrecognizable to the supposedly "free market" consumers of the twenty-first century. As we will see in the next chapter, however, from the seventeenth century onward, other hungry European states, institutions, and private individuals, grappling for status and power on their own continent and seeing the riches arriving on the shores of the Iberian Peninsula, made their own increasingly aggressive advances into the tropics.[26]

The Dutch Empire of the sixteenth to eighteenth centuries sought a foothold in the Caribbean and South America as well as South and Southeast Asia. The British Empire became the largest in history between the sixteenth and early eighteenth centuries, ruling territories on every tropical continent across a territory on which the Sun was said to never set. By the nineteenth century, the French had also established a tropical colonial empire in the Caribbean, Africa, South Asia, and Southeast Asia, and the Belgians had done so in Africa. These jostling competitors broke down former Iberian domination, and benefits became distributed across an increasingly wealthy European elite. The resulting new trading opportunities were nevertheless not driven by homogenous empires but rather by lots of individuals, including mobile merchants, gambling chancers, and food producers of all cultures. The seventeenth to nineteenth centuries saw the laying of the framework for the system of competitive, global flows of capital and labor we know today, as Europeans expanded existing inequalities and systems of exchange into the global sphere. A consolidation of profit-driven approaches to tropical environments and to people from the tropics saw haunting commercialization and racialization of enslaved labor in the form of the transatlantic slave trade. Intensifying plantation agriculture stripped forests on new scales to produce "cash crops" as cheaply and as abundantly as possible, simultaneously shaping and being driven by European and northern North American consumer demand. The result was a wealth imbalance between Euro-America and the tropics. The result was tropical nations plagued by issues of limited economic development, lagging infrastructure, and conservation challenges. The result was racial discrimination and violence that continue to afflict twenty-first-century societies. The result was a situation where we all, whether we like it or not, are responsible for what happens in tropical forests in the twenty-first century.

Chapter 11

GLOBALIZATION OF THE TROPICS

E very day we turn on the TV or open a newspaper to new
international political squabbles, economic difficulties, and
catastrophic climate change. Many tropical nations, across
the Americas, Asia, Africa, and the Pacific, are on the front lines of these
issues as they try to balance economic growth with the burning of fossil
fuels, poverty alleviation with biodiversity conservation, and nationalism with international communication and global solutions. It is in the
tropics where islands are already beginning to disappear underwater. As
global temperatures increase, ice caps melt and sea levels in the Indian
and Pacific Oceans rise at a rate of around four millimeters every year. It
is in the tropics where increasingly unpredictable climates are leading to
both droughts and floods in the same regions in the same years. It is in
the tropics where, should all tropical forests become disturbed by human
deforestation and infrastructural development, a mass extinction event
is looming for plants and animals. Some predict declines of as much as
30 percent of all tree species and 65 percent of ant species, and bushmeat
hunting and climate change provide further challenges for the dispropor-
tionate biodiversity of these regions. It is also in the tropics where many
of the fastest-growing but also the poorest populations reside, some living on less than US $1 a day, placing increasing pressures on the public

health, economies, soils, and forests of these nations. The frequency and urgency with which these shocking stories are conveyed to us, as well as their occurrence often many thousands of kilometers away, might allow us to think that they represent a recent, emerging problem that is detached from the pasts and behaviors of the rest of the world. Yet, no matter how hard some of our politicians try to convince themselves, and us, of this, nothing could be further from the truth.[1]

We have seen in Chapter 10 how the dramatic expansion of European powers, primarily from Iberia, heralded a period of new pantropical and extratropical global contacts and exchanges between the Americas, Asia, Africa, and Europe between the fifteenth and seventeenth centuries. Some recent treatments have stopped the story there, suggesting that early interactions and geographical "chances" wholly determined the rest of history as we know it. However, this ignores the ongoing colonial, imperialist, and capitalist processes that were only just beginning. Between the seventeenth and twentieth centuries, a variety of European and, later, northern North American leaders, landowners, merchants, and investors increased their attempts to control and exploit the environments and peoples of the tropics. During this period in the tropics, we see some of the earliest, but also worst, impacts of unchecked, unregulated flows of capital on human bodies, cultures, and landscapes in world history. We see the consolidation of a global, racialized trade in forced human labor. And we see the appearance of the first documented recognition that human impacts on the environment could have regional consequences for the climate. These economic and political processes were both resisted and exploited by societies and individuals from the tropics. They also eventually led to some of our modern laws on human rights and international property. However, in the end, they created a new and monstrous redistribution of wealth, away from the tropics and into the hands of the western half of Europe, the United States, and Canada. They altered tropical landscapes beyond all recognition, leaving conservation challenges for many nations today. And they sprouted the roots of racial discrimination and violence that continue within many Euro-American societies in the twenty-first century.[2]

We now turn to the latest historical, archeological, paleoecological, and anthropological research exploring how tropical forests became key witnesses to this new global order. They were home to millions of

Africans forced from their homes in West and Central Africa as local raids and European demand for labor drove the transatlantic slave trade. These enslaved individuals were also transported to tropical forests that were cut back as the increasingly monoculture sugar plantations, particularly in the Caribbean and South America, had violent, expanding impacts on Neotropical environments. Later in time, as global markets for varied tropical crops solidified, tropical forests watched as Brazil's crucial resource of rubber was planted elsewhere, as India had its own cotton sold back to it, and as farmers in Myanmar cleared entire river deltas to plant sweeping fields of rice to fuel the Western world and an ever more unequal distribution of global wealth. Landscapes previously home to tropical forests provided the tea, coffee, potatoes, and bananas that stocked increasing numbers of Euro-American kitchens. This is the story of how the world we know today emerged from the tropics, alongside the stories of the people and plants that watched it happen. It is the tale of enslaved individuals, laborers, merchants, and empires experiencing and maneuvering in a rapidly changing world of global economics and cultures and of the tropical plants that now line our cupboards and wheel our cars and bikes. It forces us to confront the origins of globalized issues in sustainability, climate change, biodiversity loss, political disputes, warfare, and racism that face not just tropical nations but our entire world in the twenty-first century.

AT THE CLOSE of the twentieth century, archeologists excavating at the Spanish colonial Hospital Real de San José de los Naturales (now San José de los Naturales Royal Hospital) in Mexico City, in preparation for the construction of a new subway line, uncovered a mass burial of human remains. Many of the discovered individuals were Indigenous people, buried after succumbing to the rampant spread of diseases that arrived with Europeans. However, scientific research published in 2020 showed that three of the individuals, dating to the very earliest phase of the hospital in the sixteenth century, had a very different origin altogether. Analysis of their preserved DNA, strontium isotope analysis of their tooth enamel (which reflects the geology of where an individual lived during the formation of the tooth), and studies of deliberate cultural dental modifications showed them to all be men, to all have been between twenty-five and thirty-five years old at death, and to all have come from sub-Saharan Africa, over 10,000 kilometers away. At another burial ground near the seventeenth-century Dutch capital of Philipsburg on the Caribbean island of Saint Martin, archeologists uncovered individuals with genetic affinities to people living in tropical Cameroon, Nigeria, and Ghana. Likewise, at British-constructed Newton Plantation in Barbados, dated to between the late seventeenth and early nineteenth centuries, disinterred individuals had strontium isotope signatures that showed they spent their childhoods in West Africa. How and why did

Figure 11.1. Map of the transatlantic slave trade between 1500 and 1870. The colonized areas of Britain, Portugal, France, Spain, and the Netherlands are shown in the Americas, as well as key locations along the Atlantic coast of Africa. The number of Africans forcibly abducted and transported to the Americas along different routes is shown. Arrows show the trade routes of different products and materials, with a focus on Britain, as discussed in Chapter 11.
Adapted from Pearson / Desai / Science / Trans-Atlantic Slave Trade Database / Emory Center for Digital Scholarship

these people move from one tropical continent to another? The answer is one of the largest, most inhumane sets of forced migrations in human history. From some of the first Africans to arrive in the Neotropics, courtesy of Iberian colonialism in the sixteenth century, to those made to work for expanding Iberian, Dutch, French, and British global economic powers two centuries later, these individuals were all firsthand witnesses to the transatlantic slave trade.[3]

As we saw in Chapter 10, although Europeans initially sought to force Indigenous populations, including the Guanches of the Canary Islands, the Taíno of the Caribbean, and the Tupí in northeastern Brazil, to work their newly conquered lands in the Neotropics, the impacts of devastating diseases left them with a demand for other sources of labor. This was particularly the case given desires to extract mineral wealth and to transform tropical landscapes into monoculture plantations of crops, such as sugar, to cheaply stimulate and satisfy the growing demand back in Europe. Sugar, in particular, necessitated not only the widespread clearance of trees for planting and fuel for processing but also a large workforce that could ensure efficient harvesting and conversion into molasses. On the Atlantic islands of São Tomé and Príncipe, off the west coast of Africa, Europeans determined that enslaved Black Africans abducted from tropical portions of the continent could, thanks to early exposure, resist tropical diseases like malaria and yellow fever that were creeping into the Americas and destroying European migrant workers and Indigenous populations alike. Although enslaved Africans had been arriving on the shores of the Americas since shortly after Columbus's arrival on Hispaniola, the expansion of sugar plantations across the Caribbean and Neotropics throughout the sixteenth and seventeenth centuries and the unification of the Spanish and Portuguese thrones in 1580 saw demand for and extraction of African labor rise progressively. Between 1550 and 1650, Iberian ships had already transported over half a million enslaved Africans to colonies in the Americas. Although slavery had existed in various forms in societies around the world until this point, this new, European-driven trade saw two significant shifts. First, enslavement became a global endeavor, based on violence against both Black African bodies and tropical landscapes. Second, although it drew upon existing economic structures and political conflicts, the earnings and wealth resulting from the increasingly dispassionate, mechanistic, profit-driven exploitation of this expanded trade in enslaved labor flowed primarily into the hands of Europeans.[4]

Working with sugarcane was notoriously brutal, requiring constant attention, heavy labor, tropical lumber, and dangerous refining through boiling and distillation. Add to this the abuse, malnutrition, and unsanitary conditions forced upon enslaved individuals by plantation owners. The demands placed on the new arrivals to Neotropical shores saw many enslaved peoples worked to death within a matter of years. Yet this trade was only in its infancy. By the seventeenth century, sugar plantations established along the Portuguese-controlled coast of Brazil, such as at Pernambuco and Bahia, produced most of the sugar exports to Europe—something that other European nations sought to change. The Dutch West Indies Company, after a failed attempt to take control of this emerging "sugar coast," worked its way into the lucrative sugar business by other means. Namely, it invested in the dedicated shipping of enslaved individuals from Africa to the Caribbean to work on the plantations of others. By the end of the seventeenth century, the British Empire had colonized a number of islands in the Caribbean, including Jamaica, Antigua, Saint Kitts and Nevis, and Barbados, that the Spanish had largely neglected due to their lack of ore resources. Here, the British established colonies, not to settle or to control but rather to extract, rolling out the sugar plantation as a concept based on chattel slavery across tropical landscapes to make maximum profit. The French founded a colony on Saint-Dominigue (now Haiti) based on a similar economic system. Although British people launched slave raids in West Africa as early as the sixteenth century, in the seventeenth and eighteenth centuries the British Royal African Company, as well as a number of French companies, sought greater control of plantation workforces and challenged Iberian and Dutch monopolies by increasingly specializing in a trade in enslaved Africans. At the same time, racist theories, supported by prominent philosophers, served to make associations between concepts of enslaved labor and Black African bodies particularly rigid in English social and cultural discourse. By the eighteenth century, the "triangular trade," as it became known, not only had enveloped the entire continent of Africa in its demand for labor but also saw African individuals shipped to European colonies and business ventures in the Americas, South Asia, and the Pacific as part of economic extensions that spread well beyond sugar plantations (e.g., mining and other crop monocultures) and the Atlantic Ocean—all to meet an increasingly coordinated, global flow of investment that had become disturbingly mechanical. Other European countries, including Denmark and Norway, became involved. So did an independent Brazil in 1822, as well as other American-based trading polities. Smaller private merchants, alongside the wealthy elite and large companies that had thus far dominated coordination of the trade and ownership of plantations, also played an increasingly large role. These merchants ensured a flow of wealth into cities like Liverpool, London, Bristol, Birmingham, Glasgow, and Nantes, managing the shipping of cloth to African merchants in exchange for enslaved individuals, the shipping of enslaved individuals on to the Americas, and the return of sugar, as well as cotton, tobacco, and other plantation goods. The human cost of these increasingly formalized and racialized flows of capital is shocking. By the time an independent Brazil became the last country in the Americas to ban slavery in 1888, anywhere between 10 million and 20 million Africans had been forcibly uprooted from their—primarily tropical—homes and families and transported across the Atlantic.[5]

Many of them died before they even stepped ashore. The eighteenth-century Portuguese shipwreck of the *São José–Paquete de Africa*, uncovered by underwater archeologists off the coast of Cape Town, is a haunting graveyard of iron bars and copper nails that shackled more than four hundred trapped African men, women, and children on their way to Portuguese-controlled Brazil until stormy waters took them to their deaths. Firsthand written accounts by freed Black African enslaved individuals such as Olaudah Equinao (Gustavus Vassa) and Ignatius Sancho highlight both the terrors faced in capture and transport and the racist and capitalist treatment by Europeans selling people as commodities. As Sancho wrote, "Look around upon the miserable fate of almost all our unfortunate colour . . . [S]ee slavery and the contempt of those very wretches who roll in affluence from our labour."[6] Although the degree remains debated, the transatlantic slave trade certainly impacted West and Central African societies and polities over the long term. Using historically recorded and estimated numbers of enslaved individuals to show how this trade resulted in depopulation and severely imbalanced ratios of men to women, African historians such as Walter Rodney and Joseph Inikori have argued that it was a major factor, alongside droughts and famine, in changing demographics in sub-Saharan Africa between the seventeenth and nineteenth centuries. Similarly, the increasingly active meddling of European powers in African politics and sponsoring of local slave raids to obtain captives for trade played an important role in further stimulating a long history of ethnic and religious violence in West and Central Africa, leading to the movement of populations further into mountain or forest settings, preferences for isolated villages as a key form of safety from attack, and the creation of political and cultural fault lines still exploited today by modern terrorist groups such as Boko Haram in the tropical forests of southern Nigeria and northern Cameroon.[7]

Traditional approaches to the transatlantic slave trade have often framed Africa as a land of passive victims, perpetuating colonial stereotypes that still depict the continent as isolated and impoverished. Much of this is a product of the fact that most of the historical records used to understand West and Central Africa at the time of European involvement were written by Europeans or Arabic traders and scholars. Meanwhile, as seen in the recent controversy of the "Benin Bronzes," preserved material culture highlighting the diversity and wealth of precolonial African states was removed from the continent by European colonizers. Nevertheless, the last thirty years of historical revisionism has presented a more active African role in an increasingly globalized world. As in Europe and the Mediterranean, capture and trade of enslaved individuals was prevalent in many parts of Africa at the time of European arrival on its shores. Trade across the Saharan caravan routes and through the Red Sea or Indian Ocean saw many enslaved peoples traded with Muslim powers from the first millennium AD. As we saw in Chapter 10, from the beginning, the Portuguese relied upon African political goodwill and traditional marriage networks to access slaves who were already being captured and traded throughout the region as part of warfare and regional demands for labor and prestige. In the fifteenth century, thriving local kingdoms, such as those of Kongo and Benin, actively exploited or resisted new European contacts, setting prices, limiting the numbers of enslaved people taken, and making economic and religious demands. Indeed, through the transatlantic slave trade, European merchants and captains were only likely to succeed if they tailored the goods they brought with them (e.g., weapons, cloth, pipes) to the luxury tastes of their regional partners, with whom they often formed close, personal relationships. Africans meanwhile incorporated and used Europeans in their geopolitics, as demonstrated by the fact that European traders built their houses alongside African elites and merchants at sites such as Elmina in Ghana and Savi in Benin. Even when Europeans did actively take part in raids in the interior, they relied on allies and could quickly succumb to a changing of political allegiances. This was amply demonstrated when Queen Nzinga (Njinga Mbande) of the Ambundu Kingdoms of Ndongo and Matamba (modern-day Angola) shifted from acting as an ambassador for the Portuguese in the region to being their staunch enemy after aligning with the Dutch to further the independence and stature of her own realm.[8]

Just as we should not reduce Africa and its diverse states and polities to a passive clay molded by Europeans, we should also not forget the individual examples of resistance to the transatlantic slave trade. Increasingly recognized numbers of escapees, slave mutinies on Atlantic voyages, and rebellions across the Americas, including, notably, in Haiti, highlight the ongoing powerful, determined resistance of Africans and their descendants to horrendous conditions between the sixteenth and nineteenth centuries. Free "maroon" settlements of former enslaved peoples, and sometimes also Indigenous communities, developed across the Caribbean, Central and North America, and South America and, as in the famous case of the Quilombo dos Palmares of Brazil, could provide staunch militant resistance to imperial powers. Away from these more dramatic events, Dr. Alicia Odewale of the University of Tulsa has also examined historical records and material remains left behind in the ground to study the everyday experiences of enslaved communities living in the Danish-controlled, Christianized, and sugar-oriented urban center of Saint Croix in the Caribbean. These communities negotiated and resisted their condition through their daily activities and use of mundane items such as pottery. At the tropical urban site of Saint Croix, enslaved communities made significant use of undecorated and handmade ceramic wares, alongside European pottery, as part of a growing reliance on skills developed within the enslaved community for daily survival. At Saint Croix and elsewhere in the Caribbean, enslaved communities also coordinated informal trading systems that saw tobacco, pipes, pottery, and glass exchanged between communities as part of the negotiation of new identities and connections. As Alicia puts it, "Even under the immense structural power of imperialism and slavery, enslaved communities still expressed their own power and transformative capacity."[9] Beyond slavery, numerous diasporic African voices are also now beginning to be heard as merchants, voyagers, investors, and crucial workers in the formation of globalized, and later industrialized, cultures across Africa, the Americas, and Europe. There were far more Africans who were captured and brutally transported to the Americas than there were Europeans who arrived on those shores prior to the nineteenth century. In many parts of the Caribbean and South America, they outnumbered Europeans by as much as 25:1, and people of African heritage still make up a considerable portion, or even the largest portion, of populations in many Caribbean and South American countries today. It is no exaggeration to say that the mixed cultures, demographies, economies, and societies of the new globalized world were built upon their shoulders.[10]

Nevertheless, it is impossible to ignore the global wealth and status discrepancies that emerged as a product of the transatlantic slave trade. Historians, exploring records of income, trade, and demography, have powerfully argued that the capital gains made by European and northern North American plantation owners, merchants, and families directly and indirectly involved in the trade provided the key financial bases for the Industrial Revolution and the development of wealth and infrastructural imbalances that went on to underpin the modern world economy as we know it today. Technological innovations in a number of sugar colonies have, themselves, been seen as a part of the dawning of an "industrial age." Analysis of historical nutrition has also shown that cheaper, more readily available, energy-rich plantation sugar between 1600 and 1850 increased public health across British society to such an extent that it spurred on the formation of an industrial working class. Even following the abolition of slavery in the United Kingdom in 1807 and its colonies in 1833, the British government paid out massive amounts of money to companies, landowners, and individuals with investments in the slave trade in repayment for losses of "free captive labour." Europeans did not *need* to take part or invest in this trade, and there was opposition back home. Yet many individuals, such as Edward Colston, who's legacy has left a mark on a number of institutions across the city of Bristol today, joined initiatives such as the Royal African Company and invested in the transport of enslaved individuals between continents. In Europe and northern North America, as we go about our daily routines, it can be easy to forget just how much of our streets and lives the resulting profits have touched and how many of our predecessors could so quickly take part in this flow of forced labor and capital. Capital accumulated from the transatlantic slave trade flowed into our hospitals, schools, churches, and universities through charitable donations. The people who earned wealth from and took part in the trade are honored in street and building names as well as with statues. Many innovations in insurance that characterize the financial sector of our economies today also find their disturbing origins in the transatlantic slave trade, as investors sought financial safety nets for both cash crops and human cargo. The racialized abuse of Black African bodies that was formalized in the transatlantic slave trade also undoubtedly left an ingrained legacy of inequality and discrimination that continues to influence North and South American and European societies and political conflicts into the twenty-first century.[11]

Back in the tropics, although some African elites and merchants made profits from the trade, the benefits of sugar and the global flow of capital did not reach African shores in the same way. As we have seen, it disrupted the demographic and political situation in much of the continent with lasting effects. From the eighteenth and nineteenth centuries, subsequent, sustained conquest of different parts of West Africa also resulted in brutal warfare, the dismantling of traditional land-use and social systems, and the theft of heritage that remains locked up in European and North American museums. Profits were also not realized on the sites of many of the industrial plantations where enslaved African labor ended up. Places such as São Tomé and Príncipe and a number of Caribbean islands were left as some of the poorest countries anywhere in the world following eventual abandonment by Europeans, brutally shaved of their tropical forests. Nearly all of the plants and animals on the Caribbean island of Nevis are now introduced or invasive species, with the majority of their native counterparts now extinct as the island was nearly deforested within the first fifty years of sugar plantation establishment. While roads might have been built to help support the movement of this valuable crop, the British made no attempt to introduce critical infrastructure such as schools for education or farms for nutrition. Dr. Justin Dunnavant of Vanderbilt University is exploring how the ecology and even the shoreline of the island of Saint Croix have changed as a result of these processes and their legacies. Widespread clearance of trees began with sugar plantations, and there is clear evidence of lasting soil degradation. Meanwhile, the island's coral shorelines have been literally and visibly mined away, not only for building materials to support the forced labor of enslaved individuals but also to clear a pathway for the tax-free smuggling of rum. The expansion in global production of this popular beverage, enjoyed straight and in cocktails, was often associated with existing or former sugar plantations in various parts of the Caribbean. More of us should take heed of the unequal global inheritances left behind as a product of flows of capital during the transatlantic slave trade as we all too often watch environmental degradation, poverty, political turmoil, discrimination, and ongoing discrepancies in pay and opportunities, both at home and in the tropics, from the comfort of our armchairs.[12]

SUGAR WAS, OF course, not the only high-value product at the heart of the transatlantic slave trade. Nor was it the only crop that saw the tropics increasingly caught up in a globally imbalanced exploitation of landscapes and labor between the seventeenth and twentieth centuries. Today, cotton quite literally puts the shirts on our backs. In the seventeenth century, however, cotton cloth from India provided a major luxury good that Europeans gave to African merchants and rulers in exchange for enslaved individuals. In turn, expanding growth in extensive plantations in the southern United States was supported by the brutally extracted labor from these people once they had been forcibly transported from their home continent. Cotton's impact on labor and environments in the tropics did, however, reach still further than this new Atlantic economy. Between the early sixteenth and the early eighteenth centuries, the Mughal Empire, which extended across India, Pakistan, and Bangladesh, was perhaps *the* main global center of cotton textile production. As writer Meena Menon states, "Farmers across the Indian subcontinent grew many different varieties of cotton as part of mixed plots alongside food for subsistence." These farmers were part of an intimate system with ginners, spinners, and weavers that produced high-quality cotton textiles sought after as far afield as West Africa and Japan. This relationship was broken by eighteenth-century British imperialism and the dramatic change to the growth of cotton as a cash crop for global export. Although the Dutch had already been importing Bengali textiles, the consolidation of the British East India Company in South Asia greatly increased the availability of these products in Europe in the eighteenth century, allowing cotton to replace wool as the wardrobe maker of choice. Following conquest of Bengal in 1757, British agents used their political and economic power to transform the international cotton markets. The investment of India's existing cotton wealth in English domestic development and technological advances that outstripped Indian hand-spun cotton, and the restriction of Indian cotton imports to Europe, catapulted British factories to the top of the global textile market during the Industrial Revolution. Attempting to make up for a loss of reliable raw cotton from North America following the War of Independence, the British set up experimental farms in India to grow American cotton varieties that were suited to the British textile mills. Demand from these textile mills led to British encouragement of cotton monoculture instead of more sustainable mixed cropping, taking valuable productive land away from food production, forcing huge numbers of Indian cotton spinners and weavers to seek alternative incomes, and sending the diverse indigenous short staple cotton varieties into decline. This resulted in a major food crisis during famine years. Diminished local production and competition from European exports also saw Indian consumers buying finished cotton textiles and even yarn back from Britain, sometimes made using the very cotton exported from India with a significant price markup! India is today the second-largest exporter of raw cotton anywhere on the planet and the largest cultivator of cotton. It remains renowned for the creation of muslin and for printed and dyed traditional textiles, but its handloom industry has been marginalized. The case of cotton is a further example of how unequal global flows of capital and economic infrastructure developments between Europe and

Figure 11.2.   Hand ginning of cotton in India by a Ponduru woman.
Meena Menon

the tropics imposed themselves on both landscapes and labor. Strikes within the Indian textile industry as part of the independence movement of the early twentieth century further illustrate the role of this crop in the manifestation of colonial power dynamics and economic inequalities.[13]

Tea is a go-to feel-good drink across much of Europe, Asia, and North America today. It is also another crop that had a big impact on tropical Asian forests and their inhabitants between the seventeenth and nineteenth centuries. Originating somewhere at the intersection between South, Southeast, and East Asia, extensive tea cultivation began in China around the eighth century AD. It reached Europe only in the seventeenth century, courtesy of Dutch and Portuguese ships. Here, the ruling classes quickly gained a predilection for the drink, particularly in England and, later, North America, and it became the preeminent import from China. In South Asia, the Singpho people of Assam and Myanmar had, however, also already been growing their own varieties of tea. The expansion of British capital into India, following conquests and political negotiation with local rulers throughout the nineteenth century, drew on this local knowledge and planted tea widely across the state of Assam and, later, other states, including Bengal and Orissa (now Odisha). Widespread production and easy access through imperial connections soon made tea cheaper and more easily available to all social classes in England, at the same time as it was impacting tropical landscapes and farmers in South Asia. Tea became a major plantation crop in Sri Lanka, for example. Between 1883 and 1897 tea plantations covered an estimated additional 20,000 acres every year in the Sri Lankan Highlands. Rural villagers lost vast swathes of agricultural land, and their traditional, mobile swidden approaches to farming among the lowland and misty montane tropical rainforest in the region were vastly curtailed, as were the extents of these forests and their ecosystems. Similar changes occurred back in India, where workers on tea plantations had few rights and received poor treatment within the social systems of racial stigma under the British Raj. The Indian and Sri Lankan tea industries remain strong international and domestic sources of income for those nations today, with much of Indian- and Sri Lankan–produced tea now consumed by those countries' own citizens and with production increasingly in the hands of smallholders. Nevertheless, the degree to which foreign interests have impacted the wealth, forests, and lives of local farmers, past and present, should not be forgotten.[14]

The coffee that wakes us up in the morning and keeps us going through days of constant Zoom meetings has no happier history. Originally native to Ethiopia, cultivated coffee arrived in Europe, via the Middle East, during the seventeenth century—warming lawyers, politicians, and philosophers in London, the nobility of Germany, and the financial sector of Amsterdam. The Dutch, recognizing that coffee plants thrived in the tropics, from 1699 cleared rainforests to make way for their cultivation in Java—a region where coffee is still grown commercially to this day. Coffee was soon exported for growth in the Caribbean. Supported by enslaved labor, the tiny French colony of Saint-Dominigue (today's Haiti) came to dominate the global coffee market in the eighteenth century. However, by the time the Haitian Revolution of 1796 to 1804 saw self-liberated enslaved people create their own independent nation, no one would have known it. Thanks to tropical deforestation and soil erosion, the French left it one of the poorest nations and most degraded landscapes anywhere in the world. Although coffee continued to be produced on the tropical Caribbean islands of Cuba, Puerto Rico, and Jamaica, by the middle of the nineteenth century Brazil dominated coffee production. Again, supported by a trade in enslaved individuals that continued until the late nineteenth century, Brazil was able to produce coffee in massive quantities and at a low cost. Plantations also expanded across Dutch-held Sumatra, Bali, Sulawesi, and Timor in the late nineteenth century. Soon coffee, previously the preserve of the wealthy, was being drunk regularly among the working classes of northern North American and European society. Heavy debate surrounds the local benefits of growing coffee. On the one hand, profits aided Brazil's emergence as a large, independent economy. On the other hand, long-term continued use of slavery, the exploitative use of cheap labor, including children and marginalized Indigenous groups, inequality, poverty, the buying up of smallholdings by larger international companies, and deforestation have offered few sustainable benefits to coffee-growing regions in the tropics and presented problems for tropical environments extending into the twenty-first century—something to think about next time you take a sip.[15]

Rubber is one of the most important tropical products around the world today. From balloons to airplane tires, from condoms to waterproof shoes, from rubber balls to supermarket conveyor belts, from diving gear to hospital tubing—you can barely go anywhere or do anything today without coming across this remarkable material. Natural rubber is made from latex, which oozes from certain plants when they are wounded. The only commercially viable sources, however, are the tropical forest plants of *Hevea brasiliensis*, a tree that grows wild in the Amazon Basin, and *Landolphia owariensis*, a wild vine that grows in West and Central Africa. Indigenous populations used the former across the Neotropics to produce clothing, boots, balls, storage carriers, and toys long before any Europeans got there. In fact, Europeans did not realize the benefits of the material until the late eighteenth century, before economic production of rubber products began in the nineteenth century following the patenting of the vulcanization process. The rise of automobiles and other rubber-reliant inventions drove the subsequent "rubber boom," which, between 1879 and 1912, spurred on an expansion of profit-making enterprises across the Amazon Basin of Brazil and Bolivia to feed growing global markets, particularly in Europe and northern North America. In the Upper Amazon, the resulting wealth galvanized the development of cities such as Manaus, where per capita income became the highest in Brazil. It had the first electric streetlights anywhere in the country. Native pests meant that attempts to grow rubber on plantations in Latin America repeatedly failed, leading to a scramble of entrepreneurs seeking to make profits in an industry reliant on the tapping of dispersed wild trees in the Amazon Basin. Indigenous groups often became caught up in the race for profit. Some, such as the Mundurukú, actively engaged with the boom, moving in search of wages and European products. Others, however, were forced into labor agreements that effectively amounted to slavery and were abused and killed by so-called rubber barons. A study of living gigantic Amazon nut (Brazil nut) trees in the vicinity of the city of Manaus by Victor Caetano Andrade, whom we met right at the beginning of this book, has even shown that these trees grew more poorly as Indigenous populations and their traditional management effectively disappeared in the face of the pressures of rubber extraction in the nearby rainforests.[16]

European and North American hands went on to shape the expansion of rubber into the commercial product we know it as today, as well as its impacts on labor and environments in the tropics. In 1876 Sir Henry Wickham oversaw the transport of 70,000 seeds of *H. brasiliensis* to Kew Botanical Gardens in the United Kingdom. From there, the British Malayan and Dutch Sumatran rubber cartels promoted the planting of rubber in Malayan tea plantations in Southeast Asia, eventually securing their own direct access to the resulting rubber and profits in the early twentieth century. Away from native Amazonian pests, the trees excelled, and plantations were rapidly expanded. Although the addition of rubber into smaller plots could benefit local smallholders and entrepreneurs, intensification of plantation land use and growing pressures from European and northern North American companies (particularly those with interests in the rapidly expanding automobile industry, such as Michelin) resulted in a widespread loss of local biodiversity and, frequently, the exploitation of Indigenous and indentured migrant labor under conditions that were little better than enslavement. Today, Thailand, Indonesia, Malaysia, India, and China are the top-five leading global producers of rubber, using a tree from the other side of the tropical world and far surpassing Brazil. Though the economic benefits are clear, monoculture growth of rubber has undoubtedly left its scars on the Southeast Asian landscape. A more horrific European intervention in the global rubber market occurred when Leopold II of Belgium, seeking his own colonial empire, established the "Congo Free State" in what is now the Democratic Republic of the Congo in the late nineteenth and early twentieth centuries. Seeking high returns to rival other European entities during the rubber boom, Leopold's state forced local populations living in the rainforest into service, killing them, razing their villages, and cutting off their hands if they refused, in order to exploit the local latex-producing vine of the region. The atrocities observed and documented by appalled European observers almost certainly contributed to the collapse of the local population in just a few decades. Based on records of rubber production and the mortality rate, an estimated one person in the Democratic Republic of the Congo died for every ten kilograms exported between the 1880s and 1923. With the advent of electricity and two world wars, Western nations and business interests continued to grapple for access to rubber throughout the twentieth century, with little regard for local working conditions. Not for nothing has historian Professor John Tully of Victoria University called rubber "the devil's milk."[17]

In the above I have attempted to distill incredibly complex, and often contentious, processes in order to highlight a few key examples where crops and products we take for granted today were at the heart of the rise of global markets between the seventeenth and twentieth centuries, with major impacts on people and environments in the tropics and economic and political echoes that still reverberate today. Even while Western imperialism and control over new worldwide flows of capital framed the all-too-often unequal outcomes, it is important to remember the huge variety of ways in which local governments, merchants, smallholders, consumers, and Indigenous populations actively negotiated and challenged the introduction and exploitation of these economically valuable products. One example serves to highlight this local agency. Oil palm is currently one of the most prominent tropical plantation crops. As we saw in Chapter 7, it was long consumed across West and Central Africa and, between the seventeenth and nineteenth centuries, was produced to feed captives sold into the transatlantic slave trade. With the abolition of the trade in 1807 in the United Kingdom (and its colonies in 1833), however, it increasingly rose to prominence as an export crop. Nevertheless, in many parts of the continent its entire production process lay in the hands of African producers, and the palm crop was ill suited to models of plantations built on enslaved labor. Attempts to introduce mechanization repeatedly failed to boost productivity, with farmers in Sierra Leone, Nigeria, and Senegal seeing no advantages. Local ecological knowledge and the benefits of growing the crop within mixed subsistence plots meant that groups with little wealth and their own labor could easily produce for export, including a group of enslaved individuals near Lagos who ultimately raised enough to buy their freedom. A similar situation is evident in the failure of the Portuguese to establish oil palm as a commercial crop among the Atlantic coast rainforests of Bahia, Brazil. Instead, oil palm, until today, is grown here within "subspontaneous" polycultural groves, underpinned by centuries of African agroforestry, that contribute to subsistence and local economic well-being. Continued European attempts to control the palm oil trade eventually stimulated colonial conquest of many African states and also saw the crop exported to Southeast Asia on a plantation-based model. Yet, ultimately, smallholders across the tropics retained some share of profits throughout. Interestingly, today oil palm is still heralded as a solution for rural, smallholder development in Africa, South America, and Southeast Asia, though, as we will see in Chapter 13, it is not without its challenges.[18]

IT WAS NOT just these more classically "luxury" crops that became part of a growing desire to control the plants, workforces, and environments of the tropics within increasingly unequal, globalized economic systems between the seventeenth and twentieth centuries. In Europe, growing subsistence reliance on, and monoculture planting of, the potato between the eighteenth and nineteenth centuries would have major economic, demographic, and environmental consequences some distance from the equator. Arriving from the cool montane forests and grasslands of the Andes, this tropical tuber was initially ignored in favor of traditional reliance on grain. Yet poor harvests in the face of the Little Ice Age, price fluctuations, frequent famine, and public unrest gave the potato its moment to shine. Extolled by agricultural economists, scientists (notably Frenchman Antoine-Augustin Parmentier), and even rulers (namely the Prussian "Kartoffelkönig" Frederick the Great), from the second half of the eighteenth century it became widely planted by smallholders and large landowners alike. By the early nineteenth century, the potato was part of the rolling out of the European "Agricultural Revolution," which involved the healthy application of fertilizer, such as seabird guano mined from its original home in Peru, and industrial ploughing methods. The result was a significantly greater and more reliable European food base—one credited with, at least in part, catalyzing population increase and even the beginnings of the Industrial Revolution in England. The result was a monoculture par excellence. Unlike the original potato fields back in the Andes, this new form of food production exchanged genetic variability for uniformity, increasing vulnerability to disease, with disastrous consequences. The Great Famine of Ireland (1845–1849) saw the decimation of this nation's new, homogenous food staple, thanks to the arrival of the potato blight from the Americas. Exacerbated by the callous negligence of the British imperial government in failing to provide alternative food sources, the resulting famine led to widespread suffering and a reduction in Ireland's population through death and emigration. Irish emigration stimulated by the Great Famine formed a significant part of the total of 10 million people of Irish ethnicity who left the island from the start of the eighteenth century, desperately seeking new labor opportunities in the face of often significant abuse. To this day the number of people living on the island of Ireland has still not recovered to prefamine levels. This catastrophe warned of the dangers of this new form of farming—one that heralded the beginning of cycles of pesticide experimentation and application, in which most monoculture agricultural fields remain locked around the world in the twenty-first century.[19]   •

Back in the tropics, imperialist control over land use and a capital-driven demand for the more efficient growth of tropical subsistence crops also had consequences for local food security and environmental change around the equator. Taking the example of rice, historical ecologist and archeologist Professor Kathleen Morrison of the University of Pennsylvania, alongside her colleague Dr. Mark Hauser, has described the creation of imperial "rice bowls" to serve growing globalized exchange networks and capital investment. In the case of the Americas, as we saw in Chapter 10, the knowledge of how to grow rice, as well as rice itself, likely took initial hold as a product of African agency. However, during the eighteenth century the organization of its production changed significantly. Plantations in primary areas of production, such as South Carolina, Georgia, and Brazil, created by European settlers and worked by enslaved African labor, became essential to providing food for Europe and northern North America and, particularly, to sustain the valuable cotton, tobacco, and sugar plantations of the Caribbean. This created interregional relationships of nutritional dependency, making local farmers and plantation communities less self-sufficient and less secure in the face of natural hazards, such as hurricanes, that could destroy lives and wipe out vital arteries of food sourcing. In South and Southeast Asia, a lively market of regional rice exchange already existed in precolonial times. However, with increasing British political and economic influence, export-driven rice production increased significantly. This was particularly the case in Burma (now Myanmar) following British annexation in 1852. The entire tropical coastal delta of the Irawaddy River was transformed and deforested into a managed landscape focused on rice growth, for local consumption but also primarily for significant export to India and Europe. As imperial structures forced a move away from local food security toward wider markets of demand, local populations became increasingly reliant on global commodity prices, imported food, and often unsympathetic governance. The result was the destruction of local biodiversity, inequality, declining flexibility for smallholder land use, and potential famine. As Kathleen puts it, in both the Atlantic and Indian Ocean cases, "the production and movement of rice around and outside of the tropics in the eighteenth to the twentieth centuries is one example of imperial imbalances of power, hunger, and risk" that have left an economic and environmental legacy in global economic relationships between the western half of Europe and northern North America and the tropical nations of the Caribbean, Asia, Africa, and the Pacific in the twenty-first century.[20]

The bananas we put in our baby food, smoothies, pies, and ice cream, or simply enjoy raw, provide a more recent example of a tropical staple food turned from a local source of calories into an agroindustrial complex at the mercy of global consumer demand in the nineteenth and twentieth centuries. We saw in Chapter 7 how this tropical forest tree crop was first domesticated in New Guinea, being introduced into the Americas by the Spanish in the fifteenth and sixteenth centuries before spreading widely among local farmers in lowland tropical forests. In the late nineteenth century, however, it became a plantation crop. In the 1870s, US railroad promoters in Costa Rica experimented with widespread planting of bananas as a cheap, nutritious way to feed their workers. Realizing the potential for mass export of bananas to the United States of America, multinational corporations (such as the United Fruit Company, which formed in 1889) began to buy up land in nations such as Costa Rica, Panama, Honduras, Ecuador, and Colombia. In the latter case, Colombian planters had actually already developed their own export networks to New York. Nonetheless, the United Fruit Company, in buying land and railroads, forced local farmers to sell to them at low prices. Moreover, in order to spread risk in terms of the potential for diseases, soil erosion, and climatic events hitting their crops, these multinational corporations frequently bought up and cleared a vast amount of tropical land simply so that it would be ready should planting need to extend to new areas to maintain profits. The result was removal of local farmers from their smallholdings, the forcing of local farmers and planters into wage labor, and an increasing reliance on imported foodstuffs. These international corporations held such economic sway and such a tight grip on access to foreign markets that they could even make and shape the politics and economies of entire tropical nations. The "Banana Wars" saw the US government exert active imperialistic political influences on various nations in the Caribbean, Central America, and South America in order to secure access to tropical production and trade. United States troops backed up the security of the United Fruit Company's operations in Honduras on more than one occasion, leading to long-term political instability. The seizing of Cuba and Puerto Rico from Spain in 1898 also saw the United States attempt to forcibly maintain beneficial economic flows of production and capital for itself at the expense of entire nations.

Similarly, the United States intervened to support Panama in its independence from Colombia, enabling the former to secure permission to
construct and control the Panama Canal (which opened in 1914) and
thereafter to dominate trade in the region. Many countries became almost completely reliant on certain key crops as a result of economic and
political interventions by the United States, leading to unequal distributions of agricultural land, infrastructure produced solely to meet the
needs of external capitalist enterprises, immense wealth inequality and
low wages, degradation of tropical forest environments at the expense of
plantations, and political conflict.[21]

Away from crops, the increasingly intensive, capital-driven pressure
on tropical landscapes between the seventeenth and twentieth centuries
is perhaps most vividly seen in how tropical timber and tropical land
themselves became the objects of a "rapacious capitalism." Mahogany
furniture is a common feature of stately homes and historical museums
in Europe and northern North America, venerated for its elegant craft-
work and exotic, workable wood. Yet the trade in this rainforest timber,
driven by growing luxury Euro-American demand and profit-hungry
merchants between the seventeenth and nineteenth centuries, involved
the exploitation of enslaved African labor in "woodcutting gangs." It also
saw waves of "harvesting bonanzas" crash upon entire ecosystems, from
the Caribbean island of Cuba to the Central American coast of Honduras. The "commercialization of the forest" occurred a little later in the
Philippines, intensifying as US imperialism replaced Spanish colonial
rule and sought to determine how much wealth it could extract from the
landscape. From 1898 until the mid-twentieth century, a moving timber
frontier, using railways and machines, began to cut lower-quality hard-
woods indiscriminately, to meet Japanese, Chinese, North American,
and European demand. Beyond being a product in its own right, timber
could also be cleared to "liberate" the land below for more "profitable"
uses. In Australia, from the nineteenth century onward, sheep and cattle
pastoralism drove intensified expansion across many parts of the continent by European settlers, and in 2010 Australia was the third-largest
dairy exporter in the world. In the late nineteenth and early twentieth
centuries, farmers began to clear significant portions of the "wet frontier" of Australia as they sought to supply increased global and regional

demand for animal products, particularly in Europe and other parts of Australia, exploiting British imperial connections. Expecting that fertile soils must lie under these heavily forested regions, larger landowners, as well as small-scale farmers who rented clearance contracts, undertook significant clearance of the semitropical eucalypt forests and wet rainforests of Queensland, including, notably, on the Atherton Tablelands. However, these farmers often instead reaped poor grazelands or wetlands, inevitable soil erosion, difficulty exporting dairy from isolated settings, and a perilous existence at the mercy of government controls and global prices. Moreover, the murder and forced relocation of the Indigenous Traditional Owners of these tropical forest landscapes, such as the Jirrbal, led to a suppression of ecological knowledge and traditional land management that, as we will see in Chapter 13, still plagues the threatened biomes of this region to this day. These examples simply scratch the surface of the increased deforestation and land conversion that took place across the tropics as governments, landowners, and companies sought to maximize the benefits of growing regional and global markets of production and consumption.[22]

New global circulations of capital and products and demands for tropical land conversion also continued to place pressures on workforces and migrants living in the tropics. The abolition of slavery did not necessarily mean that the movement of labor into and around the tropics became entirely voluntary, however. Even after formal abolition, enslaved Africans could still be forced to enter multiyear "apprenticeships" during which they worked long hours for no pay. With that practice finally stopped in 1838, landowners looked elsewhere for labor. In the early nineteenth century, many thousands of men from tropical South China and the Pacific Islands were coerced into contracts that nearly amounted to slavery as they performed frequently fatal work in the guano mines of Peru that fertilized the "Agricultural Revolution" of Europe. From the early nineteenth to the early twentieth centuries, historical records demonstrate that more than 1.5 million laborers from southern India moved to plantations in British, French, and Dutch colonies in the Caribbean; South Africa; Indian Ocean islands like Réunion, Mauritius, and Sri Lanka; Myanmar; and Fiji. Although supposedly voluntary, the contracts of "indentured servitude" that formed the basis of this system exploited sectors of the Indian caste system with low social status, offered poor working conditions, and often involved significant coercion. In the case of Sri Lanka, the poor treatment of arriving Tamil-speaking labor on British tea plantations in the highlands, which local Sinhalese farmers had refused to work, may have contributed to ethnic conflicts that have since confronted this tropical island. An often forgotten part of Germany's colonial history is the role it played in the movement of nearly 200,000 people as indentured laborers from the Solomon Islands, New Guinea, and the Banks and Torres Islands, among other regions, to plantation work or other forms of labor in Australia, Fiji, Samoa, New Caledonia, and other parts of New Guinea between 1863 and 1919. Potentially nearly one-quarter of these individuals died during their contracts. Migrating Chinese investors, landowners, and laborers were a significant force in the twentieth-century planting of rubber in parts of Southeast Asia. In the late nineteenth century, Italian laborers migrated to coffee plantations in Brazil, while in the early twentieth century the arrival of Filipino migrants as fruit pickers in Hawai'i formed the beginning of a labor export market for a tropical nation in which over 10 million members of its 85 million population live and work abroad, often facing heavy discrimination. Although this list does not come close to accounting for all of the individuals, families, and communities that moved, willingly or forcibly, as a result of growing global demands for production, capital, and consumption, it illustrates the scale on which these processes have left their mark on demographic and cultural diasporas across and beyond the tropics right up to this day.[23]

FROM THE SEVENTEENTH through to the twentieth century, the tropics, and their people and environments, were at the core of new, global flows of capital, demands for labor, and exploitations of land. While the varied examples I have dealt with have, given their unique temporal, social, economic, cultural, and political contexts, all been the subject of their own in-depth historical, anthropological, and archeological studies, together, they highlight how people, plants, and landscapes from the tropics experienced and also pushed back against the formation of economic systems that majorly shape the world in which we live today.

Throughout the seventeenth to the twentieth centuries, and indeed even before, flows of investment were based on a "taproot of imperialism."[24] This European ideal of exploiting new lands and new sources of labor to harness more profits ultimately birthed the concept of "capitalism," its critiques, and theories of modern global economics in the nineteenth century. From the racialized transatlantic slave trade to the exploitation of indentured workers and migration of various, often stigmatized ethnicities, landowners, companies, and even individuals in quest of profitable investments sought out labor from the tropics as a cheap commodity. Meanwhile tropical landscapes were shaved so that they might be used to produce crops or resources that fueled not local or regional needs but rather global flows of wealth and demand as part of a new "world system." Societies and existing regional politics and economies in the tropics were involved, and individuals resisted in various ways. Overall, however, European and, later, northern North American control of the direction and output of investment on a global scale left legacies of inequality, cultural and political prejudice, racism, and land use that saw certain sectors of Western society enriched and diasporic arrivals, as well as their tropical nations, impoverished. They also left a blatant mark on tropical environments. When the famous German naturalist Alexander von Humboldt explored the Spanish plantations of Lake Valencia in Venezuela in the nineteenth century, he noted, "When forests are destroyed, as they are everywhere in America by the European planters . . . the springs are entirely dried up."[25] In doing so, he was one of the first European observers to suggest a link between conversion of tropical land for new, profit-driven uses and a change in local and even regional climatic conditions.[26]

It might be hard for us to see the origins of our own twenty-first-century economics and politics in this early capital-driven violence against landscapes and labor in the tropics. Yet, taking the example of the transatlantic slave trade and monoculture agriculture with which we began this chapter, we can track how a so-called plantation logic still continues to shape global economic markets, the regulation or unregulation of the rights and treatment of workers, international politics, infrastructural developments, structural racism, and environments within and beyond the tropics in the twenty-first century. Dr. Alex Moulton of the

University of Tennessee, Knoxville, and his colleagues have elegantly described how much of the Caribbean, for example, has witnessed three consecutive eras of agricultural management for capitalist accumulation. The "plantation era" refers to the imposition of chattel slavery as a "productive force," which combined desires for biological control of tropical forest landscapes with brutal disciplining of and control over enslaved bodies. The second era pertains to the nineteenth- and early-twentieth-century emergence of independent states such as Jamaica, Haiti, and the Dominican Republic, which saw governments focus on the productivity of rural populations, deforestation and expansion of usable land, and the modernization of farming. Legacies of colonial infrastructure, however, often meant this involved succumbing to significant foreign debt as well as somewhat patronizing, even if well-meaning, perspectives that the "Global South" was in need of "development." The third era, from the mid-twentieth century to the present day, saw some governments of tropical nations and the ongoing diffusion of global market capitalism promote internal competition between entrepreneurial farmers seeking to outdo each other and leave behind traditional agricultural approaches. It has also seen European and northern North American leaders politicize "help" and "relief" for Caribbean communities in the face of disasters (made all the more frequent by climate change driven by the emissions from their own countries), praising the "resilience" of local people while completely ignoring the colonial injustices that have left them the most at risk. As Alex describes, "These three phases highlight the ongoing threads of colonialism, racism, and capitalism operating on tropical landscapes and labor that have channeled a 'plantation logic' into the present." While this is just one part of the tropics, we should not kid ourselves that Euro-American histories can in any way be disconnected from the fate of the tropics, its inhabitants, and the inequalities that shape global politics and economics more broadly in the twenty-first century.[27]

The twentieth century saw many tropical nations previously under the yoke of colonialism gain independence, make broad improvements in living standards, and receive increasing investment in their economies. Nevertheless, the global and national wealth disparities set in place by the processes seen in this chapter and Chapter 10, the growing economic power of multinational corporations, the need for infrastructural development following centuries of brutal extraction, and the continuation of industrial farming systems enforced under imperialism have shaped the current threats faced by tropical forests around the globe that we will look at in the remainder of this book. Between 1980 and 2000, 2,275,320 km^2, 2,590,620 km^2, and 8,520,310 km^2 of tropical forest were lost from Central Africa, Southeast Asia, and the Amazon-Orinoco Basins, respectively, in the face of logging, land conversion, infrastructural projects, and expanding settlements. The tropics as a whole lost a total of 17,364,780 km^2 of forest cover during this period. That is an area nearly double the size of Europe. Rapidly growing economies in Latin America, South Asia, and Southeast Asia increased their rate of fossil fuel burning, which, alongside industrial activities in Europe, Canada, and the United States that had continued apace since the beginning of the nineteenth century, as well as the feedback resulting from the loss of tropical forest cover, contributed to a rise in CO_2 emissions and a 0.6°C global surface temperature increase over the course of the twentieth century. Habitat loss and climate change meant that the high levels of biodiversity found in the tropics suffered, dramatically affecting global species numbers. Between 1950 and 1987 an estimated 50,000 species of plants, birds, mammals, reptiles, amphibians, and insects were lost as a result of tropical extinctions.[28]

Of course, differences in colonial history, environment, demography, cultures, and political and economic structures and decisions meant that the twentieth century saw each tropical region and nation develop its own unique conservation challenges, priorities, and strengths. Even within each region, different social groups were impacted differently as the growing grip of capitalism on the tropics drove increasingly stark inequality. The widespread appearance of favelas in Brazil as a product of the movement of rural populations to growing, industrialized urban centers during the 1970s is one example. Notably, the majority of the population living in these favelas are descendants of Black Africans, highlighting ongoing wealth disparities and discrimination that reach back to the colonial period and a racialized trade in enslaved labor. Another example is the growing poverty and marginalization of Indigenous populations and the difficult choices they must make between isolation and becoming enmeshed in new forms of settlement, wage labor, and political and social structures. The racial fault lines and traumas set up by the historical processes seen above and in Chapter 10 also began to play out on European and northern North American shores in the twentieth century. The systemic racism, segregation, discrimination, and violence facing Black people in the United States, a lasting legacy of the trade in human lives of the fifteenth to nineteenth centuries and the racist ideologies surrounding it, resulted in enduring injuries made visible in the rise of the civil rights movement of the 1950s and 1960s. In the United Kingdom, from the 1940s, immigrants arriving from tropical colonies in the Caribbean to seek work met with discrimination in the search for housing, and they, as well as ongoing arrivals from the Asian and African nations of the Commonwealth, faced racial profiling, prejudice, and poverty that resulted in their murder by white supremacists, precarious living standards, and retaliatory riots such as those in Broomielaw, Glasgow, in 1919 and Brixton, London, in 1985. Black people living across northwestern and Mediterranean Europe in the latter half of the twentieth century faced similar issues.[29] They still do.

The twentieth century also witnessed a growing understanding that the plight of tropical forests was not just a regional problem but a *global* one. Rampant tropical deforestation and losses of species diversity became symbols of environmental movements around the world, as many ecologists drew attention to the imbalanced amount of planetary biodiversity that would be lost should these habitats disappear. Yet, as we saw right from the beginning of this book when we met the first plants and forests to take root on Earth in Chapters 1 and 2, tropical forests are not just globally significant in the context of their amazing roll calls of plant and animal species. As we discovered then and will explore further in the next chapter, they also play a crucial role in the regulation of precipitation, flow of groundwater, soil formation and stability, and exchange of gases with the atmosphere, on local, continental, and planetary scales. Chapters 13 and 14 will explore whether or not European and northern North American nations and their citizens are willing to accept the fact that modern conservation tensions stem from their colonial and imperial encounters with and attitudes toward the tropics—and whether they are willing to do something about it. Regardless, however, the exploitation of tropical forests that has expanded throughout the twentieth and into the twenty-first centuries has impacted not just environments and human livelihoods in the tropics but also, increasingly, the climates and atmospheres of all of us. Determining when and how this process began may offer important lessons as we try to deal with environmental feedbacks of our own making today. To do so, we must reckon with the increasingly urgent concept of the "Anthropocene"—the period when humans began to have a dominating impact on earth systems.[30]

Chapter 12

A TROPICAL "ANTHROPOCENE"?

Since the beginning of this book, the broad, ordered periods of time, or epochs, that we have been using (e.g., Cambrian, Carboniferous, Pleistocene, Holocene) to split up our journey through the tropics have been identified on the basis of physical rock characteristics and the presence or absence of certain fossil plant or animal communities. Distinct boundaries between these periods are often related to major events such as climate shifts, volcanic and tectonic activity, or, as we saw in Chapter 4, an extraterrestrial impact. However, the pace and extent of twenty-first-century human impacts on the Earth's surface, its rock layers, and climate have led some scientists to argue that we have now entered a completely new era of geological history, the "Anthropocene," one whose boundary is set by us. Combining the Greek word for "human" with the standard suffix denoting a period in geological time ("-cene"), this term signifies that we are now living in an epoch when human activities are playing an equal, or even dominant, role to natural forces in the operation of so-called earth systems. "Anthropocene" has become an increasingly popular term, not only thanks to artistic and philosophical musings on what the so-called end of nature might mean for all of us but also because the Anthropocene Working Group (AWG) has been increasingly pushing for its official acceptance as a division of geological time by the International Commission on Stratigraphy in order to highlight the sheer force our species now wields in the context of the deep time of Earth's history. Geologically, such a status requires agreement on a clearly visible global "spike," such as the biological and chemical fallout caused by the asteroid impact that doomed the reptilian dinosaurs at the Cretaceous-Paleogene boundary. Attempting to do the same for the Anthropocene, the members of the AWG, headed by Jan Zalasiewicz at the University of Leicester, have narrowed in on plastic particles produced during an expansion of manufacturing of synthetic materials or radionuclides produced by the testing and dropping of the devastating atomic bomb during the twentieth century, which turn up in marine and lake sediments and caves around the world. Yet, while it is undeniable that human impacts on planetary systems since the twentieth century are on a scale never seen before in human history, to truly understand their origins and predict their outcomes, we quite literally need to dig deeper than a single line of plastic or radiation.[1]

As Professor Nicole Boivin, the director of my own institute in Jena, puts it, "As a process that involves humans, it is essential that geologists include social scientists, specialists in the study of human-environment interactions, to properly come to terms with what the 'Anthropocene' is and where its origins lie." Our species has shaped the Earth for much of its existence, and to limit focus to postindustrial time frames is to miss a long history of human relationships with the natural world. By including archeological, historical, and paleoecological perspectives, we can search for earlier potential roots of the Anthropocene epoch, from the increased burning of fossil fuels following the Industrial Revolution of the late eighteenth and nineteenth centuries to the origins of agriculture thousands of years ago. We can also move past a desperate search for a single, formal epoch and place the geological timescale to one side. Using the anthropocene—with a small *a*—to denote a framework, we are able to explore the long arc of human impacts on environments and the dynamic and often imbalanced human influences on earth systems that have led us to where we are today. Tropical forests provide perhaps *the* ideal setting in which to do just that. From our exploration of their origins in Chapter 1, we have seen that tropical forests, since their first appearance on our planet over 300 million years ago, have been a crucial part of earth systems. They built on the work of the first plant life to make the atmosphere breathable for life on land. They have regulated the distribution of the Sun's energy and also shaped where and how much rain fell in a given area. Their roots stabilized the planet's surface and created soils. They housed some of the first complex ecosystems to inhabit the planet and shaped some of the most marveled-at periods in the evolution of life on Earth. They remain perhaps the most significant biomes at the heart of anthropocene-derived changes seen around the world in the twenty-first century. As even local and regional human alterations to these environments can combine to have global feedbacks, what better place for us to look than tropical forests for the origins of human interactions with earth systems?[2]

We will now turn to the latest ecological and earth science evidence that shows how tropical forests remain keystone environments for various earth systems today. They are host to the planet's most varied plant and animal communities and protect them from natural hazards, they are key to regional and global precipitation, they curate the soils of entire river basins, and they interact with the atmosphere. Using the archeological and historical background developed throughout the book to this point, we will then investigate the changing interaction of human societies with tropical forests and the potential scale of its consequences. We will explore how the expansion of water buffalo and rice in the tropics 6,000 to 3,000 years ago may have resulted in visible greenhouse gas emissions, how the spread of farming and associated tropical deforestation in Africa might have influenced regional and continental soil erosion, and how the murder and disease inflicted upon huge Indigenous populations across the Americas at the time of European contact may have resulted in a regrowth of forests so significant that it changed the amount of CO_2 in the atmosphere. We will then move to the colonial and imperial land use that followed and its legacy for perspectives on the "Anthropocene," and indeed conservation, that can too often fail to acknowledge the wealth and power imbalances that have produced current sustainability tensions in the tropics. This chapter in no way suggests that our current impacts on tropical forests are "precedented" or "OK." Rather it highlights how our actions depart in both scale and severity, how we have become depressingly detached from the consequences of our choices, and how politicians and scientists must bear this long history of planetary influence and social exclusion in mind if they are to design effective conservation and policy.[3]

THOSE OF YOU who regularly watch nature documentaries on television or make wildlife encounters a key part of your summer holidays will not be surprised to learn that tropical forests are ecological hotspots. They are home to over half of the world's plant and animal species. Their lofty trees and bustling undergrowth, as well as their vast array of crawling, walking, swimming, and flying life-forms, make up a staggering third of the Earth's land-surface biological productivity. To put this into perspective, more species of ant can be found scurrying on a single tree in the Amazon Basin than across the entirety of the British Isles. Similarly, tropical forests can have up to 1,000 species of tree per hectare, as opposed to around a dozen for temperate forests. The huge diversity of tropical and subtropical forests we first met in Chapter 2—ranging from humid lowland evergreen rainforests to semi-deciduous tropical forests with a pronounced dry season, from acidic peat swamp forests to cool montane tropical forests—means that each and every tropical forest ecosystem has its own characteristic roll call of plants, insects, mammals, amphibians, reptiles, and birds that are often strung together in vital, intimate relationships. Certain key species can have massive influences over forest regeneration and health—for example, seed-dispersing primates, bats, and birds the world over play a critical role in tropical forest ecosystem dynamics, which we gained a deep-time perspective on in Chapters 3 and 4. As world-renowned tropical forest ecologist Professor Yadvinder Malhi of the University of Oxford puts it, "In the context of the biosphere, tropical forests, in many ways, have the most to offer. However, they also have the most to lose." These environments are currently facing some of the worst habitat loss and threats to their wildlife in the twenty-first century. Given the huge proportion of planetary biodiversity tropical forests host, such losses will undoubtedly echo around the planet.[4]

Lord of the Rings aficionados will be delighted to know that tropical forests are also gaining something of an Ent-like reputation for the way they manage and secure the land on which they grow. Belowground, grasping root systems anchor soils in place and shape the flow of water through the ground around them. Aboveground, significant canopies and leaves regulate the flow of rain to the floor, further stabilizing the earth. Together, these attributes enable tropical forests to quell rushing floods and to stem the movement of land. Coastal mangroves and forest plantations were even shown to have protected parts of Sri Lanka against the full force of the 2004 Indian Ocean tsunami; elsewhere, in their absence, the impacts were much worse. Tropical forests additionally host unique microorganisms that rapidly break down dead organic material, processing almost the entirety of falling leaf litter within a year, to keep up an important supply of nutrient-rich soil. These microorganisms, roots, and decaying vegetation even filter the water passing through and over them, removing contaminants for their inhabitants. A team of scientists, led by ecological modeler Dr. Nadja Rüger of the German Centre for Integrative Biodiversity Research, working on an island in the middle of the Panama Canal, has shown that long-lived pioneer tree species such as mahogany and Amazon nut (Brazil nut) can represent nearly half of the biomass in a given tropical forest. Meanwhile, studies of large-diameter trees in tropical forests more widely show the major control they have on carbon storage and cycling in forests. As a result, "many of these services to the geosphere, and a series of earth systems, can be attributed to the tropical 'Treebeards' out there," as Nadja puts it.[5]

The contributions of tropical forests to earth systems also reach far into our skies. Their large leaves and extensive canopies provide the perfect transfer platforms for water back into the atmosphere. Tropical forests are responsible for one-third of the evapotranspiration (evaporation of water from leaves) around the globe at any one time. Once in the air, the pollen, fungal spores, and compounds unleashed by tropical forests bind this moisture together to form clouds and, thus, life-giving rain. Climate scientists, using a sandbox technique similar to that of Tim Lenton, whom we met when exploring the first planetary impacts of plants in Chapter 1, have shown that this evapotranspiration is incredibly important to the water cycle (or hydrosphere) on both local and regional scales. If one were to completely deforest the Amazon Basin, precipitation could decrease by an average of 324 millimeters per year locally (or as much as 640 millimeters in some estimations, which would be more than the entire average annual rainfall in and around Newcastle-upon-Tyne), while also leading to reductions more widely across South America. Similarly, deforestation of the Congo River basin in Central Africa would not only reduce rainfall by approximately 50 percent in the immediate region but could also lead to less precipitation in West Africa along the Guinea Coast. Modeled impacts of tropical deforestation even extend beyond the tropics, thanks to global climate circulation systems. Complete deforestation of the Amazon Basin could potentially lead to declines in rainfall in midwestern and northwestern portions of the United States. Meanwhile, models of deforestation in the Congo Basin simulate reduced rainfall in Mexico, the United States, the Arabian Peninsula, southern Africa, and southern and eastern Europe.[6]

The cloud-forming canopies of tropical forests are not just significant for their role in the water cycle; they can also quite literally act as parasols in the regulation of local, regional, and global temperatures. Due to the fact that the tropics are exposed to energy from the Sun more prominently than any other parts of the planet, the shade and moisture provided by the water-filled air skirting above these forests not only cools the ground beneath them but actually impacts the distribution of temperature across the Earth more widely. Add to this that the forests of the tropics store nearly one-quarter of the world's land-based carbon, with peat swamp forests acting as particularly bulky carbon "sinks." As tropical forests perform a staggering 34 percent of the planet's photosynthesis, if they disappeared, not only would this trapped carbon be released into the atmosphere, but less CO_2 would be absorbed into the biosphere in their absence. Together, these greatly increased emissions would trap more of the Sun's energy and exacerbate global warming still further. Although interactions between changes in landcover, "albedo" (how much of the Sun's rays are reflected back into the sky, determining how much the surface warms), and CO_2 shifts are highly complicated, a series of modeling studies, as well as direct observations, have highlighted the potential impacts of deforestation on regional and global temperatures. Complete deforestation of the Amazon Basin, Central Africa, and Southeast Asia is expected to produce regional warming of anywhere between 0.1°C and 3.8°C, 0.5°C and 2.5°C, and approximately 0.2°C, respectively.

Meanwhile, pantropical deforestation would also lead to an increase in average temperatures globally by as much as 0.7°C, a total that would double the warming observed since the mid-nineteenth century and the intensified industrial burning of fossil fuels. This would undoubtedly lead to further polar ice melt, climatic instability, and the demise of various species of plants and animals within and beyond the tropics. This is not to mention the fact that most of the coal already burned since the start of industrialization, as we saw on our search for some of the first trees on Earth in Chapter 1, originated from the earliest tropical forests of the Carboniferous period in the first place.[7]

Tropical forests are also often described as the "lungs of the Earth." Although they do not actually supply 20 percent of the planet's oxygen, they remain entwined in regional and global air quality. Without forests, landscapes become dustier. Meanwhile, the loss of their contributions to evapotranspiration increases the likelihood of fires. These impacts would only be exacerbated should their deforestation, or the draining and removal of tropical peat, occur as a result of deliberate burning to begin with. In 2019, Southeast Asian forest fires showed how widespread burning can have drastic impacts on the air quality across a sizable area, with blazes in Indonesia argued to be behind smog descending on Malaysia, Singapore, and Vietnam and even forcing school closures in these areas—although, it should be noted, Indonesia disputed that it was the only country at fault. A lack of air filtering from forest cover, increasing airborne dust, and rising fire frequency, alongside deforestation-linked emissions of CO_2, nitrous oxide, and methane, will eventually impact the air we breathe, and perhaps not only in densely populated tropical areas. If not our own breathing directly, alterations to tropical forests will certainly interfere with our food supplies. Human populations in Africa, Asia, the Pacific, and the Neotropics are already facing challenges to food security as increasingly unpredictable climates impact their harvests. Furthermore, given that those of us in the Western world often rely on vast international networks, thanks to the colonial processes reviewed in Chapter 11, tropical and subtropical climate changes will also soon affect the coffee we import from Brazil and Indonesia and the rice we obtain from Asia. The potential global climate changes wrought by dramatic declines in tropical forests could even damage agricultural growth of staple crops across the United States, China, and India, with disastrous impacts for the large human populations of these regions. Tropical forests therefore provide perhaps the clearest case of twenty-first-century human entwinement in earth systems. We can see how we are all reliant on these environments and are all, in some way, linked to them. So when did this begin? And how have past interactions with tropical forests shaped the potentially perilous position we now find ourselves in, where the loss of these environments will be felt all over the world?[8]

CHAPTER 6 DEMONSTRATED how tropical forests were repeatedly occupied by late Pleistocene humans potentially as early as 200,000 to 100,000 years ago and certainly by 80,000 to 45,000 years ago. But could these early hunting-and-gathering groups have somehow left lasting impacts on these environments and thus earth systems? One way they might have done so is through removal of the megafaunal inhabitants of tropical forests. Conventionally classed as animals heavier than forty-four kilograms, many of these giants faced extinction around the world during the late Pleistocene, at a time when *Homo sapiens* was occupying new areas of the planet. In South America, including the Amazon Basin, a staggering array of weird and wonderful megafauna disappeared during this time, such as giant sloths, the elephant-like gomphothere *Stegomastodon*, and the rhino-sized *Toxodon*. Similarly, in Chapter 8, we saw how humans have been linked to later megafaunal extinctions in the Caribbean and Madagascar. Phylogenetic analysis of the late Pleistocene and early–middle Holocene extinctions of a series of unique, large mammalian lineages, notably within South America, has been argued to show that this represented the demise of 2 billion years' worth of evolution. Not only that, but because megafauna are so crucial to tropical forest ecosystems, through the dispersal of large seeds and fruits and the disturbance of forests with the swinging of their bulk, some scientists have suggested that if, indeed, humans brought about their demise, this could be the first example of our species impacting not only the biosphere but also the carbon cycle, as their absence hindered the regrowth of large trees. Other cases of potential early human impacts on earth systems come in the form of deliberate burning and maintenance of forests in

Borneo in Southeast Asia, the New Guinea Highlands in Near Oceania, and Queensland in Australia. Ethnographic study of Indigenous populations demonstrates how widespread and long-lasting such activities can be in altering and managing tropical ecosystems through the maintenance of forest boundaries, open patches, and clear forest floors. Nevertheless, it can be difficult to definitely and directly attribute human activity to these changes, and in the three "megafaunal" cases mentioned, it seems unlikely that our ancestors were the sole factor in their tropical decline. Moreover, in general, the overall earth systems influences of these activities would, while visible, also seem to be relatively localized.[9]

More continental and global-scale human impacts on tropical forests and earth systems may have begun with the emergence of farming practices. We saw in Chapter 7 how a number of our favorite domesticated plants and animals have their origins in the tropics, including as part of some of the earliest examples of cultivation in the archeological record. Meanwhile, we witnessed how the entrance of agricultural systems from extratropical areas could have major environmental repercussions. One of these latter cases has actually even been implicated in possible globally recognizable preindustrial "greenhouse gas" emissions. We are used to thinking of the rear ends of cows as significant methane producers but perhaps less so rice. Nevertheless, where rice grows in flooded fields, or rice paddies, the water stops oxygen from entering the soil, leaving it a hotbed for methane-producing bacteria. When William Ruddiman became the first scientist to suggest that the beginnings of the anthropocene could actually be found with the origins of agriculture approximately 8,000 years ago, he suggested that, combined with CO_2 emissions from rampant agricultural deforestation in Eurasia, methane emissions due to prehistoric rice expansion across Asia produced a greenhouse gas effect so large that it actually prevented the onset of a glacial cycle. Other scientists have linked the expansion of ruminating livestock into sub-Saharan Africa, monsoonal India, and central China, as well as the expansion of irrigated rice and water buffalo throughout Southeast Asia, to later, significant methane emissions between 6,000 and 3,000 years ago. These models do not yet even include detailed estimates of the potential CO_2 emissions caused by prehistoric deforestation across the entire tropics, as a product of both locally developed and externally introduced farming innovations. Although precisely and directly linking human activities to global atmospheric records remains challenging, and reconstructing the resulting climatic feedbacks is even more so, these examples illustrate how prehistoric farming activities in the tropics may have already interacted with planetary systems.[10]

The impacts of past tropical food producers on regional biospheres and geospheres can be a little easier to discern than those of their planetary counterparts, particularly in island ecosystems. The introduction of the domesticated pig and the Pacific rat as humans expanded across Micronesia and Polynesia are cases in point, resulting in observable soil erosion and the mass extinction of thousands of bird species, respectively (Chapter 8). Scientists believe that the arrival of swidden farming on another Pacific island, Guam, led to a complete environmental transition from forest to "savannah." Such changes did not only occur on islands, however. In the Amazon Basin, cultivation of Indigenously domesticated plants involved significant landscape modifications to produce raised fields. Furthermore, the active selection and promotion of economically useful species by past human societies has shaped the modern genetics and distribution of rainforest trees across this vast region to this day. Working at Lake Caranã within a forest reserve in the Brazilian Amazon, paleoecologist Dr. Yoshi Maezumi of the University of Amsterdam and her colleagues were able to demonstrate that 4,500 years of forest management not only showed an enrichment of edible plant species represented in the sediment record of the lake but had also left its mark on the composition of the surrounding forests. Nonetheless, as Yoshi puts it, "the widespread role of humans in ancient deforestation and burning associated with these early tropical farming systems, as well as the legacy of human land use on biodiversity, soils, rainfall, and temperature, remains to be better-resolved through the comparison of detailed, well-dated multiproxy records." The same is true across Southeast Asia, southern China, much of tropical Africa, and most of the island environments we have visited on our tropical tour, where current investigations are often limited to local comparisons and the integration of archeological and past environmental data into regional earth systems models remains a pressing area for future research.[11]

The impacts of food-producing societies on tropical environments and earth systems were, however, only to increase and, as Chapter 9 showed us, between 2,500 and 500 years ago, a variety of populous urban settlements, states, and empires had emerged in many of these regions. LiDAR and other remote sensing technologies have proven particularly powerful for revealing the sheer scale of the past populations buzzing across the tropics. In the Neotropics, in North and Central America, the Classic and Postclassic Maya occupied extensive urban landscapes, planted crops, and manipulated forests. Recent assessments suggest as many as 11 million people may have been present in the region in AD 800. Estimates of the deforestation associated with these populations have now been factored into climate models to suggest that Mesoamerican urban societies reduced precipitation by 5 to 15 percent in the region, amplifying droughts and potentially contributing to their own issues of sustainability in dry, lowland regions. This is not to mention the vast hydrological and agricultural modifications made by the Triple Alliance (Aztec Empire) across the Valley of Mexico and out into their area of imperial control. Turning further south, the Inka Empire ruled over approximately 20 million people, building massive terracing and irrigation networks that deforested mountains and hillsides. Inka rulers were continuously combatting soil erosion through the planting of trees and regulating the use of wood. Meanwhile, according to recent demographic estimates, another 8 million to 20 million people among the "garden cities" of the Amazon Basin, through significant earthworks, occupation mounds, and networks of causeways, were by five hundred years ago permanently shaping forest composition, dynamics of regrowth and succession, and perhaps even overall forest cover aboveground, as they simultaneously changed the nutrient contents and fertility of soil below. Although evaluations of their population sizes based on census data, later colonial records, and archeological records remain hotly debated, these precolonial Neotropical societies, additionally including the precolonial populations of the Caribbean, left their mark not just on tropical forests but also on the regional and continental biosphere, geosphere, hydrosphere, and, perhaps, as we will see shortly, the Earth's atmosphere. Notably, however, these populations, even in urban systems, were often dispersed across the landscape, practicing a combination of crop cultivation, tree management, and wild animal use on land and in rivers, reducing the overall per capita pressure on each parcel of land at any given time. As a result, despite some local and regional challenges, a population approaching that of Europe was broadly being successfully sustained in the Neotropics by the time of contact with Europeans.[12]

Turning to Asia, the world's largest preindustrial urban center had documented impacts on the dry tropical forests of Greater Angkor in Cambodia. Studies of lake cores and the urban infrastructure itself have shown that human deforestation across this vast artificially constructed drainage basin gradually led to soil erosion that contributed, alongside climatic variability, to rulers eventually abandoning the region as their capital. Limited paleoenvironmental investigation means that we have almost zero understanding of how the significant states and empires of West and Central Africa, southern China, elsewhere in Southeast Asia, and Sri Lanka and the Indian subcontinent impacted their surrounding tropical forest environments and thus earth systems, though current representations are almost certainly underestimates. In contrast to perspectives often put forward in conservation, ecology, and earth science circles, then, these environments were not a blank canvas for what came next. Instead, many millennia of increasingly intensive manipulation meant that human societies were already locked into earth systems feedback mechanisms with plant and animal communities, soils, and perhaps even local and regional climates. Some scientists have suggested that such examples are, to some extent, anecdotal and incomparable to what happened next, and they might be right—not necessarily only because these examples were on a smaller scale or because they were inherently less impactful when compared to the dominating impacts of humans on the tropics and earth systems today. Instead, while precolonial tropical societies modified their environments, often with far-reaching consequences, they clearly retained a certain flexibility to adapt. We have seen that, with some notable exceptions, elites, merchants, and, most particularly, farmers and horticulturalists in these societies could see what was happening around them, change where they planted and what they exploited, modify landscapes, shift their political centers and structures, or even simply move on. The viselike grip of proto-global and, later, global economic, political, and social structures increasingly swept these options from the board, particularly for marginalized, oppressed groups. This period set us on a course for the present day, where some of the people most impacted by anthropocene-related changes to tropical forests are also the least able to change their practices and protect their livelihoods.[13]

THE SIZE OF human populations and extent of their modifications across tropical forests by 1491 is dramatically illustrated by what happened as their worlds collided with Europe. We saw in Chapter 10 how Alexander Koch and his colleagues have used models of existing population estimates from historical documents, settlement areas, and other sources to estimate that within 150 years after European arrival, staggering numbers of Indigenous people across the Neotropics had died as a result of diseases arriving from across the Atlantic. However, he and his team did not stop their research there. Assuming a given area of land use by these Indigenous populations in different parts of the Neotropics, they sought to work out what would happen if these human societies were suddenly largely removed from tropical landscapes. They found that the estimated additional carbon uptake, as new forest growth rapidly occurred across the Americas in their absence, would have made up a significant amount of the seven-parts-per-million decline in global atmospheric CO_2 concentration recorded in polar ice cores between the late sixteenth and early seventeenth centuries. This decline reduced global temperatures by 0.15°C. This might not seem like a lot, but it actually contributed to the coldest part of the Little Ice Age from approximately AD 1600 to 1650, when harvests failed across Europe and ice-skating on frozen lakes became a source of artistic inspiration. More detailed archeological-, paleoecological-, and historical-based models of changing land use are necessary to confirm the scale of this "reforestation." However, for Alexander, the current balance of evidence suggests that "the 'Great Dying' of the Indigenous Peoples of the Americas, particularly within the Neotropics, initiated by European colonization led to a human-driven global impact on earth systems, with serious feedbacks occurring far away from the tropics, long before the Industrial Revolution."[14]

Earth scientists Simon Lewis and Mark Maslin have even suggested that the "Great Dying" and the earth systems feedbacks that ensued represent the most significant global marker for the "Anthropocene" as an actual epoch—separating what went before from the beginning of the modern "world-system." Certainly, from a biosphere standpoint, the first global trade networks that joined the American tropics to Africa, China, and Europe transferred not just microbes on a scale never before seen but also plants and animals. Lewis and Maslin, for example, in a prominent paper published in the journal *Nature* have suggested that pollen from maize in European lakes and marine sediments and phytoliths from bananas in lake records in Central and South America, from 1600 onward, represent clear stratigraphic markers for the globalization of human foods and cuisines initiated by the so-called Columbian exchange. The same could be said of archival and archeobotanical records of sweet potato in the Philippines, manioc in Africa, wheat in Mexico, potatoes and tomatoes in Europe, and sugarcane in the Caribbean as the tropics sat at the center of a process of changing landscapes and mealtimes across the world. Bones of horses, cows, goats, and pigs, as well as less-intentionally moved rodent stowaways, appearing for the first time in archeological sites in many parts of the tropics offer further potential markers preserved in sediments. These animals could have major impacts on tropical biodiversity, vegetation cover, plant community structure, and the stability of soils as part of a "swift, ongoing, radical reorganization of life on Earth without geological precedent."[15]

This "reorganization" also included new ways of appropriating, using, and perceiving tropical landscapes that arrived with Europeans. We saw in Chapter 10 how, by the second half of the seventeenth century, after just a century of mining in the region, Spain had already received 16 million kilograms of silver from the Neotropics—tripling the amount that had formerly existed in Europe! This could leave deep scars in regions such as the silver mines of Potosí, where labor was forced into an intensive extractive exercise. Here, by 1600, trees were stripped, domesticated and wild animals disappeared, and the soils of entire mountain slopes had turned to loose gravel. Meanwhile, mercury seeped into the soils and air. Likewise, excavation was so intense at the gold deposits in the Brazilian Highlands that scientists visiting the mining zone in the early nineteenth century assumed that the formerly lush, forested area had always been a barren plateau. A similar disregard for sustainability in favor of profit is visible in the form of increasingly widespread plantation agriculture. Europe's insatiable lust for sugar had already eaten up approximately 5,000 km² of Brazil's now threatened Atlantic rainforest by the middle of the seventeenth century—a territory nearly the same size as the entirety of Trinidad and Tobago. We find a comparable pattern on British-, Dutch-, and French-owned sugar plantations in the Caribbean, Dutch spice plantations in Southeast Asia, and French vanilla plantations in Madagascar. Beyond cash crops, globalized economic systems even began to appropriate tropical landscapes for their own everyday food and materials. For example, running out of wood on Mediterranean shores, the Spanish and Portuguese increasingly turned to the lumber resources of the Philippines and Cuba and of South India, respectively, to build their fleets. The exact scale of tropical deforestation initiated by this lust for a "cheap nature," to be used and profited from, currently remains poorly known. However, given that, as we saw in Chapter 11, concerned observers had already postulated a connection between Neotropical plantation agriculture and aridity as early as the nineteenth century, significant earth systems feedbacks seem highly likely.[16]

The expanding tendrils of this "Age of Capital" have also been implicated in one of the most prominent anthropocene processes identified in human history, the onset of the Industrial Revolution in northwestern Europe. While this happened away from tropical landscapes, the source of capital for the investors, entrepreneurs, factory owners, and merchants who drove it was in many cases the wealth, productivity, and human lives forcibly taken from, or moved across, the tropics at an increasingly merciless pace between the seventeenth and twentieth centuries. Privileged, controlled access to energy-rich sugar grown using chattel slavery, rice planted across entire tropical river basins for the main purpose of export, and potatoes and fertilizer to develop local agricultural efficiency meant that the tropics also experienced a rerouting of calories. Opportunities to grow food were taken away from the inhabitants of the tropics to fuel an increasingly nourished society in first Britain and then other nations in western Europe and, later, northern North America. As such, the western European imperial appropriation of tropical plants, land, and labor can be seen as a significant factor in the onset of mass factory-based, capitalist industrial production, expansion of effective transport infrastructure of railways and steamships, and insatiable burning of fossil fuels that left significant, globally recognizable increases of atmospheric CO_2 in Arctic ice cores and records of methane, nitrates, ash, and burning in ice and lake records from the nineteenth century onward. Suggested by some as possibly representing the true beginnings of the "Anthropocene" as a formal epoch, industrial- and transport-based emissions remain some of the most significant culprits behind human-induced global warming in the twenty-first century.[17]

The processes described above are not just important for our search for the gradual intensification of anthropocene phenomena over time. They have also been important for deconstructing the idea of a definitive, unifying, geological, scientific "Anthropocene." Studying the colonial and imperial processes that swept across the tropics, anthropologists, historians, and historical geographers have suggested that searching for a single spike, in either the Industrial Revolution or later twentieth-century events in Europe and northern North America, obscures the long-term processes of inequality on which twenty-first-century impacts on the natural world are founded. Global history between the fifteenth and twentieth centuries shows that the "Anthropocene" is not a product of humanity as a whole; nor has it impacted all of humanity equally. Instead, as Janae Davis of Clark University and her colleagues have put it, "It is the product of interconnected historical processes driven by a small minority that provided the conditions for global capitalism through colonialism, enslavement, and racism."[18] Expanding the work of anthropologist Donna Haraway, they argued that the "Plantationocene" might be a more appropriate concept than the "Anthropocene." Rather than seeing the steam and factories of western Europe as the active origins of the present globalized world, we should look at how plantation systems converted foreign investment into homogenized tropical landscapes, how they forced people, plants, and microbes from all corners of the globe together, and how they were caught up in processes of Indigenous mortality and racialized capital-driven systems of enslavement. In such a framework, the forced labor of plants, landscapes, and people provided the "model and motor" for the greedy Western factory system seen by many as the embodiment of the industrial-based "Anthropocene" we contend with today.[19]

Some anthropologists and historians have also proposed the broader term "Capitalocene," highlighting how plantation systems were just one part of the origins of a global system that wrapped tropical and extra-tropical landscapes together. This view highlights how, from the onset of European colonization in the Atlantic and the Americas, tropical lands were mapped, modeled, and valued as open spaces ready to be put to work. Labor, in the form of slavery and indentured servitude, and nature alike became commodities to be accumulated, appropriated, and used. The result was a world in which environments and livelihoods in the tropics—from the silver mines of Bolivia to the coffee plantations of Haiti, from the cattle-modified landscapes of Mexico to the logged hills of the Philippines, and from the tea plantations of highland Sri Lanka to the fields of Indian farmers desperately planting cotton instead of food to earn enough money to pay heavy taxation—were heavily degraded to provide capital and power for populations, factories, and governments in the western half of Europe and northern North America. The result was a world in which even human life across the tropics became reduced to a cheap, often racially based commodity, grist for the mill for national, corporate, and public wealth on the other side of the world. This is not to say that oppressed minorities and populations were always powerless, and Chapters 10 and 11 showed us many examples of their resistance and shaping of globalized cultures, cuisines, and economic systems across the tropics. Overall, however, any analysis of anthropocene processes must clearly recognize that the current global role of the tropics in political, economic, social, and earth systems in the twenty-first century has not emerged from random, uniform human processes, and certainly not everyone has had an equal seat at the table.[20]

TROPICAL FORESTS ARE, of course, not the only environments in which the "multiple beginnings" of anthropocene processes can be investigated and observed in prehistory and history.[21] Some archeologists have focused on the origins and spread of pottery in soil deposits around the world as a "novel" marker of human behavior and waste, a process now extending back to 20,000 to 12,000 years ago in East Asia. Alternatively, domesticated animals emerging from the Near East represent the beginnings of "distinctive organic remains" that herald a new period of human impacts on the biosphere.[22] Nonetheless, tropical forests are the land-based environments that are perhaps most caught up in biological, climatic, and atmospheric earth systems and best show how even local and regional modification can have widespread repercussions. With their "keystone" position in planetary function in mind, we can imagine how the disturbance of megafauna through hunting and burning; how the expansion of domesticated animals, emissions-producing crops, and deforestation in the face of expanding and locally developed farming systems; and how the emergence of precolonial cities, states, and empires within tropical forests might all have had preindustrial impacts on earth systems ranging from the local to the global. Working within the anthropocene framework, we can see how these human insertions into earth systems may not represent a clear, global "Anthropocene" spike and may not approach the intensity or scale on which such interactions are happening in the twenty-first century. However, Chapters 6 to 9 have undoubtedly demonstrated how almost all tropical forests have a deep-time history of human manipulation, leaving lasting legacies in terms of species distributions, biodiversity, and ecosystem dynamics. We will see in the next chapter how these long histories of Indigenous management show that there are few "pristine" tropical forests, and we must shape our conservation choices and practices accordingly. In turn, the growing, vibrant archeological, historical, and paleoecological literature in tropical forests forces us to face challenging questions as to what kind of environments we actually want to preserve around the world moving forward: empty and uninhabited, or sustainably managed and used and, therefore, perhaps better protected.[23]

Tropical forests and their human, plant, and animal inhabitants are also certainly not the only environments and communities to have witnessed the impacts of colonialism, imperialism, and capitalism between the fifteenth and twentieth centuries. In 1541, England, under the rule of Henry VIII, began to extract from Irish forests and relocate Irish labor as part of colonial rule. Meanwhile, a desire for iron, copper, and silver ores had seen significant portions of central Europe laid waste by the early sixteenth century as a consequence of imperial demands. Beyond Europe, the horrific treatment of Indigenous populations and the atrocities of slavery that occurred as part of growing appropriation of nature and flows of capital were repeated in temperate North America, the nontropical portions of Australia, and southern Africa, among other places. Nevertheless, we have seen how the tropics were at the center of the intensifying global spread and flow of imperial power and capital. The mines of the Neotropics quite literally provided the money for European economic and political ventures, tropical plants and landscapes the world over provided the basis for plantation and ranching systems that amassed national and private wealth in Euro-America, and human lives across the tropics were lost, commodified, forcibly relocated, and altered forever by colonial and imperial processes. The scars left behind on our planet, in the form of barren landscapes, political and capitalist economic inequalities, racial conflict, and earth systems consequences, are profoundly apparent through the lens of tropical history. They mean that all of us are caught up in the ongoing balancing act between tropical commoditization and conservation, between necessary sought-after infrastructure and economic expansion and planetary health, and between Western demands for national and global action and the difficulties facing often impoverished tropical nations if they are to act. Only by recognizing this can we begin to develop practicable conservation solutions, involving diverse international, governmental, and local stakeholders, to the crisis now facing some of the most threatened land-based environments on the surface of the Earth.[24]

Chapter 13

HOUSES ON FIRE

I n January 2019 Swedish student Greta Thunberg declared to world leaders at the World Economic Forum, "Our house is on fire." Her words proved disturbingly prophetic as the year wore on, particularly for the tropics. Between June and August, fires swept across the rainforests of the Brazilian Amazon. Although natural fires are rare in these environments, a weakening in environmental protection regulations by the government, long-term trends of deforestation, fragmentation, and timber cutting, climate change, and linked regional aridity resulted in the highest levels of active fires and forest clearance seen in a decade. Fires in Brazil, as well as Peru, Bolivia, and Paraguay, had catastrophic consequences for Amazonian biomes, and by August blankets of smog shot into the high atmosphere, extended across both the Amazon and Paraná River systems, and adorned cities, such as São Paulo, many kilometers away. By January 2020 smoke was arriving in South America from a very different source. Natural wildfires are common in many parts of Australia during the dry season. However, between June 2019 and January 2020, the wildfires of what became known as the "Black Summer," spurred on by human-induced global climatic warming, were so bad that the resulting smog made it all the way to Chile and Argentina. Not only that, but fires even began within Australia's rainforests—environments that scientists had thought would rarely burn. As firefighter Renier van Raders put it when I met him on a recent trip to the Atherton Tablelands, "In my twenty years in this job, I have never seen these wet forests burn. This year we had two local events where the rainforest canopy was on fire." These two sets of infernos vividly illustrated the increasingly precarious state of the tropics, and indeed our entire planet, in the twenty-first century—and worrying trends in fire continued in the Amazon Basin in 2020. They also showcased the diverse social, political, economic, and environmental factors that make developing sustainable conservation and policy solutions so challenging.[1]

Scientists looking at changes in forest cover using satellite imagery taken from space between 2000 and 2012 have shown that, globally, tropical forests and woodlands are disappearing at a rate of 91,400 km^2 every year—an area roughly the size of Portugal. It is therefore not hard to see why tropical forests are frequently considered, in academia, policy, and

Figure 13.1. Photograph of Queensland's rainforest on fire in 2019. This wet forest was formerly considered to be safe from fire, but climate change and the buildup of dry vegetation on the forest floors has led to intense burning in recent years. *Renier van Raders*

Figure 13.2. The scene left behind following a fire in the rainforest in Queensland, Australia. *Renier van Raders*

the media alike, to be key battlegrounds in the fight for human sustainability. The drivers behind their disappearance can be various. In some cases, it is low-income farmers seeking to plant food to survive or cash crops to scrape together a living. In others, it is multinational corporations sponsoring widespread land conversion to maximize productivity and profits. Logging for local, national, and international timber needs represents another threat. Although selective, well-managed logging can be sustainable, poorly regulated, corner-cutting approaches and the building of roads and other forms of infrastructure can lead to forest degradation and fragmentation with impacts that can ultimately be just as bad as wholesale clearance. Beyond deforestation, the diverse plant and animal life of tropical forests faces further challenges. In Africa, nearly 5,000 tons of hunted tropical bushmeat are dragged from their leafy homes every year. Here, as well as in Southeast Asia and South America, the "catch" often includes large-bodied animals, such as primates, that, as we have seen, can have crucial roles in the overall health of forests. Atmospheric CO_2 increases and global warming represent more indirect, though no less significant, human threats for tropical forest life-forms, radically altering their distribution and dynamics. This is only going to get worse as raging fires and deforestation in the same forests release yet more CO_2 from the land into the atmosphere, creating an even hotter and less nurturing "greenhouse." In fact, the average carbon emissions produced by tropical deforestation around the world are currently higher than those produced by the entirety of the European Union.[2]

The prominent contribution of tropical deforestation to rising global emissions, as well as the local, regional, and global earth systems feedbacks that, as we saw in Chapter 12, are possible should these forests continue to disappear, has meant that the conservation and management of tropical forests has become a major environmental, economic, and political priority. This is the case not just for governments in and around the equator but, as demonstrated by the international political fallout during the Amazonian fires of 2019, also for those further to the north. And rightly so. Because by 2050 over half of the world's human population and two-thirds of its children will live in the tropics, looking increasingly intently at tropical forests to make their living and to secure their futures. The possible solutions can seem obvious from Euro-American homes. Cordon off the land like a crime scene, remove all humans to let it return to "nature," plant more trees, and encourage foreign governments to prohibit logging, hunting, and landscape conversion. However, these sometimes patronizing perspectives usually ignore two major issues. First, as we've seen, most tropical forests have not been completely "natural" for many millennia. Second, the responsibility for the current impacts of capitalism on tropical landscapes, as well as for the global wealth imbalances at the core of deforestation, hunting, and climate change crises, does not lie solely with people in the tropics. Instead, we are all, particularly those of us enjoying comfort in the Western world, liable for the fate of these threatened biomes. While this can be hard to swallow, as we will now see, only by acknowledging the above two points can we develop more successful and just conservation, economic, and political policies.[3]

GIVEN ALL THAT we have seen throughout this book, it may come as something of a surprise that some conservationists and ecologists have sought to define areas of tropical forest as "wilderness." The current rate of global species extinctions is estimated at 1,000 times its natural baseline rate as a result of twenty-first-century human planetary influences. Thanks to the sheer quantity of plant and animal species in tropical forests, a plethora of which are thought to still await discovery, conservationists have focused on these ecosystems as bastions in the global fight against species loss. Some have even suggested that by protecting vast areas of "wilderness" with low human "footprints" and high levels of biodiversity (i.e., tropical rainforests), we can secure the future of the majority of the Earth's plants and animals. Similar ideas drive the proposal that by adding more trees around the world, notably in dry tropical and subtropical areas, we can successfully counteract climate change by returning large areas of land to a "restored" state. Alternatively, ecologists have emphasized the significance of the condition of the remaining forests—namely, whether they remain "intact" or have been "degraded" by human actions. Such terminology hints at desires to reach a primeval, undisturbed condition, something clearly visible in many UNESCO designations of natural heritage areas. For example, the Wet Tropics of Australia is said to contain "relicts of the great Gondwanan forest that covered Australia . . . 50–100 million years ago."[4] Now, the maintenance of intact forests, with their characteristic canopies, stratification, and dynamics, can certainly have its benefits. These forests undoubtedly play immensely significant roles in air quality, soil stability, climate, and biodiversity protection. In fact, many of the best-protected ecoregions worldwide occur in "intact" areas of tropical forest within the Neotropics and Southeast Asia. Nevertheless, the fencing of wholly "natural" tropical areas is becoming increasingly challenging in the face of human population growth. Moreover, ecologists are recognizing that "intact" does not have to mean "pristine," while many ecological benefits also exist in secondary and "anthropogenic" forests. As we've seen from the beginning of this book, tropical forests are not static. Instead, they have been constantly on the move thanks to plate tectonics, climate change, and, later, deliberate human interventions. The exclusion of these dynamic histories, as well as their traditional human occupants, can thus misrepresent these forests, often with unintended consequences.[5]

The establishment of protective reserves can still represent a highly effective solution to tropical conservation if done in the right way. However, for many areas of tropical forest, they are simply not an effective, or even viable, option. So what can be done instead? Can we learn something from the archeological, historical, and paleoecological insights we have already gained? Bushmeat hunting, for example, is often taken to be inherently unsustainable, leading to "empty" forests. However, as we saw in Chapter 6's tour of our species' first global expansions, semiarboreal and arboreal mammals, including two primates (the gray langur and purple-faced leaf monkey) that are now endangered, were consistently hunted by bow-and-arrow-wielding human populations in Sri Lanka's rainforests between 45,000 and 3,000 years ago, without apparent negative impacts. Furthermore, even deliberate early human modifications to tropical forests to maintain clearings, tracks, and grassland biomes, dating as far back as 45,000 years ago, likely stimulated rather than hindered populations of prey such as wild boar in Borneo. Certainly, ethnographic and historical observations in Central Africa and South America document that Indigenous hunting methods and cultural restrictions see hunters move between different forest areas and different species to maintain the overall health of their prey communities. Meanwhile, a review of ethnographic research across the Amazon Basin, New Guinea, and Australia has revealed the benefits of human burning for the new growth of starch-rich plants as well as for populations of locally hunted birds and ground-dwelling and arboreal mammals. In fact, although, as discussed in Chapter 12, prehistoric "overhunting" was potentially an early human earth systems intervention, the majority of tropical animal extinctions convincingly linked directly to hunting actually postdate European colonialism. Indeed, today, the most unsustainable bushmeat hunting occurs where traditional weaponry, local ecological knowledge, and subsistence requirements have given way to heavy population pressure, "trophy" killing by tourists, or a commercial, individualistic drive to use snares and rifle technologies to maximize capture. In the latter case, this is often done to feed a rampant international illegal trade in pets, medicinal ingredients, or valuable items such as ivory. With the right conditions and practices in place, however, hunting in tropical forests, particularly as a local source of food, need not inevitably result in their falling silent.[6]

We have also seen that farming activities need not necessarily be in conflict with tropical forests. As shown in Chapter 7, these environments have yielded some of the earliest examples of deliberate cultivation anywhere in the world. Not only that, but Indigenous farming activities in the tropics reveal the possibility of very different avenues to food production than our common Euro-American perceptions of "agriculture." Archeology, paleoecology, and Indigenous knowledge are all now advocating the benefits of such "usable pasts" for improving food security in the tropics in the twenty-first century. Nigerian archeobotanist Dr. Emuobosa Orijemie of the University of Ibadan emphasizes how the archeobotanical record of the last 1,000 years in central Nigeria shows that "mixed landscapes of indigenous crops, such as oil palm, groundnuts, yams, cowpea, and pearl millet, and not monoculture" provided food security for human populations through periods of demographic expansion and climatic variability. Indeed, all around the tropics, scientists, local smallholders, and Indigenous communities are calling for a move away from the precarious world of maximum productivity encouraged by Euro-American expectations of genetically modified technologies, monoculture, and market demand and a turn instead toward the security of diverse traditional grains, legumes, farming systems, and extended family-based social networks. These practical legacies of food production are also the reason why, at the very beginning of this book, we saw villages on the banks of the Amazon Basin still sitting astride the fertile human-created Amazonian Dark Earths and combining small fields of manioc and maize with stands of wild Amazon nut (Brazil nut) trees and palms. Although many imagine agricultural fields and clearance as the prime route toward supposed "civilization" and development, the millennia of farming practices recorded in the soils of tropical forests around the world suggest very different, more sustainable routes forward for food production, particularly as the tropics and their inhabitants are increasingly threatened by climatic instability.[7]

Because 70 percent of the world's population will be living in urban settlements by 2050, and much of this will be focused on the tropics, we also need to come to terms with the fact that cities and their infrastructure will come into increasing conflict with tropical forests. Although high concentrations of people, roads, railways, airports, houses, and industry associated with Euro-American ideals of the city seem incompatible with tropical sustainability, Chapter 9 showed us that the tropical archeological record may once again offer some solutions. In particular, the "low-density agrarian-based urbanism" seen in Cambodia, North and Central America, the Amazon Basin, and Sri Lanka provide models for the low-density distribution of people across a landscape and the intermeshing of food production with urban facilities, administrations, and dwellings. Similarly, precolonial walled cities in West Africa, such as Benin City, demonstrate how well-organized systems of urban nodes, wild forest, land for cultivation, and fallow land formed to protect large populations from enemy incursions but also from resource shortages, a lack of fertile soils, and the loss of crucial forest environments. There is also renewed present-day scientific interest in the soil modifications that many of these past urban landscapes developed and relied upon to maintain their large tropical populations. Although they represent very different settings and socioeconomic contexts than their present counterparts, twenty-first-century urban planners are actively using and pushing these archeological examples as they seek to develop more sustainable models of cities that are more resistant to fluctuations in climate and natural hazards and have greater local security in terms of food and livelihood. In particular, they are highlighting how a planned mixture of "green" and urban space, as opposed to the informal developments currently seen expanding across Latin America and Southeast Asia, can provide a better model for tropical cities. Moreover, they point out that past urban examples highlight the importance of local ownership, consultation, and involvement, rather than top-down decisions and economic and political elites, in the ultimate long-term success of tropical urban sprawls.[8]

Accepting the "deep human history in tropical forests" also allows us to use archeology and paleoecology to make some specific suggestions in terms of the species and ecosystems we want to preserve or even try to reintroduce in different parts of the tropics. Take the example of planting more trees. We clearly need to take care as to *which* trees are planted, *where*, and *how*. Oil palm has been highlighted, by governments, companies, and conservationists alike, as an important route to prosperity for tropical nations and their populations thanks to a large and growing global demand. Where agricultural fields, pastures, and grasslands are covered in oil palm trees in a systematic manner, the land can capture more carbon, leading to long-term environmental and economic benefits for local populations, and despite its bad media press, compared liter for liter, oil palm threatens fewer species than coconut oil and olive oil. However, in Southeast Asia in particular, remote sensing satellite imagery has shown that much of the land converted to oil palm plantations over the last two decades has actually been native, peat swamp forest, leading to significant carbon emissions through fire, peat drainage, and loss of significant biodiversity. Archeology and history can have a significant role to play in developing more ecologically relevant solutions, highlighting the profitable and sustainable inclusion of economically valuable crops within diverse farming systems, as seen with oil palm in Africa in Chapters 7 and 11. Similarly, paleoecological research, ranging from the study of ancient plant remains in archeological sites in Hawai'i to the analysis of ancient plant DNA preserved in lake records at the high-altitude border of Uganda and Rwanda, has helped to bring past landscapes back to life. Records from before European colonialism of forests that existed under different Holocene climatic conditions can better shape "rewilding" and conservation priorities. Study of ancient animal fossils in the tropics can play a similar role in determining where and how species that have disappeared from the area, but were present in the past, might be locally and productively reintroduced. Nevertheless, ultimately, the question remains, what do we actually want to protect or restore? Are we really always trying to reconstruct a "natural" landscape that, as we have seen throughout this book, has not truly existed in many parts of the tropics for thousands of years? Or do we rather want to engage with the idea that many of tropical forests were being successfully managed for a long time before colonial and imperial forces arrived and subsequently made them the threatened biomes they are today?[9]

This question is becoming increasingly pressing as the creation of reserves by national governments can damage the rights and livelihoods of Indigenous peoples whose ancestors have occupied tropical forests

Figure 13.3.   Monoculture oil palm plantation in Borneo, Indonesia. In principle oil palm offers a valuable crop for boosting incomes and, when planted on
pastoral or open land, can even improve the carbon sequestration in the soil.
Although increasingly regulated, in the last two decades significant conversion
of rainforest and peat swamp forests to plantations has occurred, releasing massive amounts of CO_2 into the atmosphere and destroying wildlife habitats.
Douglas Sheil

for millennia. There is sometimes a narrow conservation assumption
that because industrialized, capitalist impacts on forests are devastating,
we must lock out *all* human communities. The Twa, who had lived in
the Kahuzi-Biega National Park in the eastern Congo Basin, were forcibly expelled, leading to rampant malnutrition as they were forced to
integrate into wage systems and lost their traditional sources of food
from the forest. The Wanniya-laeto of Sri Lanka have been similarly
prohibited from hunting in their traditional forest lands and now face
the gradual disappearance of their culture and even their language. In
some cases, the removal of Indigenous groups evidently impacts the forests in a negative manner. In Australia, bureaucratic obstacles hindering
Traditional Aboriginal Owners, such as the Jirrbal community, from

practicing burning and forest management on their traditional forest land can have ecological as well as cultural consequences. A reduction in the application of long-held burning practices has meant that dense undergrowth has formed in both the dry sclerophyll forests and the rainforests of the region. The land now looks "sick," as Jirrbal elder Desley Mosquito described to me. As the climate dries in the face of human-induced CO_2 emissions, dense forest floors provide the ideal kindling for increasingly frequent forest fires. Significantly, a number of initiatives by the Wet Tropics Management Authority and the Rainforest Aboriginal Peoples now support the application of traditional knowledge within ecologically protected areas. Paleoecological research in the Amazon Basin has demonstrated how drastic changes in forest burning practices between precolonial and colonial times have had a major bearing on ongoing fire frequency and intensity. A Western scientific arrogance that insists we should always keep parts of the world fenced off and "undegraded"—perhaps as a way to somehow morally offset our own growing destructive impacts—while potentially ecologically productive, can sometimes actually damage the very ecosystems it seeks to protect. Not only that, but, like the history of the last five centuries, the creation of reserves has often been undertaken in a way that continues the unjust dismissal and marginalization of the very people who have the best long-term knowledge of local ecological dynamics and thresholds, as well as the most intense experiences of integrating new social, economic, and political structures into tropical land management—in short, the people we should be supporting and asking for advice.[10]

WE NEED TO accept that colonial and imperial processes, documented in Chapters 10 and 11, have led to a series of environmental, economic, and political injustices that tie the hands of people living in the tropics today. At the most fundamental level, five centuries of ecological and land-use disruption by Euro-American powers have made tropical communities some of the most vulnerable to climate change and other conservation threats. We can trace the challenges of meeting subsistence needs in the face of rampant soil erosion, declining nutrients, and invasive species that plague islands such as Haiti, São Tomé and Príncipe,

Madeira, Mauritius, Madagascar, and the Canary Islands back to profit-driven, colonially established plantation agriculture. Similarly, the deforestation of the Philippine uplands and coasts, which has left behind eroding slopes and a vulnerability to flash flooding, is a product of centuries of extractive logging by first Spanish and later US and Japanese imperial governments. Colonial and imperial extractive demands have also impacted the wildlife in certain regions. For example, in Sri Lanka, due to massive hunts by British colonial agents in the Horton Plains Highlands, animals once crucial for seed dispersal and forest dynamics are now completely absent from mountainous regions on the island today. One nineteenth-century British hunter reported that he alone had killed 6,000 elephants. Finally, the massive amount of fossil fuels burned in the name of industrialization—mainly, to date, to the benefit the Euro-American world—is the primary cause behind the present state of the polar ice caps and global climatic systems, which has seen frightening sea level rise and extreme typhoons and cyclones batter the islands of the Pacific, Indian Ocean, and Caribbean. Given that a significant portion of capital for this process in many cases came from trade in human lives, cheap labor, and extractive profits, often from these very same tropical landscapes, this climate injustice and the current plight of these drowning nations should perhaps sicken us more than it often does.[11]

Beyond the tropical landscapes themselves, long legacies of colonialism and imperialism, by producing very different socioeconomic conditions, have also dictated the flexibility with which local communities can now even afford to act in the face of threats. This has included the suppression of local Indigenous traditional knowledge in many regions through murder, disease, and relocation. But it has also involved fundamental emerging wealth imbalances that inevitably shape local priorities. Many tropical nations in Africa are home to some of the fastest-growing, but also the poorest, populations anywhere in the world. For example, in Madagascar, more than 92 percent of the population lives on less than US $2 per day. The government announced schemes of protected areas in 2003; however, breakdown in law enforcement and political crisis has hindered their progress, opening the door to illegal logging and mining, as well as bushmeat hunting and slash-and-burn farming by desperate populations attempting to survive in a drying landscape. In the southern area of the island, twentieth-century French imperial policies that forced mobile pastoralists to adopt more "modern" sedentary cultivation of cash crops have also increased contemporary pressure on tropical forests and a lemur population that represents c. 20 percent of the world's primate species. Similarly stark wealth discrepancies and political instability pose major challenges to coordinated conservation efforts across tropical Africa as well as New Guinea, where small-scale farmers are often the primary drivers behind deforestation. Contrast this to Australia, the richest nation in the world to host tropical forest, thanks to its very different colonial history. Here, a far higher level of per capita income and the economic and social benefits offered by comprehensive access to education provide a context in which a designated Wet Tropics World Heritage area can be effectively extended and policed across 894,420 hectares in Queensland (an area eight times the size of Greater London) and profit from the arrival of 5 million tourists (from home and abroad) and income of AUS $426 million every year.[12]

We should therefore not underestimate the degree to which Western conservation ideals are something of a luxury, born of the fact that, in many ways, we have already "made it" in terms of "development" and industrialization. The centuries of colonial and imperial removal of wealth from tropical nations, with limited return in terms of local investment and infrastructure, have, by contrast, exposed governments in the tropics to tensions between demands for improvements to living conditions at the same time as conservation and climate crises are reaching a tipping point. Brazil is an increasingly wealthy member of the BRICS economic group, with low population growth and relatively high earnings. Nevertheless, its government remains under heavy pressure from its supporters to develop further. The state of Amazonas is Brazil's poorest but also contains the largest area of tropical forest cover. As Victor Caetano Andrade and I saw at the start of this book, dense rainforest necessitates taking slow boats or expensive planes to move around, within, or beyond the state. Political, economic, and social incorporation of Amazonas, and its capital Manaus, into Brazil's federal network has always been challenging and involved a relatively late history of European contact with Indigenous communities. Poor populations are desperate for access to growing markets or even basic medical supplies, as tragically witnessed during the recent COVID-19 outbreak. Better infrastructure and land conversion, either government mandated or illegal, are seen as key routes to improving local livelihoods but will also inevitably represent a major conservation threat to significant areas of tropical forest, necessitating detailed planning. Local situations such as these must be considered when international calls are made for South American nations to stop cutting down trees or when Kenya and India are told to stop burning fossil fuels, especially in the context of long histories of Western extraction, exploitation, and unequal development efforts.[13]

Multinational corporations, as well as global consumerism, also now have a viselike grip on tropical nations, thanks to centuries of profits made on the basis of poorly paid labor, conversion of tropical land, and capitalist-based approaches to agriculture and tropical landscapes. Even relatively wealthy nations, such as Australia, do not escape, and the extent of and slow response to the "Black Summer" can be, at least partially, linked to poor policies on climate change in a nation that relies on coal-mining companies to bolster its economy. Similarly, governments in South America and West and Central Africa have often leased vast areas of tropical forest to dominating corporations seeking oil and gas. In the case of "cash crops"—from coffee in North, Central, and South America to tea in Sri Lanka and from bananas in the Caribbean to soybeans in Brazil—global demands, including from ourselves, many miles away, have long driven large landowners and small farmers toward cheaper, profit-making monoculture fields. Returning to palm oil, companies relying on this ingredient for products as diverse as lipstick, pizza dough, instant noodles, shampoo, chocolate, cooking oils, packaged bread, and biodiesel, catering to growing demand from Asian and Euro-American markets and consumers, have been a major driving force behind Southeast Asian governments having poor control over, or even actively encouraging, unregulated, unsustainable conversion of tropical forests to widespread plantation systems, to the jeopardy of their endemic wildlife. Meanwhile, simultaneous rising costs of local land, as well as corporations' expansion of their own holdings, can leave the door open to the abuse of the land rights of local people, as well as rising prices that leave small-scale farmers and Indigenous populations with little option but to convert, contract, abandon, or sell their own plots. Aggressive market conditions thus threaten more mixed, more sustainable, and often more productive approaches. The expansion of cattle ranching by large land-owners across the Amazon Basin to meet regional and global meat markets has represented a similar problem in Neotropical forests.[14]

Euro-American consumerism can even dictate which areas and tropical forest environments are made conservation priorities in the first place. Reserves, whether they are 39,000 km², like the Tumucumaque National Park in Brazil, or under 2 km², like the Bukit Timah Nature Reserve in Singapore, are expensive—both to establish and to maintain. Where tropical governments are able to invest in parks or in laws pro-hibiting logging or clearance, they need there to be some form of payoff. Ecotourism—travel to protected reserves to explore wildlife—has pro-vided one way in which foreign wealth can simultaneously fund the maintenance of forests with crucial ecosystem roles while also produc-ing additional economic benefits for local communities. The problem is that the forests receiving the most attention by protective policies are of-ten those that provoke Western fascination rather than necessarily those that serve the most pressing local needs. For example, they frequently include the green homes of iconic conservation stars such as the troops of montane gorilla in Uganda's Bwindi Impenetrable National Park, the orangutans of the Kabili-Sepilok Forest Reserve in northern Borneo, the jaguars of Brazil's Pantanal Matogrossense National Park, and the casso-waries and tree kangaroos of the Wet Tropics of Australia. This can be significant, especially as these giants often occupy a critical "keystone" position in tropical ecosystems. Nevertheless, it means that governments are less willing to pay for the conservation of less traditionally glamor-ous forests, including dry tropical forests, montane rainforests, coastal mangroves, and peat swamp forests, which often have more unique taxa, are often under far greater threat, and perform crucial functions in landscape stability and buffering against natural hazards. The position of many of these forests in climatic boundary areas, such as the sides of mountains or precipitation gradients between ever-wet and seasonally dry rainfall regimes, also makes them some of the most vulnerable to increased pressures of human land use and climate change.[15]

Tropical nations and their governments are, of course, not com-pletely blameless in the conservation challenges facing the tropics.

Political corruption, crime networks, increasing nationalism, and willful neglect of science can pose real problems for tropical forest conservation. Nor does the world need any repeat of the condescending ideologies of "saving" supposedly incapable local tropical populations that served as justification for many colonial and imperial enterprises and atrocities over the past half a millennium. In fact, local voices, such as the Association of Small Island States, the Caribbean Community Climate Change Center, cooperative unions of fair-trade cultivators, Coordination of the Indigenous Organizations of the Brazilian Amazon, and, as mentioned above, Rainforest Aboriginal organizations in Australia, are showing themselves perfectly capable of expressing the existential threats that they, and all of us, face in the context of climate change, governmental development strategies, and corporate interests, as well as advocating practicable solutions. However, top-down international and even national declarations and demands are often blind to the social, economic, political, and environmental histories that can limit the power of these groups and their voices. They can also hide the role of Euro-American nations in continuing to create an imbalanced tropical world through consumer choices and demands and wealthy, exploitative businesses. If we want to truly protect tropical forests for future generations, we must not only listen to local stakeholders and Indigenous populations such as the groups mentioned above but also practically support them by putting our money where our mouths are. In this way, we can combine local, traditional ecological and economic knowledge and strategies with archeological and paleoecological insights into how tropical landscapes have evolved and been managed by humans through time, to give tropical forests, and ourselves, the best chance to fight some of the most extreme sustainability challenges to ever face humanity during its time as a widespread tropical species.[16]

THE HISTORICAL COLONIAL restrictions placed upon significant amounts of Indigenous and local knowledge, which was developed over millennia, means that the application of traditional management and adaptations in tropical forests is not always straightforward and can sometimes even have negative impacts. Here, archeology and paleoecology, as well as the revitalization of Indigenous languages and oral histories, can enable local stakeholders to engage with ancestral practices and solutions. For example, in the Caribbean, whereas postcolonial houses are built of hard materials that are expensive to replace and dangerous in the face of intensifying cyclones, hurricanes, and earthquakes, the archeological record of precolonial housing shows something very different. Semipermanent housing, built on foundations of resistant materials and structures as well as more lightweight, easy-to-replace components, was not a sign of "backwardness" but an ideal, economic solution to the everyday challenges of living on these islands. On the Atherton Tablelands of Australia, combined visits of Aboriginal elders, archeologists, and paleoecologists to previous sites of occupation such as Urumbal Pocket have enabled more thorough documentation of the nature and importance of past human land and resource use through a combination of preserved oral histories with soil traces of human food processing, plant and animal use, and records of controlled burning. Sometimes, such interdisciplinary work and collaboration can even highlight negative elements of past practices, emphasizing the need for detailed, context-specific reviews and solutions rather than blanket statements or applications. For example, archeological and environmental investigation of human-made sediment traps at the locations of Konso and Engaruka in Tanzania show that traditional approaches enabled the capture of fertile, fine-grained soils and the avoidance of field salinization. However, while these practices improved growing conditions in the valley regions, soil erosion and vegetation loss in the highlands may have resulted in ultimate abandonment and loss of livelihoods. More detailed work of this nature is essential to understand how food security, resilient settlement, and sustainable use of tropical forests and landscapes was achieved, or indeed not, prior to the arrival of domineering colonial, imperial, and capitalist forces across much of the tropics—using the past to plan for the future.[17]

Where protected reserve initiatives are possible, the involvement of local stakeholders, whether they are recognized Indigenous groups or local farmers, has repeatedly resulted in more successful outcomes than when they are ignored. In the Brazilian Amazon, Indigenous classification systems of plants and ecosystem transitions have long been at the center of attempts to maintain botanical diversity throughout the giant

130,000 km² reserve of the Kayapó and Upper Xingu people. In the Ecuadorean Amazon, Indigenous groups combine with local nongovernment organizations (NGOs) and government initiatives to promote ecotourism to individual villages as a source of additional wealth but also a way to promote tropical forest preservation and conservation. Government and administrative support for Aboriginal ranger groups, such as those of the Jabalbinna, Djabugay, and Girringun, have had significant success preserving wildlife in the Wet Tropics of Queensland. Similarly, the allocation and networking of so-called local gorilla guardians in Cameroon has enabled a better understanding of the frequency and distribution of some of the most endangered great apes in Africa, revealing ongoing illicit hunting and strengthening communication and understanding between local villages, hunters, governments, and conservation scientists in order to increase support for protective measures and action. Sometimes conservation initiatives spring completely from the actions of local interest groups. In the Mamberamo-Foja region of Papua (the Indonesian portion of the New Guinea land area), the Ijabait elders act quite literally as "guardians" of local tropical forests. They live at strategic sites, such as forest boundaries and rivers, to monitor and control human use of forest and freshwater animals. In each case, the involvement of local groups, whether through cultural and traditional concerns, social contracts, or added economic benefits, has led to the more beneficial imposition of protected areas and a reduction of illegal incursions. As conservationist Professor Douglas Sheil of the Wageningen University puts it, "The inclusion of populations living in tropical forests not only acts as a robust local barrier to incoming external threats but is also more likely to reconcile economic and subsistence concerns with conservation needs."[18]

Considering local interest groups is also essential to developing and supporting more sustainable ways of extracting resources and products from tropical landscapes. For example, Reduced Impact Logging initiatives use dedicated training for workers, effective patrolling, protection of hillside forests, and refined tree cutting to reduce ecosystem and soil disturbance. The long-term ecological and economic benefits of retaining tropical forest cover are clear. However, these programs also require short-term costs for producers, consumers, and governments.

In Bolivia, local government laws, enforcement, tax benefits, and regulated extractions have been required to encourage the effective use of certified forests. To reduce bushmeat hunting in Africa, governments and NGOs similarly need to consider the provision of protein substitutes given the current importance of wildlife to local diets. On the Yucatán Peninsula, honey production represents the main source of income for as many as 16,000 rural ethnic Maya farmers, and Mexico is the world's fourth-largest honey exporter. These Indigenous farmers produce honey traditionally, refusing to use chemicals, limiting impacts to the forest environment, and taking close care of the health of bees. Nevertheless, the cost of meeting global "organic" certification standards means that, to reap the benefits, they will require government assistance if they are not to resort to more "conventional" methods. Similarly, some of the first pioneering growers of organic coffee included Indigenous farmers practicing agroforestry in Chiapas and Oaxaca in Mexico. Their approach proved beneficial for soils, ecosystems, and even the carbon cycle, but as larger landowners have taken up organic coffee cultivation and official organic status has become harder to sustain in the face of global competition, Indigenous farmers have faced the difficult choice of whether to risk intensification or revert to cheaper alternatives. Ultimately, "trade-offs" require governments, companies, and NGOs to support local communities in the short term in order to cash in on the ecological and economic benefits of longer-term sustainable approaches to forest use within the tropics.[19]

On a planetary scale, there are also growing movements by Western nations to actively "pay into" tropical conservation. In 2019, the summit of G7 nations in Biarritz, France, agreed on a US $22 million aid package to help combat and mitigate the wildfires ravaging the Amazon. As is often the case, however, the idea of "aid" proved controversial: Brazilian president Jair Bolsonaro suggested that he would reject the money as the conditions attached amounted to colonial interference in South American agrobusiness competition. The Reducing Emissions from Deforestation and Forest Degradation (REDD+) program represents a more organized, long-term means for wealthier nations, which have often benefited from the exploitation of the tropics, to begin to pay something back to support conservation efforts in and around the equator and, in so doing, simultaneously protect local environments and economies and the global climate. Developed by the United Nations Framework Convention on Climate Change, the REDD+ program acknowledges that tropical forests should be valued not only for their extractive potential but primarily for the carbon that they store, protecting our atmosphere from further CO_2 emissions, warming, and ultimate climatic disaster. In fact, it places a financial estimate on the benefits of this vital "carbon sink." Poorer countries receive results-based payments, subsidized by wealthier nations, for reducing unregulated logging impacts, slowing overall deforestation, preventing fires, and restoring tropical forest biomes. Not only that, but REDD+ ensures that Indigenous peoples, small and large businesses, and national governments are all heard during the development and application of these initiatives across the tropics. Although the overall economic and ecological benefits are still being determined, and it has also been seen as a controversial, top-down colonial scheme potentially prone to abuses in its own right, this program currently supports sixty-four nations across Africa, Asia, the Pacific, Central and South America, and the Caribbean. Furthermore, it does, at least, represent perhaps the first explicit recognition by the international community that the responsibility for tropical forest conservation and sustainability is a global one. If we want to do something about the intensifying loss of these environments, as well as the associated earth systems repercussions, then we will all have to play (and pay) our part.[20]

Which brings us to perhaps the most important way in which we can all limit our own impacts on tropical forests: modifying our consumer habits and choices. Many of us have benefited from the globally imbalanced social, economic, and political processes of the last five hundred years, with our own relative comfort to this day relying on the resources, people, and environments of the tropics. Tropical deforestation and its causes all remain tied to our own choices and histories, whether we want to accept it or not. The effects will also, sooner or later, impact us. It is time for us all to take action—whether it is paying a premium on furniture that uses timber from the Forest Stewardship Council or studying what a purchase of "organic" coffee actually means for Indigenous-led growing initiatives; whether it is buying fair-trade chocolate or boycotting clothing companies that use poorly paid, and often child, labor from the tropics; whether it is reading in depth about the practical benefits of the next ecotourism destination for local communities and economies or taking the time to read biscuit labels to distinguish products that include sustainable palm oil versus mass-produced, uncertified palm oil; or whether it is reducing our car journeys or flights abroad, curbing emissions, and reducing the climatic impacts on the most vulnerable tropical communities as well as, eventually, ourselves. Back home, be that in a tropical city away from forest frontiers or miles and miles away on an icy, dreary morning in northwestern Europe, these products and facilities may seem to just drop into our laps and fly onto our shelves. But their origins, their production, and our empowered ability to purchase and use them all derive from the tropics and the last half millennium of Euro-American economic and political integration with, and often exploitation of, tropical forests and their communities. Our decisions matter.

IN THE WINTER of 2018, the supermarket chain Iceland released a Christmas ad very different from the cute, comfortable viewing the UK public was used to at this time of year. Produced in partnership with Greenpeace, the video shows a cartoon orangutan swinging through the bedroom of a young girl, messing up her possessions as it went. Reaching a bottle of shampoo, "Rang-tan" suddenly stops and howls, transporting the viewer back to its own, tropical home. Here, fire, destruction, and devastation are suddenly strewn across the screen, causing "Rang-tan" to flee, an ecological refugee. While people will undoubtedly question Iceland's motives and timing, as a multinational corporation, its move to ban uncertified palm oil products from its stores is a positive one. Furthermore, the video provides a beautiful artistic vision of the influence all of us can have on tropical homes many thousands of kilometers away from us merely by picking up the wrong bottle in a supermarket. Do we really want rampant flames and falling trees messing up forests and livelihoods on our behalf? Even if we cannot always see them? In a manner perhaps characteristic of the way the Western world often seeks to bury its head in the sand, the UK monitoring body Clearcast actually banned the ad from television, deeming it "too political" for the Broadcast Code of Advertising and Practice due to the involvement of the conservation organization Greenpeace. That did not stop it from receiving millions of views on social media, however. Ultimately, like that video, this chapter and those preceding it have sought to repeatedly highlight how all of us are, in some way, implicated in the fate of tropical forests and the success of their conservation. From Indigenous groups to small-scale farmers seeking to survive, from companies extracting profit to governments trying to improve the lot of their citizens, from conservation scientists campaigning for particular species or biomes to each person standing at a supermarket shelf, starting a car, or queuing for a flight, we are all dependent on the tropics for our economic and social situations and all responsible for what happens next. Is it so hard, therefore, to find a common, intense interest in their ongoing existence? They need all of us, now, more than ever. Will we answer?[21]

Chapter 14

A GLOBAL RESPONSIBILITY

I t was the middle of the night on the shores of Ponta da Castanha on the Téfé River tributary of the Amazon, toward the end of the Brazilian journey that began this book. Lying in my hammock, restless from the heat, mosquitoes, and a hard day of trekking, I felt incredibly small. Just beyond the door of the wooden hut where I, Victor, and our colleagues were sleeping stood environments that have, in some form or other, been on our planet for at least 300 million years. The specific microclimates, species, and structure of the particular Neotropical forests that swayed in the nighttime breeze outside extend back to 60 million years ago. That's older than the most significant mountain range of the South American continent, the Andes, and highlights the resilience, but also the changeability, of these remarkable environments over geological timescales. The building's open holes for windows let the incessant chattering of insects, swishing of bats, and, toward dawn, squawking of birds wash over me. I had never heard such a concentration of wildlife in such close proximity. But these noises quite literally spoke to the role tropical forests have played in the evolution of animal life on this planet—from the early reptilian ancestors of the crocodile our boat driver had seen loitering by our boat that evening to the dinosaurian ancestors of birds, from the gliding mammals of the Jurassic to the hominin ancestors of me and my companions resting inside. I had never felt more in awe of tropical forests and their contribution to the story of life on Earth than I did in that moment.[1]

These environments have often been framed as dangerous, exotic, pristine "deserts" of preindustrial human activity. But closing my eyes, I felt incredibly connected to this place. I remembered the pottery we had seen on the forest floor the day before on top of distinctive Amazonian Dark Earth soils, a solid record of the long human history of the Amazon Basin that extends back to between 13,000 and 12,000 years ago. In my mind I retraced my steps through the clusters of palms and Amazon nut (Brazil nut) trees and the shifting swidden fields of manioc and maize that stood on the other side of the village and that our host Jucelino had shown us the day before. The vegetation and forest structure around many such villages along the Amazon and its tributaries provide a living archive of the millennia of land management in these environments. When it came time to leave, I took my seat on a western European–operated aircraft. I thought about the way in which huge distances and home comforts can make those of us living in Europe feel so separate from tropical forests—but also about how colonial and capitalist processes that followed European arrival in the Amazon Basin over the last five hundred years were visible all around me; how the growth of Amazon (Brazil) nut trees Victor and I were analyzing changed as disease, warfare, murder, and ongoing marginalization caused Indigenous inhabitants to flee parts of the forest; how the entwinement of Brazil in the transatlantic slave trade helped drive the global availability of coffee in my airplane cup and the sugar I was about to put into it; and how the rising pressures to feed smallholders, multinational corporations, ambitious governments, and Western consumers were driving the deforestation and land conversion taking place before my eyes on the ground passing by below. I might be able to fly away from Brazil physically, but the historical origins of global and regional inequalities and the importance of increasingly threatened tropical ecosystems to global sustainability mean that, ultimately, I am still connected to the tropical forests of the Amazon Basin and indeed the rest of the world. We all are.[2]

Through this book I hope to have convinced you that tropical forests are a part of your history, your present livelihood, and your future security.

I hope that next time you turn on a nature documentary, you feel a little closer to the miraculous plant and animal life tropical forests have to offer. I hope that the next time you watch a film with an explorer cutting through a tangle of tropical vines, you are less surprised to witness the incredible achievements of preindustrial human societies rise out of the "jungle" before them. And I hope that the next time you see a news report about the poor public health of Indigenous populations, racial tensions, multinational corporation mining, deforestation, and rampant fires in the tropics, you do not look away. Instead, I hope you keep watching. I hope you feel a responsibility. And I hope you begin to take time to research and explore the ways in which you might be able to make a difference. Remarkably, despite our complex global social and communication systems, humans often struggle to empathize with issues unless they are on their doorstep. Yet Chapters 12 and 13 have shown that the environmental, political, and economic threats facing populations living in the tropics will, through the connection of tropical forests to earth systems, lead to changes in biodiversity and climate the world over. Not only that, but participation in a colonial and capitalist system that has exploited the tropics for the last half a millennium means that we all have some obligation to help obtain justice in fights against the global economic and infrastructural inequality, rampant profit-driven tropical deforestation and land-cover change, marginalization of Indigenous rights and knowledge, and ongoing racial violence and systemic discrimination that are still visible across almost all of Europe and northern North America. Yet "hope" at this stage won't do us much good. So I will now make one last attempt to show you what is at stake. Let's look at what is happening and what is predicted to happen to (1) tropical forests and their species, (2) to human inhabitants of the tropics, and (3) to the global climatic, economic, social, and political situation as a consequence of tropical forest disappearance. Perhaps by seeing what we all stand to lose, we can gain a new urgency in negotiating a more just, resilient future for the tropics.[3]

STUDIES OF TWENTY-FIRST-CENTURY forest loss based on global yearly satellite images at high levels of resolution suggest that deforestation is continuing unabated, despite growing awareness of its planetary impacts. The majority of the world's deforestation is happening in the tropics, where total forest loss increased by over 2,000 km^2 (an area about two-thirds the size of Rhode Island) every year in the first decade of this century. Tropical rainforests, such as those of the Amazon Basin, accounted for nearly a third of global forest-cover reductions, while tropical dry forests in South America and moist deciduous forests in Africa saw annual increases in total deforestation of 459 km^2 and 536 km^2 a year, respectively. Southeast Asia also saw a particularly large expansion of forest removal over this period, with Indonesia alone showing an increase in *annual* losses from 10,000 km^2 to around 20,000 km^2 (so that an area of forest roughly equivalent to the size of New Jersey was being removed by 2010). This highlights the pantropical nature of rampant deforestation in the twenty-first century. Some predict that, if nothing is done, by 2100, Amazonia, the Congo Basin, West Africa, the Caribbean, and Wallacea will retain just half of their forest cover. Meanwhile, the tropical Andes, Madagascar, the Atlantic forests of Brazil, and New Guinea will have just 26, 23, 26, and 18 percent of their present-day forests still standing.[4]

It is not just complete deforestation that is a concern but, as we saw in Chapter 13, also the less obvious degradation of those forests left standing. A recent satellite study found that forest degradation occurred across the same area of forest in the Amazon Basin as complete deforestation did between 1995 and 2017. Since 2017 alone, approximately 1,036,800 km^2 of Amazonian tropical forest area has been disturbed in some form (an area four times the size of the United Kingdom). One of the main causes of disturbance is wood harvesting and export, and early twenty-first-century extraction rates from selective logging are particularly high for Guyana (13 m^3 of wood per hectare) and Indonesia (34 m^3 of wood per hectare). However, the causes behind this damage vary in different parts of the tropics, and the use of wood for fuel is by far the largest contributor to forest disturbance in much of tropical Africa. Forests change even further as tropical cities expand. In Dar es Salaam, Tanzania, a capital likely encountered by any of you seeking safari holidays in eastern Africa, researchers found that urban demand for fuel production and high-value timber left a clear impact on the surrounding tropical forests that extended many kilometers away from the city. Similarly, the highest rates of forest degradation in a studied region of southeastern Mexico are, unsurprisingly, currently found in regions with the highest human population density. Even those tropical forests that are left unshaven in the face of human activities are going to have to come to terms with a growing human presence.[5]

Perhaps the most emphatic of forest disturbances is fire, and twenty-first-century tropical forest fires are only going to intensify. Studying rates of charcoal accumulation in sediment sequences across the tropical forest environments of the Amazon, as well as the Brazilian Atlantic coast, researchers have concluded that the fire frequency and intensity seen in these areas today have no precedents throughout the 11,700-year span of the Holocene. In fact, based on current "worst-case" scenarios of changing land use and forest fragmentation in the Brazilian Amazon, models predict that the probability of monthly fire occurrence will increase by 73 percent by between 2071 and 2100, even before expected changes in climate are factored in. This is not just a South American problem, and crucial peat swamp forest ecosystems in Africa and Southeast Asia are increasingly "in the line of fire." During even relatively mild climatic conditions in 2002, nearly three-quarters of the area impacted by fire in Borneo was peat swamp forest, and peat fires are now becoming common in every dry season in this part of the world. The next decade could see a total loss of peat swamp forest ecosystems across Southeast Asia as a product of increased fires and permanent human land-use changes. Observations of fire frequency across Australia during the early twenty-first century suggest that weekly bushfire occurrences will also only further impact the tropical forests of the Wet Tropics in coming years.[6]

Human-induced climate change will undoubtedly exacerbate the rates of fire, disturbance, and tropical forest structural change. It is becoming increasingly apparent that a temperature rise of between 1.5°C and 2°C from preindustrial levels will be enough to "tip" the planet into a completely new climatic and environmental state. Ice collapse and melt has already advanced beyond expected levels, and recent estimates suggest that the entirety of the Arctic ice cap will melt in summer by 2050. Although growing economies in Latin America, South Asia, Southeast Asia, and Africa, in particular, have contributed to growing carbon emissions, tropical nations still only contribute around 20 percent of the global total. Yet they are still taking the brunt of the damage. Average surface temperatures increased by an average of 0.7°C around the tropics between 1961 and 2018. While this may not seem like a lot, compared to around 1.5°C for a region such as Scotland during the same period, normally low intra- and interannual temperature variation in the tropics means that smaller changes there can have bigger impacts on climate systems, environments, and species than larger changes in temperate regions. Atmospheric CO_2 is already at levels only seen 4 million years ago in the Pliocene, and as climate modeler Tim Lenton, whom we met in Chapter 1, has pointed out, we are rapidly on our way to levels seen during the "hothouse" state of the Eocene when temperatures were 14°C higher than those of the preindustrial period. One might argue that this could actually be good for tropical forests, given their proliferation at that time, which we explored in Chapter 2, but given rising human pressures and the fact that climate change will also result in an increasing unpredictability of rainfall in many tropical regions, they will almost certainly be placed into an equal state of disarray as the rest of the planet, should these trends continue.[7]

In fact, scientists have shown that a decrease in forest cover of anywhere between 20 and 40 percent or a rise of 4°C in temperature would be enough to begin a permanent transition to savannah environments in most of the central, southern, and eastern Amazon. Alongside rises in fire frequency, this would undoubtedly spark an end to some of largest extents of tropical forest biomes anywhere in the world. Changes in climate and tropical forest habitat are also placing enormous pressures on the rich animal species of these environments. Insects, amphibians, and reptiles can all only tolerate small variations in temperature and will face increasing challenges in the decades to come. Birds have been demonstrated to struggle in the face of fragmentation, and isolated patches of Amazonian forest smaller than 1 km^2 can lose nearly half of their bird species within a period of just fifteen years. Tigers in India, forest elephants and gorillas in Africa, and orangutans in Borneo are all considered particularly vulnerable to habitat loss, and their protected areas may soon become too small to sustain viable populations. Overhunting is also a growing issue, especially given an ongoing, profit-driven global trade in tropical animals and an increasing number of people seeking adequate protein in the tropics. Using a database of 3,281 records of how hunting has affected mammals around the tropics, researchers have found that about half of the existing intact tropical forest areas and 62 percent of wilderness areas are now, at least in part, already devoid of large mammals, with particularly large declines expected in West and Central Africa and Southeast Asia. Human introduction of invasive species will also continue to reconfigure tropical plant and animal communities, particularly in the face of habitat change. Invasive plants can suppress native species seeking to regrow. Meanwhile, exotic animals often tip the balance of predator-prey relationships. For example, the widespread distribution feral cats in the Wet Tropics of northeastern Queensland, which prey on often endangered, endemic small mammals, is a cause for considerable concern.[8]

IT GOES WITHOUT saying then that tropical forests are up against it. So too are their human inhabitants, according to the latest "State of the Tropics" report. In 2020, almost half the world's total human population lived in the tropics. And by 2050 over half of the world's population will call the equatorial regions their home. The future of our species as a whole undoubtedly relies on what happens here in the years to come. Many of these people will also increasingly live in cities. By 2050, 6.3 billion of the world's human population will live in an urban area. The tropics account for over half of this planned increase and have witnessed growing rural-to-urban migration between the late twentieth and early twenty-first centuries, with over 1.5 billion people currently living in tropical cities. Growing economies and the social, commercial, and health opportunities provided by urban networks are driving these trends, particularly in Southeast Asia. Indonesia's urban population, for example, grew from 32 million in 1990 to nearly 150 million by 2018. Nearly three-quarters of the populations of Central America, South America, and the Caribbean are already living in urban areas. Just under one-quarter of the tropical urban population lives in slums with incredibly poor living conditions, which have characterized regions experiencing rapid urbanization since the nineteenth century. These informal settlements are broadly declining, thanks to rising wealth and settlement-improvement programs, and often offer access to diverse income opportunities and cultural connections. Nevertheless, ensuring the sustainability and inclusivity of cities in the tropics, including moves toward "green space" integration, is a major United Nations concern in the context of environmental conservation and human living standards.[9]

Growing populations will require growing access to food. In Chapters 10 through 13, we followed an intensifying conflict between tropical forests and expanding agricultural, pastoral, and extractive land use. The tropics are home to 39 percent of the planet's agricultural land area, and while this land is used to support food security for the region's growing population, it also continues to feed the desires of the export market, as it has since at least the seventeenth century. Cropland in tropical countries expanded by 48,000 km^2 per year between 1999 and 2008, with Nigeria (10,259 km^2), Indonesia (5,826 km^2), Ethiopia (5,405 km^2), Sudan (5,227 km^2), and Brazil (4,205 km^2) showing the greatest areas of annual expansion. Yet expansion of agricultural production has not necessarily expanded the tropical food base. Islands in the Pacific and Indian Ocean have actually seen a reduction in agricultural land due to pressures to expand tourist and urban infrastructures. Similarly, the rising proportion of agricultural land used for biofuels or cash crops for export may benefit local incomes but not food security. Foreign demand has also primarily driven increases in meat production in the tropics (a rise of 76 percent between 2010 and 2017), with Brazil, India, and Australia acting as particularly significant meat exporters to China, the United States, Japan, Europe, and the Middle East. The economic benefits of these changes may not be reflected in local tropical sustainability. These problems are particularly acute because remote sensing and mapping models of soil erosion predict that tropical Africa, South America, and Southeast Asia will see increasingly poor, unstable soils over the course of the twenty-first century. The inhabitants of the tropics are going to face increasingly difficult decisions between exploiting imbalanced economic export markets to improve incomes and developing more productive, sustainable subsistence agricultural systems to meet their own nutritional requirements.[10]

Extreme poverty is defined as "severe deprivation of basic human needs, including food, safe drinking water, sanitation facilities, health, shelter, education and information."[11] In positive news, globally, 1 billion fewer people were living under these conditions in 2015 than in 1990.

Nevertheless, the majority (85 percent) of the remaining more than 100 million of the world's population experiencing extreme poverty live in the tropics. Population growth, conflicts, and a complex interplay of political and economic factors mean that in some tropical nations the proportion living under such conditions is actually increasing. Just five nations—Nigeria, the Democratic Republic of the Congo, Ethiopia, India, and Bangladesh—house 50 percent of the human population living in extreme poverty the world over. In terms of "moderate poverty," where basic needs are met, but just barely, and people earn less than $3.20 per person per day, the tropics have also seen much more modest improvement: 1.3 billion people live in moderate poverty in the tropics, double the number for the rest of the world. As we saw in Chapter 13, poverty is a major issue for tropical forest conservation and sustainability, as populations and governments seek to balance demands for food, water, and income with environmental impacts. The prevalence of poverty across the tropics is widely recognized as a product of colonial processes that continue to shape economies based on agricultural systems with large, poorly paid workforces, infrastructural imbalances, a reliance on foreign investment, and colonially drawn boundaries that can catalyze political and cultural conflicts. However, this has not stopped nations and companies in Europe and northern North America from continuing to exploit the situation, while simultaneously releasing more than their fair share of emissions into the atmosphere, which will again disproportionately affect the livelihoods of tropical residents.[12]

Climate change is already beginning to alter and destroy the lives of many people in the tropics. Climate models show that human-induced warming had already increased the likelihood of extremes in precipitation and temperature occurring in the tropics by 2010, and extreme weather events are likely to only become more acute and more frequent as the twenty-first century continues. Changing precipitation and water runoff modeled under global warming scenarios of 1.5°C to 2.0°C indicate increased flooding in South Asia, Southeast Asia, and the western Amazon, with some studies suggesting that fifteen of the twenty most impacted nations will be in tropical or subtropical regions. Many of the same regions, as well as other areas in the subtropics and tropics, will face increasingly frequent and intense drought periods over the course of the twenty-first century. These changes can have major impacts on human populations. For example, in Malaysia rice makes up 98 percent of current cereal production. It is estimated that for each 1°C rise in temperature, rice yields will decline by 10 percent, while wet rice systems may become increasingly unsustainable in the face of growing drought conditions. Meanwhile twenty-first-century flash floods and landslides have resulted in mass destruction and loss of life in cities such as Vargas in Venezuela. Growing storms, cyclones, and hurricanes are also battering the tropics more frequently. In 2017, Hurricane Maria killed over 3,000 people and caused $91.61 billion worth of damage in the northeastern Caribbean, and this was just one storm in a hyperactive Atlantic hurricane season. Rising sea levels represent a slower but eventually more permanent climate-based change to tropical livelihoods. Estimates project that an increasing intensity and frequency of storms and floods, sea level rise, and drought and desertification will together create around 200 million climate refugees worldwide in the next fifty years and perhaps as many as 700 million (nearly one-tenth of the world's current population).[13]

Tropical deforestation is also causing a number of public health issues. Tropical forest edges have proven to be a major petri dish for pandemics. Roughly 1.7 million viruses exist in mammals and birds around the world at any one time, though a minute fraction is known and properly described. Tropical forest edges provide the greatest opportunity for humans and their close animal companions to come into contact with these significant epidemiological reservoirs. Deforestation and forest fragmentation, for logging or infrastructural development, create longer and more ragged forest edges, where growing numbers of humans and domestic animals become interspersed with populations of wild animals, priming a large human host for viral outbreaks. Hunting for bushmeat or medicinal trade further expands opportunities for contact. In fact, deforestation and human exploitation of tropical wildlife have resulted in many of the major disease outbreaks of the last two decades. The 1980s emergence of HIV has been linked to bushmeat hunting in the Democratic Republic of the Congo. Outbreaks of the incredibly deadly Ebola virus in West Africa, which killed over 11,000 people between 2013 and 2016, have been linked to a single human contact with a bat colony. Bats are thought to be particularly prominent reservoirs of disease and feed closer to human settlements when their forest homes are disturbed. They are also thought to be the most probable original carrier of the SARS-CoV-2 virus that causes COVID-19, which likely began as a product of human-wildlife interactions in the humid subtropics of China. It is predicted that ongoing deforestation and fragmentation in the tropics in the twenty-first century will only increase the probability of highly virulent diseases making their way into human populations.[14]

Another challenge facing tropical communities seeking more sustainable livelihoods is the loss of Indigenous and traditional knowledge. The tropics are home to a staggering amount of human cultural diversity, including 80 percent of all living languages. As we have seen, local communities around the tropics have long histories (and indeed prehistories) of recognizing ecological thresholds and dynamics, performing sustainable agroforestry and farming practices, and incorporating new crops, technologies, and settlement patterns into their tropical environments. For example, Ifugao communities in the Cordillera Mountains of the Philippines have land rights laws, upland cultivation practices, knowledge of soil and water conservation, and forest stand management strategies, collectively known as *muyong*, that are critical for forest maintenance and biodiversity protection. Indigenous languages can be particularly crucial for passing on knowledge: people speaking the Indigenous Huastec language in western Mexico show more diverse and evenly shared knowledge of tropical plant use than Spanish-speaking Indigenous populations. However, Indigenous populations are currently some of the most marginalized groups within the tropics, facing higher rates of disease, higher child and adult mortality, and significant social disadvantages. Their languages are also disappearing, as young Indigenous people are integrated, often forcibly, into national capitalist economies and education systems. These losses represent a major threat, particularly in light of the fact that Indigenous populations and other local smallholders are crucial to the development of balanced, sustainable approaches to tropical forest use and conservation. This is especially the case given that over one-fifth of the significant portion of the world's carbon stored in tropical forests lies on land belonging to or claimed by Indigenous communities.[15]

YOU MAY FEEL that this concluding chapter has so far resembled the nightmare international newsreel you are fed up with scrolling through on your smartphone, tired of hearing coming over the airwaves, and thoroughly done with watching on your television. But I am asking you to keep reading: your life is irrevocably intertwined with tropical forests. Even if you struggle to place yourself in the shoes of inhabitants living at the "ground zeroes" of tropical forest sustainability, you cannot get away from the fact that, before long, these problems will have made their way into your back garden. Increasingly seasonal, unstable soils and watersheds will reduce the productivity and availability of cash crops (e.g., rubber, oil palm, cocoa), staple crops (e.g., rice), and meat (e.g., sheep, goat, cow) in the tropics, reducing the size of exports making their way to reliant European, North American, and now also Middle Eastern nations. Furthermore, tropical forest deforestation and degradation will lead to continental and global changes in precipitation, temperature, and agricultural growing conditions. Finally, since the Carboniferous period, tropical forests, alongside the oceans (and, later, Northern Hemisphere forests), have played a major role in capturing carbon and regulating the degree to which global carbon emissions actually translate into rising atmospheric CO_2. It is possible that higher temperatures could bring about faster tropical growth and more rapid fixation of CO_2. However, as tropical deforestation and degradation, including through more frequent fire outbreaks, continue throughout the twenty-first century, particularly in swamp forest areas, the very real prospect of these environments becoming carbon *sources* rather than carbon sinks is a truly terrifying thought. The United Nations imperative to limit increases in global average temperature to below 2°C above preindustrial levels will drift increasingly out of reach. What happens in the tropics in the next century will shape what happens on our entire planet, be in no doubt about that.[16]

The predicted loss of tropical forest plant and animal species also represents a truly global problem, given their enormous contribution to planetary biodiversity. Assuming a complete loss, degradation, and alteration of all primary tropical forests by the year 2225, scientists have calculated that there will be a global species richness decline of approximately 44 percent for ant species, 30 percent for dung beetles, and 20 percent for trees. For the coming two centuries, predicted extinction rates for different global species groups vary between around 200 and 2,000 extinctions per million species per year, all relating to the loss of habitat in the tropics. As Professor Xingli Giam of the University of Tennessee has pointed out, these rates are between 2,000 and 20,000 times higher than we might expect under natural geological and biological conditions. Furthermore, they are "two or more orders of magnitude higher than extinction rates associated with four of the five previous mass extinction events," including those of the Devonian, Ordovician, Permian, and Triassic.[17] In fact, they are comparable to the rate predicted for the Cretaceous-Paleogene mass extinction that wiped out the dinosaurs. Put together, then, estimates project that the deforestation and degradation of tropical forests, not even mentioning the additional, likely contributions of human-induced climate change, will result in a sixth planetary mass extinction event at the end of the next two centuries. Just think about that for a second. The events unfolding around the equator today as we go about our daily lives are predicted to have the same, or greater, impact on the world's biodiversity than geologically known catastrophes that took life on Earth to the very edge of the cliff face.[18]

The public health crises facing the tropics mentioned above have, in the past decade, also now very blatantly blanketed the world. Cases of the Ebola virus in Texas, in the United States, and Glasgow, in the United Kingdom, in 2014 showed the potential for this deadly disease to be carried beyond the African tropics, and outbreaks continued in the Democratic Republic of the Congo into 2020. The extratropical transmission seen for Ebola so far, however, pales in comparison to that seen as part of the ongoing COVID-19 pandemic. By the end of January 2020, the SARS-CoV-2 virus had spread around much of China, which rapidly went into a state of lockdown. Although recent studies suggest it may have made it to Europe as early as the end of December 2019, certainly by March 2020 the virus was transmitting between people in Europe and North America, with Italy and the United States becoming new centers of the outbreak. At the end of January 2021 there were more than 96 million globally documented cases of COVID-19, with more than 2 million deaths. Although many countries have managed to reduce cases, and a number of vaccines are now available, as I write this book, there is no rapid end in sight. The emotional, psychological, and economic damage caused by death rates, understaffed hospitals, abandonment of work-places, the closure of schools, the decline in domestic and international travel, and reduced access to infrastructure and facilities is expected to be dramatic and long-lasting. Nevertheless, while this virus has already changed our lives forever, more, perhaps deadlier pandemics linked to human–wild animal contact in the tropics are likely to emerge as the twenty-first century wears on.[19]

Legacies of colonialism and imperialism met in Chapters 10 and 11 will also continue to shape the face of global economics, politics, and sustainability in the twenty-first century. Gross domestic product (GDP) has increased across the tropics between 1990 and 2018, but so too has its gap with the rest of the world. Paychecks sent home by diasporic workers make up significant proportions of the GDP of many tropical nations. Furthermore, debt has increased across the tropics since 2013, placing tropical nations in a precarious position, particularly when ex-ternal events such as pandemics require rapid responses. These issues might have traditionally seemed "far away" to you. However, they present challenges we will all have to deal with. Climate change is expected to have a particularly intense impact on the growing, impoverished inhabi-tants of the tropics, resulting in increasing numbers of "climate migrants" that nations across Europe, North America, and Australasia will have to contend with. To take one example, at the current pace of sea level rise, the people forced to migrate away from Bangladesh by the end of the century will outnumber all current refugees around the world. Western European and northern North American nations will also have to man-age the challenge of reducing global carbon emissions (the majority of which they are responsible for) while respecting and offsetting the needs of tropical nations to grow and attain living conditions that many of their Northern Hemispheric counterparts have enjoyed for so long after half a millennium of colonialism.[20]

THE ABOVE MAY suggest that the situation is bleak. But it is necessary to underscore the urgency of the situation at hand. This does not mean that there is no room for positivity. The deep human history of tropi-cal forests revealed by this book highlights the way in which societies have repeatedly adapted to change and challenges in the tropics. Indeed, this ecological flexibility may even define our species. However, we have somewhat lost our way. Global consumer demand, wealth inequalities, and political priorities continue to exert pressure on some of the habitats we can least afford to lose. This book has revealed two key avenues out of this dangerous cycle. First, whether it is Pleistocene hunter-gatherers encountering giant sloths in the Amazon or food producers experimenting in highland New Guinea, pioneering precolonial voyagers to uninhabited Pacific islands or water-tank-building urban dwellers in Sri Lanka, intimate on-the-ground knowledge of these ecosystems characterizes successful adaptations to tropical forests. Only if local smallholders and Indigenous populations are properly consulted and empowered, supported by detailed archeological, paleoecological, and historical records of past human management, can more beneficial, adaptable ways of living with tropical forests be enacted. Second, we must acknowledge that some of the greatest pressures facing tropical forests and their inhabitants today are a product of long-term colonial and imperial extraction that have enriched nations and social groups in Europe and northern North America and degraded tropical forest environments and local possibilities for economic growth. Moving forward requires that global consumers and governments recognize that they too must bear responsibility for what happens in these environments.[21]

Fortunately, there are some signs that both of these avenues are beginning to be recognized and supported. The African Forest Landscape Restoration (AFR100) Initiative was founded in December 2015 as a country-led effort to restore 100 million hectares of degraded and deforested land across Africa by 2030. Thirty countries within the African Union have signed on to the scheme, and Ethiopia's innovation and technology minister famously declared that the country had planted more than 350 million tree seedlings in just twelve hours as part of the program in 2019. Importantly, AFR100 was developed with local outcomes as a top priority. It not only plants trees but also promotes sustainable agroforestry, the intercropping of trees with other forms of land use, and erosion control. This means that broader continental and global needs for forest cover and the production of a "Great Green Wall" to push Sahara desertification back are balanced with needs to enhance local food security. Farmers working in Burkina Faso and Niger have, for example, through experimental modifications to traditional agroforestry, water retention, and soil management practices, converted between 200,000 and 300,000 hectares of land, which produce 80,000 tons of food per year. Such initiatives also offer the opportunity to build more equitable futures in the region. For example, women and girls make up more than 50 percent of the active agricultural labor force across the tropics. Consultation with them when planning forest-restoration projects ensures the wider sharing of educational, infrastructural, ecological, and land-ownership benefits across those societies living within tropical forest environments.[22]

As we saw in Chapter 13, a growing number of initiatives around the tropics are also seeking to actively combine archeological, paleoecological, historical, and traditional knowledge to construct more equitable, effective conservation plans—policies that integrate smallholders and Indigenous groups in the sustainable management of forests rather than forcing them out. Where local smallholders in western Ghana have been involved and supported in reforestation initiatives, they have planted a significant number of trees on the International Union for Conservation of Nature's Red List of Threatened Species as part of their tree plantations, acknowledging and conserving the most vulnerable portions of their surrounding ecosystems. We have seen how, in Australia, Rainforest Aboriginal People are receiving increasing support for their application of traditional, historically documented burning to landscapes and their ongoing stewardship over endemic plant and animal diversity. Similarly, collaboration with Indigenous Kayapó and Upper Xingu groups produced one of the largest protected tropical forest areas in the world at 130,000 km^2 in Brazil, as well as effective policies for biodiversity management. Indeed, prior to the rise of Brazil's recent nationalist government, the country had begun to stem the tide of deforestation, recording a dramatic reduction between 2000 and 2012 that had significantly offset rises elsewhere around the planet. Further support and realistic assessment of the priorities and capacities of local communities interacting with tropical forests are essential. Colombia and Costa Rica have provided one such route of support, introducing taxes on carbon emissions to fund more sustainable practices by landowners, such as reforestation and mixed agroforestry.[23]

Nonetheless, given the legacies of colonialism and imperialism, it is also essential that extratropical governments and consumers recognize the role they have to play in bolstering change. Global deforestation and fragmentation should be greatly reduced, according to predictions, if the international Reducing Emissions from Deforestation and Forest Degradation (REDD+) scheme, which pays tropical nations to reduce emissions from these activities, is successful. Some scenarios and payment plans are expected to lead to the preservation of between 75 and 98 percent of current forest cover across the tropics, though care must be taken that such schemes do not come to mimic further Western colonial dictation. Many oil palm–using companies, seeking to retain access to high-value international markets, have begun to voluntarily sign on to the sustainability standards of the Roundtable on Sustainable Palm Oil (RSPO) established in 2004. Obtaining RSPO certification necessitates consideration of impacts on soil and water sources and air quality, as well as plans to mitigate damage to biodiversity. Where this certification system has been applied, livelihoods have improved and negative environmental impacts have been reduced. Your power as a consumer to shape sustainability has been highlighted by the fact that European countries such as the Netherlands and Belgium and giant multinational corporations such as Unilever and Carrefour have, as of 2020, committed to using only palm oil ingredients with 100 percent RSPO certification in their food and cosmetic products. The international pressure (and support) must be kept up, however, as smallholders often lack the means or incentives to adopt RSPO guidelines, and noncertified markets, particularly in Asia, remain open for business.[24]

We should also perhaps find some optimism in how figureheads such as Greta Thunberg and Vanessa Nakate have led waves of school strikes and protests around the world to demand urgent, united political and economic action on environmental justice and climate change. Furthermore, we are potentially at the threshold of a moment in which Euro-American governments are finally forced to reckon directly with centuries of colonialism and imperialism that have left scars of global inequality and legacies of racial prejudice within their own borders. The year 2020 saw the inequality and discrimination that have plagued European and northern North American interactions with the tropics since the fifteenth century boil to the surface. The killing of George Floyd by Minneapolis law enforcement on May 25, 2020, saw the Black Lives Matter (BLM) movement soar to national and international prominence. An estimated 15 million to 26 million took part in protests across the United States to protest against this brutal killing, as well as the deaths of Ahmaud Arbery and Breonna Taylor in February and March of the same year, respectively. It became one of the largest movements in North American history. National and international support for BLM also expanded well beyond the United States, with protests organized across Australia, Canada, Denmark, France, Germany, Japan, New Zealand, and the United Kingdom. We have also seen a rise in demands to deal with the symbols and systems left behind by long-term northern North American and European colonial abuses, as well as the toppling of statues, including that of Edward Colston whom we met in Chapter 11, of individuals known for their racist views, participation in slavery, and contribution to the murder of Indigenous populations. We are finding multinational corporations and academic institutions making commitments to the production of more equitable working environments and societies. Admittedly, much of this still remains superficial or "tokenistic," and a number of politicians and members of our societies remain resistant. But are we perhaps at the cusp of a major social reckoning—one that acknowledges and addresses the centuries of marginalization, persecution, and exploitation of certain sectors of society and that sees European and northern North American nations accept the role they have played in the political, social, and economic situation of the nations of the tropics and the challenges facing them today?[25]

BY NOW YOU have seen how tropical forests reach into almost every recess of your life. They have shaped the bouquets you pick up for your loved ones, the scents and displays of your gardens, and, for those of you who celebrate Christmas, the trees at the center of your living room. They have also played a major role in the evolutionary origins of the ants running along your paving slabs, the birds singing outside your window, many of the stars of your local zoo, and, of course, you, your families, your friends, and the entirety of our species. Tropical forests have also

grown into your kitchens, providing the rice in your cupboards, the crucial ingredients in many of your medicines, the chicken in your oven and
the eggs in your omelet, the sweetcorn on your barbeque, the coffee in
your cup, and the chocolate you treat yourself to when no one is looking.
They have also provided the rubber for your bikes and cars, your ability
to erase mistakes, your Wellington boots, and the insulation surrounding
almost anything with an electric wire in your home and workplace. Tropical forests are all around you. Sadly, so too are the legacies of centuries of
imbalanced colonial and capitalist interactions between the western half
of Europe and northern North America and the tropical world, molding
the racial tensions, the economic inequalities, the political battlegrounds,
and a lot of the social and cultural marginalization in the societies in
which you live. These are not distant, exotic environments on the other
side of the world. Rather, through an entangled prehistory and history,
they have found their way into your homes—no matter where you are.

So, what are you going to do about it? A number of conservation
policies are being enacted across the tropics. Trees are being planted to
stabilize landscapes and extract carbon from the atmosphere. Protected
areas are being developed and enforced to stem the tide of deforestation, degradation, and biodiversity loss. And human populations in
the tropics are beginning to be given the resources to apply many of
the "usable pasts" we have explored to develop more sustainable, often
Indigenous-led approaches to farming, agroforestry, and economic productivity that integrate forests into new forms of land use. However, those
on the front lines of this key fight for sustainability cannot be left to take
a stand alone. They need economic initiatives that redistribute resources
from richer nations and richer sectors of tropical societies. They need
global consumers to pay for sustainably produced tropical products, consider the footprint of their tropical tourism, and hold poorly regulated
multinational corporations to account. They need global voters to see
national interests as global interests, to recognize the mistakes and legacies of colonial pasts and the need to fight for fairer societies around the
world. They need the inhabitants of Europe and northern North America, alongside China, to cut their significant per capita contributions to
global CO_2 emissions in order to stop further environmental change and
to give them a buffer to undertake their own, more sustainable economic

development. Every single one of you could, today, begin to play at least a small part in every single one of these things. And, as this chapter should have shown, it does need to be today. You might act because this book has shown you that you have an affinity to and a responsibility for tropical forests. You might act because you actually already felt a moral or empathetic desire to help others and just needed some direction as to how. Or you might act because you have seen that, if you don't, climate change, declining food sources, economic catastrophe, political instability, mass migration, and an explosion of pandemic diseases will very soon be knocking at your own door.[26]

APPENDIX:
TROPICAL FORESTS
IN GEOLOGICAL TIME

Era	Period		Epoch	Start (Millions of years ago)	Events
Cenozoic	Neogene		Anthropocene?		■ Industrial revolution
			Holocene	0.0117	■ European colonialism and the 'Columbian exchange'
					■ Massive pre-industrial urban centres (e.g. Greater Angkor)
					■ Origins of food production and farming practices
		Quaternary	Pleistocene	2.58	■ *Homo sapiens* on every continent apart from Antarctica by end of period
					■ Specialized tropical forest use by populations of our species
					■ Evolution of *Homo sapiens* in Africa
					■ Dispersal of first hominins outside of Africa to Southeast Asia tropics (*Homo erectus*, *Homo floresiensis*)
					■ Glacial cycles become more intense – Milankovitch cycles
			Pliocene	5.33	■ Emergence of the genus *Homo* and stone tool use
					■ Hominin evolution continues in Africa (e.g. 'Lucy')
					■ Climatic variability, glacial and inter-glacial cycles
			Miocene	23.04	■ First hominins in Africa (e.g. 'Ardi')
					■ C_4 grasses expand in different parts of the tropics
					■ Drying leads to tropical forest retreat
					■ Initial gradual increase in aridity
					■ Warm climate with extension of tropical forests and great apes across Eurasia
	Paleogene	Tertiary	Oligocene	33.9	■ Mammals evolve in response to changing environments, horses increasingly begin to graze and get bigger
					■ 'Mega-bats' evolve
					■ Increasing climatic swings between cooling and drying and warming and wetting
			Eocene	56.0	■ First simian primates in Africa at end of period
					■ Perissodactyls (ancestors of horses, tapirs and rhinos) initially very successful, until ungulates expand as climates cool
					■ Tropical and sub-tropical forests extend from Europe to South America
					■ Rapid warming to middle 'optimum' prior to cooling and forest retreat
			Paleocene	66.0	■ Explosion of mammals in terms of size and diversity, including early small horses
					■ First true tropical rainforests (e.g. Cerrejón, Titanoboa), legumes
					■ Warmer and wetter climate

Era	Period	Start (Millions of years ago)	Events
Mesozoic	Cretaceous	143.1	■ Period ends with mass extinction in the face of an asteroid ■ Mammals continue to evolve ■ Debated link between dinosaurs and early flowering plant evolution, dispersal and diversification (angiosperms) ■ Angiosperms emerge and expand ■ The remainder of Gondwana breaks up further into Madagascar, India, Australia and Antarctica
	Jurassic	201.4	■ Dinosaurs living within gymnosperm-dominated forests and with some plants perhaps adapting to dinosaur feeding (ginkgo, monkey puzzle trees) ■ First mammals (*Juramaia*) ■ Dinosaurs diversify, clear emergence of three separate lineages (Theropoda, Sauropodomorpha, Ornithischia) ■ Pangaea divides into Laurasia (north) and Gondwana (south), Gondwana in turn begins to break up with Africa (excluding Madagascar), separating from the rest
	Triassic	251.9	■ First dinosaurs make it to the tropics by the end of the period ■ First gymnosperms (conifers, ginkgo, cycads) appear and dominate forests that slowly re-emerge ■ Forests nearly completely absent ■ Arid and unstable tropics
Paleozoic	Permian	298.9	■ Mass extinctions in sea and on land at the end of the period ■ Supercontinent of Pangaea forms ■ Reptiles proliferate and diversify into different lineages ■ 'Rainforest collapse' of Carboniferous forests
	Carboniferous	359.3	■ First reptiles emerge ■ Extensive tropical swamp forests lead to the formation of massive coal fields that later drove the industrial revolution
	Devonian	419.0	■ Period ends with glaciation and extinction ■ First 'arboreal' forms – *Eospermatopteris* ■ First proto-gymnosperms and true trees – *Archaeopteris*, more soil formation and deep rooting systems ■ First amphibians and insects
	Silurian	443.1	■ First fossil body forms of 'higher' vascular plants (xylem, phloem, true roots)
	Ordovician	486.9	■ Ice Age at end of period ■ First direct evidence for green land plants, enhanced weathering
	Cambrian	538.8	■ First non-vascular land plants estimated based on genetic clocks ■ 'Cambrian explosion' and thriving marine ecosystems

ACKNOWLEDGMENTS

This book is indebted to the many generations of ecologists, biologists, earth scientists, botanists, zoologists, and conservationists who have made their careers revealing the remarkable secrets of our "jungles." These pages are built on their labors of love for environments that they have seen increasingly threatened as the decades have ticked by. Since the first chapter, I have argued that tropical forests have been generally marginalized in studies of human history. Nevertheless, I also hope to have highlighted the work of archeologists, paleoanthropologists, anthropologists, historians, and paleoecologists who have persistently bucked this trend—endeavoring to recenter these habitats in discussions of the evolution of our species, the emergence of food production, preindustrial urban settlements, and the impacts of European colonialism on landscapes and people. This book is a tribute to them. This book should also have demonstrated that advancements in "science" are far from the only sources of knowledge about tropical forests. Indigenous voices have long emphasized the ecological, cultural, and economic importance of these environments, not just as stewards of their own lands but also as key advocates for global human sustainability. Governments and scientists often simply haven't listened closely enough. Or even listened at all. I am incredibly grateful to the Indigenous peoples I have had the great fortune to meet and work with, including, specifically, Uruwaruge Heenbanda and Uruwaruge Wanniya-laeto of the Wanniya-laeto community in Sri Lanka and Desley Mosquito, Barry Hunter, and Gerry Turpin of the Jirrbal, Djabugay, and Mbabaram Aboriginal groups, respectively, in

Australia. Indigenous groups and their traditional knowledge must be acknowledged and celebrated instead of ignored or exploited. They offer some of the best hopes for sustainable human use and management of these environments around the world today.

In an era when science is, quite rightly, facing increased scrutiny over its ethics, work environments, and treatment of researchers, I have been immensely privileged to be able to call upon some of the most supportive mentors during my explorations of tropical forests. My two PhD advisors at the University of Oxford set me off on my first encounters with these dynamic environments, and I have never looked back since. Mike Petraglia introduced me to Sri Lanka and its remarkable prehistory. Julia Lee-Thorp gave me the methodological tools to explore it further. Looking back, I am sure I was an incredibly irritating student. But hopefully all the red pen and the constant knocking on their doors was worth it. I will also always be immensely grateful for my first advisor at the same institution, Peter Mitchell, without whom, I can categorically say, I would be neither an academic nor an archeologist. I would also, of course, have never even made it into my first tropical forest without the support and kindness of my local collaborators in Sri Lanka: Dr. Siran Deraniyagala, who put the island on the map of Pleistocene archeology; Dr. Nimal Perera, who cemented it there; and Dr. Oshan Wedage, who is one of the truest, most loyal friends anyone could ever ask for and an incredibly talented researcher. Thanks to them, the future of the discipline in this part of South Asia is in very good hands indeed. I also owe an immense debt of gratitude to my current employer, Nicky Boivin, who has been the most encouraging, understanding, and inspirational of mentors as I have gone on to expand my research around the tropical world. Thank you for the opportunity and resources to pursue my passion as a career. I will never forget how incredibly lucky I am, and this book would not exist without your support.

While horror stories of bad collaborations can often ring around the conference circuit and coffee breaks, in the context of tropical forest research, I have been constantly left stunned by the kind encouragement, input, and assistance provided by my colleagues—from across the world and across disciplines. This book was no different, and the quotes and images included are a permanent testament to the generosity of my colleagues, as well as to the diversity of voices, from PhD students to professors, from historians to earth scientists, that are invigorating our understanding of the importance of tropical forests to life on Earth. A huge thank-you to each and every one of the people who suggested references, answered my annoying questions, and read through different parts of this work: Victor Lery Caetano Andrade (Preface), Tim Lenton, Silvia Pressel, and Luke Meade (Chapter 1), Carlos Jaramillo (Chapter 2), Emma Dunne, Leonardo Salgado, and Paul Barrett (Chapter 3), Zhe-Xi Luo, Tyler Lyson, and Gina Semprebon (Chapter 4), Yohannes Haile-Selassie, Tim White, Sarah Feakins, and Kira Westaway (Chapter 5), Eleanor Scerri, Oshan Wedage, and Sue O'Connor (Chapter 6), Tim Denham, Dolores Piperno, José Iriarte, and Umberto Lombardo (Chapter 7), Scott Fitzpatrick, Monica Tromp, and Kristina Douglass (Chapter 8), Lisa Lucero, Damian Evans, and Eduardo Neves (Chapter 9), Alexander Koch, Noel Amano, Amanda Logan, and Grace Barretto-Tesoro (Chapter 10), Alicia Odewale, Ayushi Nayak, John Hemming, John Tully, Kathy Morrison, Åsa Ferrier, Meena Menon, Justin Dunnavant, and Alex Moulton (Chapter 11), Nicky Boivin, Yadvinder Malhi, Nadja Rüger, Yoshi Maezumi, and Janae Davis (Chapter 12), and Renier van Raders, Emuobosa Orijemie, Desley Mosquito, and Douglas Sheil (Chapter 13). The views expressed are my own, and they should in no way be blamed for any mistakes made. However, they have also helped make this book what it is.

While the other colleagues who have helped me in writing this book and pursuing this line of enquiry are too many to name individually, I would like to particularly thank my local collaborators, beyond Sri Lanka, who have welcomed me into their countries and research environments and taught me more than I can ever repay. These include Grace Barretto-Tesoro and Francis Gealogo in the Philippines, Letícia Morgana Muller, Hilton Pereira, Eduardo Neves, Eduardo Tamanaha, Carolina Levis, and Charles Clement in Brazil, Oscár Solis in Mexico, Emuobosa Orijemie in Nigeria, Åsa Ferrier, Alice Buhrich, Desley Mosquito, Barry Hunter, Gerry Turpin, Richard Cosgrove, Simon Haberle, and Janelle Stevenson in Australia, Ravi Korisettar in India, and Mahirta Mahirta in Indonesia. It is their work and their efforts, alongside the Indigenous groups and stakeholders that many of them work with (or are a part of), that are cementing tropical forests as landscapes of cultural and natural heritage of both national and global significance. I would also like to thank the often unnamed administrative and support staff that make all of our research a possibility in the first place. For me, that is Anja Schatz, Anja Hannewald, Ellen Richter, Dorit Wammetsberger, Graeme Richardson, Daniela Gütsch, Beate Kerpen, Anna Pallaske, Michelle O'Reilly, Hans-Georg Sell, Christian Nagel, Gerd Kusserow, Thomas Baumman, Thomas Melzer, and Thomas Brückner at the Max Planck Institute for the Science of Human History. It is Stefanie Schirmer, Ulrike Thüring, Dovydas Jurkenas, Mary Lucas, Erin Scott, Elsa Perruchini, Sara Marzo, Bianca Fiedler, and Jana Ilgner in our laboratories. It is Anneke van Heteren (Zoologische Staatssammlung München), Jacques Cuisin, Violaine Nicolas-Colin, Géraldine Véron, Joséphine Lesur, and Christine Lefèvre (Muséum National d'Histoire Naturelle), and Malcolm McCallum and Tom Gillingwater (University of Edinburgh) for their support in the sampling and repatriation of museum collections. It is also all of the visitors of museums, journalists who feature our findings, peer reviewers, journal editors, members of national governments and administrative sectors that process fieldwork permits, and teachers, as well as the taxpayers, voters, and politicians who support our work around the world.

It is perhaps not said enough that, while leading a research group comes with a lot of responsibility and stress, it also comes with immense benefits. Chief among those is the ability to supervise gifted students and postdoctoral researchers from a variety of backgrounds. It has been an immense privilege to interact with the talented and driven individuals who have already, in some way, passed through my lab: Victor Lery Caetano Andrade, Ayushi Nayak, Maddy Bleasdale, Rebecca Jenner Hamilton, Bob Patalano, Max Findley, Jillian Swift, Anneke Janzen, Alicia Ventresca-Miller, Noel Amano, Asier García-Escárazaga, Eleftheria Orfanou, Oshan Wedage, Michael Ziegler, Phoebe Heddell-Stevens, Patxi Perez Ramallo, Neha Dhavale, Celeste Samec, Verónica Zuccarelli, Xueye Wang, Courtney Culley, Óscar Ricardo Solís Torres, Letícia Morgana Muller, Giulia Riccomi, and Clara Boulanger. I have learned an immense deal from all of you, and it has been a privilege to, in some small way, be involved in all of your exciting research, which I am sure will be setting agendas in archeology and paleoecology, including within the tropics, for many years to come. As well as many of the individuals listed above, I am also grateful to a number of people who have, at various points in my career, forced me to put work to one side and (usually) talk about something else for a change. Thank you especially to Tom Boulton, Jonny Rollings, Max Mewes, Christoph Brückner, Christoph Klose, Gerd Gleixner, Florian Ott, Rob Spengler, Andrea Kay, Julien Louys, Gilbert Price, Katerina Douka, Tom Higham, Brian Stewart, Sam Challis, Luíseach Nic Eoin, Rachel King, Mark McGranaghan, Ed Peveler, Abi Tomkins, Olly Beeley, Adam Besant, Vaughan Edmonds, Dan McArthur, Andy Baldock, Alison Crowther, Cosimo Posth, Mathew Stewart, Huw Groucutt, and Eleanor Scerri for the good company and for the beer.

This book has been an immense pleasure to write, though it has not been all plain sailing. I am incredibly grateful to my editors, especially Connor Brown in the United Kingdom and Eric Henney in the United States, for their enthusiasm and patience in crafting this into a final product that was hopefully readable enough for those of you that kindly picked it up! I am also indebted to copy editors Annie Lee and Jennifer Kelland for their detailed, patient reading of the manuscript. Thank you also to the arts team at Viking, Penguin Random House in the United Kingdom for turning some of my scribbles and ramblings into effective artwork that brings the story of tropical forests to life. The same goes to those fantastic creators of the "paleoart" used in different parts of the book: Mark Klinger for his fantastic drawing of *Juramaia sinenesis*, Jay Matternes for his *Ardipithecus ramidus* picture, Velizar Simeonovski for his elephant bird reconstruction, and the other artists whose work I have credited in these pages. Without these images, it would have been that bit harder to describe the long journey of tropical forests and their inhabitants. I would, of course, not have even got anywhere near the editing or illustration stage if it were not for my incredible agents, Joanna Swainson and Thérèse Coen, at Hardman & Swainson. Thank you for having faith in this project, for reading every page, and for negotiating contracts and payments. A first-time author of a trade book could not have asked for a better experience. Thank you also to the "unseen editors," particularly

Noel Amano, Max Findley, and Rebecca Hamilton, who took the time out to provide extraordinarily helpful comments on a number of different parts of the book (and a number of different times!).

The research discussed in this book would also not have been possible without the generous funding and support I have received from various agencies throughout my career so far. They include the School of Archaeology at the University of Oxford, St Hugh's College at the University of Oxford, the Clarendon Fund, the Santander Foundation, the Natural Environmental Research Council, the Boise Fund at the University of Oxford, the PAGES and Society for African Archeologists, the Bundesministerium für Bildung und Forschung, the Beutenburg Campus, the National Geographic Society, and the European Research Council. I am also particularly grateful to the Max Planck Society for providing the funding for my position, my laboratory equipment, and my research initiatives throughout the tropics. Research funding, particularly for the social sciences and humanities, is often some of the first to be cut in the face of economic crises, growing nationalism, and, most recently, pandemics. The European Research Council has prominently recently faced cuts to its budget, for example. While there are undoubtedly many things more worthy of financial support in trying times, it is this funding that enables us to explore questions like those investigated in this book. Without it, none of it could have happened. I am also fully conscious of the fact that many of the same historical processes and inequalities explored in Chapters 10 and 11 have, ultimately, resulted in my having a privileged platform on which to produce this work. There is an urgent need to further support research initiatives in the tropics that are led and driven by talented researchers from the tropics, a number of whom we have heard from in these pages.

Finally, beyond mentors, colleagues, friends, and funding, it is ultimately family that allows a book like this to be written. And I have one of the most supportive that I know. My parents, Julia and Neil, provided the trips to museums, archeological sites, and nature reserves that sparked off my interest in the human past. My brother, Tom, put up with it. Only with their financial and loving support could I have taken the first step on the academic ladder, studying archeology and anthropology at university. Every book that I asked for, no matter how boring it sounded, they bought. My grandmothers, Kay and Margaret, must also take credit here, and I get the feeling that they are still watching over me as I travel through my own "jungles." Kindness makes a huge difference in this life, and I have never seen such kindness in a person as I have in both of them. I hope that, in some small way, I carry that irrepressible desire to help people forward. In my aunty, Ali, I also have one of the most caring, supportive, and loving relatives possible. Nowadays, it is my partner, Jana, and our children, Rhys, Ida, and Livia, that are the core foundations of my tropical endeavors. They welcome the colleagues and friends who arrive on our doorstep from all over the world (though they do often get the benefits of good food!). They tolerate the time I spend away on fieldwork in the tropics and at conferences with other forest enthusiasts. They also let me find the time to write. As Livia says, "In diesem Haus leben Chaos und Liebe" (in this house live chaos and love). No truer words were ever spoken. Thank you for everything and all my love to you all. Thank you also to a wonderful acting "mother-in-law," Bettina Ilgner, whose heart has no bounds and whose care for her family gets them through the tough, as well as the good, times more than she will ever know.

SOURCES AND NOTES

PREFACE

1. Domínguez-Rodrigo, M. 2014. Is the "Savanna Hypothesis" a dead concept for explaining the emergence of the earliest hominins? *Current Anthropology* 55(1): 59–81; Mellars, P. 2006. Why did modern human populations disperse from Africa ca. 60,000 years ago? A new model. *Proceedings of the National Academy of Sciences of the United States of America* 103: 9381–9386; Meggers, B. J. 1971. *Amazonia: Man and Culture in a Counterfeit Paradise*. Wheeling: Harlan Davidson; Turnbull, C. 1961. *The Forest People: A Study of the Pygmies of the Congo*. New York: Simon & Schuster; Watson, J. E. M., et al. 2018. The exceptional value of intact forest ecosystems. *Nature Ecology & Evolution* 2: 599–610; Ghazoul, J. 2015. *Forests: A Very Short Introduction*. Oxford: Oxford University Press.

2. Spracklen, D. V., Arnold, S. R., Taylor, C. M. 2012. Observations of increased tropical rainfall preceded by air passage over forests. *Nature* 489: 282–285; Malhi, Y. 2010. The carbon balance of tropical forest regions, 1990–2005. *Current Opinions in Environmental Sustainability* 2: 237–244.

3. Curry, A. 2016. "Green hell" has long been home for humans. *Science* 354: 268–269; Pimm, S. L., Raven, P. 2000. Extinction by numbers. *Nature* 403: 843–845; Roberts, P., et al. 2021. Reimagining the relationship between Gondwanan forests and Aboriginal land management in Australia's "Wet Tropics." *IScience* 24: 102190. doi: 10.1016/j.isci.2021.102190.

CHAPTER 1: INTO THE LIGHT— THE BEGINNING OF THE WORLD AS WE KNOW IT

1. Vandenbrink, J. P., Brown, E. A., Harmer, S. L., Blackman, B. K. 2014. Turning heads: The biology of solar tracking in sunflower. *Plant Science* 224: 20–26; Appel, H. M., Cocroft, R. B. 2014. Plants respond to leaf vibrations caused by insect herbivore chewing. *Oecologia* 175: 1257–1266; Hughes, S. 1990. Antelope activate the acacia's alarm system. *New Scientist*. September 29. www.newscientist.com/article/mg12717361-200-antelope-activate-the-acacias-alarm-system; Yuan Song, Y., Simard, S. W., Carroll, A., Mohn, W. W., Sen Zeng, R. 2015. Defoliation of interior Douglas-fir elicits carbon transfer and stress signalling to ponderosa pine neighbours through ectomycorrhizal networks. *Nature Scientific Reports* 5. doi: 10.1038/srep08495; Wohlleben, P. 2016. *The Hidden Life of Trees: What They Feel, How They Communicate: Discoveries from a Secret World*. London: William Collins.

2. The official dates of geological time periods change as new stratified records are found and dating resolution increases. Throughout this book, where dates are provided for an officially ratified geological period (e.g., the Cambrian), I have followed those given in Volumes 1 and 2 of Gradstein,

F. M., Ogg, J. G., Schmitz, M. D., Ogg, G. M. 2020. *The Geologic Time Scale 2020*. Amsterdam: Elsevier. This is the standard reference used by many paleontologists and geologists and is regularly updated as new information is obtained. Where an unofficial division is mentioned (e.g., the "early" Cambrian) the date refers to that given in the specific scientific literature under discussion in that section of the book or is otherwise specified in a note. In the case of the "early" Cambrian, I have combined the time period discussed in Paterson et al. (2019) and Collette and Hagadorn (2010), both referenced below in this note, with the new dates given for the start of the Cambrian (538.8 million years ago) and the boundary in the middle of the Cambrian between Epoch 2 and the Wuliuan (509.0 million years ago) in *The Geologic Time Scale 2020*. *Paragraph sources:* Paterson, J. R., Edgecombe, G. D., Lee, M. S. Y. 2019. Trilobite evolutionary rates constrain the duration of the Cambrian explosion. *Proceedings of the National Academy of Sciences of the United States of America* 116: 4394–4399; Collette, J. H., Hagadorn, J. W. 2010. Three-dimensionally preserved arthropods from Cambrian Lagerstätten of Quebec and Wisconsin. *Journal of Palaeontology* 84: 646–667; Beck, H. E., et al. 2018. Present and future Köppen-Geiger climate classification maps at 1-km resolution. *Scientific Data* 5: 180214. doi: 10.1038/sdata.2018.214; Lenton, T. M., Daines, S. J., Mills, B. J. W. 2018. COPSE reloaded: An improved model of biogeochemical cycling over Phanerozoic time. *Earth-Science Reviews* 178: 1–28; Mills, B. J. W., et al. 2019. Modelling the long-term carbon cycle, atmospheric CO2, and Earth surface temperature from Late Neoproteozoic to present day. *Gondwana Research* 67: 172–186.

3. Mills, B. J. W., et al. 2019. Modelling the long-term carbon cycle, atmospheric CO2, and Earth surface temperature from Late Neoproteozoic to present day. *Gondwana Research* 67: 172–186; Lenton, T. M., Daines, S. J., Mills, B. J. W. 2018. COPSE reloaded: An improved model of biogeochemical cycling over Phanerozoic time. *Earth-Science Reviews* 178: 1–28; Bergman, N. M., Lenton, T. M., Watson, A. J. 2004. COPSE: A new model of biogeochemical cycling over Phanerozoic time. *American Journal of Science* 304: 397–437.

4. Ruhfel, B. R., Gitzendanner, M. A., Soltis, P. S., Soltis, D. E., Burleigh, J. G. 2014. From algae to angiosperms—inferring the phylogeny of green plants (Viridplantae) from 360 plastid genomes. *BMC Evolutionary Biology* 14. doi: 10.1186/1471-2148-14-23; Berner, R. A. 2006. GEOCARBSULF: A combined model for Phanerozoic atmospheric O_2 and CO_2. *Geochimica et Cosmochima Acta* 70: 5653–5664; Shimamura, M. 2016. Marchantia polymorpha: Taxonomy, phylogeny and morphology of a model system. *Plant and Cell Physiology* 57: 230–256.

5. Rubinstein, C. V., Gerrienne, P., de la Puente G. S., Astini, R. A., Steemans, P. 2010. Early Middle Ordovician evidence for land plants in Argentina (eastern Gondwana). *New Phytologist* 188: 365–369; Retallack, G. J. 2020. Ordovician land plants and fungi from Douglas Dam, Tennessee. *The Palaeobotanist* 68. https://cpb-us-e1.wpmucdn.com/blogs.uoregon.edu/dist/d/3735/files/2020/09/Retallack-2020-Ordovician-land-plants.pdf; Edwards, D., Feehan, J. 1980. Records of Cooksonia-type sporangia from Late Wenlock strata in Ireland. *Nature* 287: 41–42; Renzaglia, K. S., Nickrent, D. L., Garbary, D. J., Duff, R. J. T. 2000. Vegetative and reproductive innovations of early land plants: Implications for a unified phylogeny. *Philosophical Transactions of the Royal Society of London B: Biological Sciences* 355: 769–793.

6. Clarke, J. T., Warnock, R. C. M., Donoghue, P. C. J. 2011. Establishing a time-scale for plant evolution. *New Phytologist* 192: 266–301; Puttick, M. N., et al. 2018. The interrelationships of land plants and the nature of the ancestral embryophyte. *Current Biology* 28: 733–745; Morris, J. L., et al. 2018. The timescale of early land plant evolution. *Proceedings of the National Academy of Sciences of the United States of America* 115: E2274–E2283.

7. Morris, J. L., et al. 2018. The timescale of early land plant evolution. *Proceedings of the National Academy of Sciences of the United States of America* 115: E2274–E2283; Puttick, M. N., et al. 2018. The interrelationships of land plants and the nature of the ancestral embryophyte. *Current Biology* 28: 733–745.

8. Sheldrake, M. 2020. *Entangled Life: How Fungi Make Our Worlds, Change Our Minds, and Shape Our Futures*. London: Random House; Popkin, G. 2019. "Wood wide web"—the underground network of microbes that connects trees—mapped for the first time. *Science*. May 15. doi: 10.1126/science.aay0516; Steidinger, B. S., et al. 2019. Climatic controls of decomposition drive the global biogeography of forest-tree symbioses. *Nature* 569: 404–408; Taylor, T. N., Remy, W., Hass, H.,

Kerp, H. 1995. Fossil arbuscular mycorrhizae from the Early Devonian. *Mycologia* 87: 560–573; Rimington, W. R., et al. 2018. Ancient plants with ancient fungi: Liverworts associate with early-diverging arbuscular mycorrhizal fungi. *Proceedings of the Royal Society B: Biological Sciences* 285. doi: 10.1098/rspb.2018.1600; Field, K. J., Pressel, S., Duckett, J. G., Rimington, W. R., Bidartondo, M. I. 2015. Symbiotic options for the conquest of land. *Trends in Ecology and Evolution* 30: 477–486.

9. Reiny, S. 2016. Carbon dioxide fertilization greening Earth, study finds. NASA. April 26. www.nasa.gov/feature/goddard/2016/carbon-dioxide-fertilization-greening-earth; Lenton, T. M., Crouch, M., Johnson, M., Pires, N., Dolan, L. 2012. First plants cooled the Ordovician. *Nature Geoscience* 5: 86–89; Lenton, T. M., et al. 2019. Climate tipping points—too risky to bet against. *Nature* 575: 592–596.

10. Lenton, T. M., Crouch, M., Johnson, M., Pires, N., Dolan, L. 2012. First plants cooled the Ordovician. *Nature Geoscience* 5: 86–89.

11. Kotyk, M. E., Basinger, J. F., Gensel, P. G., de Freitas, T. A. 2002. Morphologically complex plant macrofossils from the Late Silurian of Arctic Canada. *American Journal of Botany* 89: 1004–1013; Petit, R. J., Hampe, A. 2006. Some evolutionary consequences of being a tree. *Annual Review of Ecology, Evolution, and Systematics* 37: 187–214; Lenton, T. M., Daines, S. J., Mills, B. J. W. 2018. COPSE reloaded: An improved model of biogeochemical cycling over Phanerozoic time. *Earth-Science Reviews* 178: 1–28; Bergman, N. M., Lenton, T. M., Watson, A. J. 2004. COPSE: A new model of biogeochemical cycling over Phanerozoic time. *American Journal of Science* 304: 397–437.

12. Goldring, W. 1927. The oldest known petrified forest. *Science Monthly* 24: 514–529; Stein, W. E., Mannolini, F., VanAller Hernick, L., Landing, E., Berry, C. M. 2007. Giant cladoxylopsid trees resolve the enigma of the Earth's earliest forest stumps at Gilboa. *Nature* 446: 904–907; Stein, W. E., Berry, C. M., Hernick, L. V., Mannolini, F. 2012. Surprisingly complex community discovered in the mid-Devonian fossil forest at Gilboa. *Nature* 483: 78–81.

13. Stein, W. E., et al. 2020. Mid-Devonian *Archaeopteris* roots signal revolutionary change in earliest fossil forests. *Current Biology* 30: 421–431.e2; Retallack, G. J., Huang, C. 2011. Ecology and evolution of Devonian trees in New York, USA. *Palaeogeography, Palaeoclimatology, Palaeoecology* 299: 110–128; Meter-Berthaud, B., Scheckler, S. E., Bousquet, J.-L. 2000. The development of *Archaeopteris*: New evolutionary characters from the structural analysis of an Early Famennian trunk from southeast Morocco. *American Journal of Botany* 87: 456–468; Guo, Y., Wang, D-M. 2011. Anatomical reinvestigation of *Archaeopteris macilenta* from the Upper Devonian (Frasnian) of South China. *Journal of Systematics and Evolution* 49: 590–597.

14. Morris, J. L., et al. 2015. Investigating Devonian trees as geo-engineers of past climates: Linking palaeosols to palaeobotany and experimental geobiology. *Palaeontology* 58: 787–801; Averill, C., Turner, B. L., Finzi, A. C. 2014. Mycorrhiza-mediated competition between plants and decomposers drives soil carbon storage. *Nature* 505: 543–545; Algeo, T. J., Scheckler, S. E. 2010. Land plant evolution and weathering rate changes in the Devonian. *Journal of Earth Science* 21: 75–78; Lenton, T. M., Daines, S. J., Mills, B. J. W. 2018. COPSE reloaded: An improved model of biogeochemical cycling over Phanerozoic time. *Earth-Science Reviews* 178: 1–28; Isaacson, P. E., et al. 2008. Late Devonian–earliest Mississippian glaciation in Gondwanaland and its biogeographic consequences. *Palaeogeography, Palaoeclimatology, Palaeoecology* 268: 126–142.

15. Ghazoul, J., Sheil, D. 2010. *Tropical Rain Forest Ecology, Diversity, and Conservation.* Oxford: Oxford University Press.

16. Cleal, C. J., Oplustil, S., Thomas, B. A., Tenchov, Y. 2009. Pennsylvanian vegetation and climate in Variscan Euramaerica. *Episodes* 34: 3–12.

17. Thomas, B. A., Clearl, C. J. 2017. Arborescent lycophyte growth in the Late Carboniferous coal swamps. *New Phytologist* 218: 885–890; Wilson, J. P., et al. 2017. Dynamic Carboniferous tropical forests: New views of plant function and potential for physiological forcing of climate. *New Phytologist* 215: 1333–1353; Prestianni, C., et al. 2015. Early seed plants from Western Gondwana: Paleobiological and ecological implications based on Tournaisian (Lower Carboniferous) records from Argentina. *Palaeogeography, Palaeoclimatology, Palaeoecology* 417: 210–219; Retallack, G. J., Germanheins, J. 1994. Evidence from paleosols for the geological antiquity of rainforest. *Science* 265: 499–502; Wright, V. P. 2018. An early carboniferous humus from South Wales preserved by marine hydromorphic entombment. *Applied Soil Ecology* 123: 668–671; Lenton, T. M., Daines, S. J., Mills,

B. J. W. 2018. COPSE reloaded: An improved model of biogeochemical cycling over Phanerozoic time. *Earth-Science Reviews* 178: 1–28; Puttick, M. N., et al. 2018. The interrelationships of land plants and the nature of the ancestral embryophyte. *Current Biology* 28: 733–745; Edwards, D., Kerp, H., Hass, H. 1998. Stomata in early land plants: An anatomical and ecophysiological approach. *Journal of Experimental Botany* 49: 225–278; Duckett, J. G., Pressel, S. 2018. The evolution of the stomatal apparatus: Intercellular spaces and sporophyte water relations in bryophytes—two ignored dimensions. *Philosophical Transactions of the Royal Society of London B: Biological Sciences* 373: 20160498; Ruszala, E. M., et al. 2011. Land plants acquired active stomatal control early in their evolutionary history. *Current Biology* 21: 1030–1035.

18. Garrouste, R., et al. 2012. A complete insect from the Late Devonian period. *Nature* 488: 82–85; Harrison, J. F., Kaiser, A., VandenBrooks, J. M. 2010. Atmospheric oxygen level and the evolution of insect body size. *Proceedings of the Royal Society B: Biological Sciences* 277: 1937–1946.

19. Meade, L., Jones, A. S., Butler, R. J. 2016. A revision of tetrapod footprints from the Late Carboniferous of the West Midlands, UK. *PeerJ* 4: e2718. doi: 10.7717/peerj.2718.

20. Meade, L., Jones, A. S., Butler, R. J. 2016. A revision of tetrapod footprints from the Late Carboniferous of the West Midlands, UK. *PeerJ* 4: e2718. doi: 10.7717/peerj.2718.

21. Meade, L., Jones, A. S., Butler, R. J. 2016. A revision of tetrapod footprints from the Late Carboniferous of the West Midlands, UK. *PeerJ* 4: e2718. doi: 10.7717/peerj.2718.

CHAPTER 2: A TROPICAL WORLD

1. Whitmore, T. C. 1998. *An Introduction to Tropical Rainforests* (2nd edition). Oxford: Oxford University Press.

2. Maslin, M. 2005. The longevity and resilience of the Amazon rainforest. In Y. Malhi, O. Phillips (eds.). *Tropical Forests and Global Atmospheric Change*. Oxford: Oxford University Press. Pp. 167–182; Morley, R. J. 2000. *Origin and Evolution of Tropical Rain Forests*. Chichester: John Wiley and Sons; Tabor, N. J., Poulsen, C. J. 2008. Palaeoclimate across the Late Pennsylvanian–Early Permian tropical palaeolatitudes: A review of climate indicators, their distribution, and relation to palaeophysiographic climate. *Palaeogeography, Palaeoclimatology, Palaeoecology* 268: 293–310.

3. Corlett, R. T., Primack, R. 2011. *Tropical Rain Forests: An Ecological and Biogeographical Comparison*. London: Wiley-Blackwell; Basset, Y., et al. 2019. Arthropod diversity in a tropical forest. *Science* 338: 1481–1484; Campos-Arceiz, A., Blake, S. 2011. Megagardeners of the forest—the role of elephants in seed dispersal. *Acta Oecologia* 37: 542–553; DryFlor et al. 2015. Plant diversity patterns in neotropical dry forests and their conservation implications. *Science* 353: 1383–1387; Walker, R., et al. 2014. The conservation of the world's most threatened tortoise: The ploughshare tortoise (*Astochelys yniphora*) of Madagascar. *Testudo* 8: 68–75.

4. Cascales-Miñana, B., Cleal, C. J. 2014. The plant fossil record reflects just two great extinction events. *Terra Nova* 26: 195–200; Barnosky, A. D., et al. 2011. Has the Earth's sixth mass extinction already arrived? *Nature* 471: 51–57; Morley, R. J. 2000. *Origin and Evolution of Tropical Rain Forests*. Chichester: John Wiley and Sons.

5. Cleal, C. J., Opluštil, S., Thomas, B. A., Tenchov, Y. 2009. Late Moscovian terrestrial biotas and palaeoenvironments of Variscan Euramerica. *Netherlands Journal of Geosciences* 88: 181–278; Montañez, I. P., et al. 2007. CO_2-forced climate and vegetation instability during Late Paleozoic deglaciation. *Science* 314: 87–91; Benton, M. J., Tverdokhlebov, V. P., Surkov, M. V. 2004. Ecosystem remodelling among vertebrates at the Permian-Triassic boundary in Russia. *Nature* 432: 97–100; Cascales-Miñana, B., Cleal, C. J. 2014. The plant fossil record reflects just two great extinction events. *Terra Nova* 26: 195–200.

6. Linkies, A., Graeber, K., Knight, C., Leubner-Metzger, G. 2010. The evolution of seeds. *New Phytologist* 186: 817–831; Looy, C. V., Brugman, W. A., Dilcher, D. L., Visscher, H. 1999. The delayed resurgence of equatorial forests after the Permian-Triassic ecologic crisis. *Proceedings of the National Academy of Sciences of the United States of America* 96: 13857–13862; Schneebeli-Hermann, E., et al. 2012. Palynology of the Lower Triassic succession of Tulong, South Tibet—evidence for early recovery of gymnosperms. *Palaeogeography, Palaeoclimatology, Palaeoecology* 339–341: 12–24.

7. Frohlich, M. W., Chase, M. W. 2007. After a dozen years of progress the origin of the angiosperms is still a great mystery. *Nature* 450: 1184–1189; Doyle, J. 2012. Molecular and fossil evidence on the origin of angiosperms. *Annual Review of Earth and Planetary Science* 40: 301–326; Morley, R. J. 2011. Cretaceous and Tertiary climate change and the past distribution of megathermal rainforests. In M. B. Bush, J. R. Flenley, W. D. Gosling (eds.). *Tropical Rainforest Responses to Climatic Change.* Berlin: Springer-Verlag. Pp. 1–34; Silvestro, D., et al. 2021. Fossil data support a pre-Cretaceous origin of flowering plants. *Nature Ecology and Evolution.* doi.org/10.1038/s41559-020-01387-8; Feild, T. S., et al. 2004. Dark and disturbed: A new image of early angiosperm ecology. *Paleobiology* 30: 82–107; Davis, C. C., Webb, C. O., Wurdack, K. J., Jaramillo, C. A., Donoghue, M. J. 2005. Explosive radiation supports a mid-Cretaceous origin of modern tropical rain forests. *American Naturalist* 165: E36–E65; Boyce, C. K., Jung-Eun, L. 2010. An exceptional role for flowering plant physiology in the expansion of tropical rainforests and biodiversity. *Proceedings of the Royal Society B: Biological Sciences* 485: 1–7.

8. Ghazoul, J. 2016. *Dipterocarp Biology, Ecology, and Conservation.* Oxford: Oxford University Press; Hu, S., Dilcher, D. L., Jarzen, D. M., Taylor, D. W. 2008. Early steps of angiosperm-pollinator coevolution. *Proceedings of the National Academy of Sciences of the United States of America* 105: 240–245; Duperon-Laudoueneix, M. 1991. Importance of fossil woods (conifers and angiosperms) discovered in continental Mesozoic sediments of northern equatorial Africa. *Journal of African Earth Sciences* 12: 391–396.

9. Wing, S. L., et al. 2009. Late Paleocene fossils from the Cerrejón Formation, Colombia, are the earliest record of neotropical rainforest. *Proceedings of the National Academy of Sciences of the United States of America* 106: 18627–18632.

10. Wing, S. L., et al. 2009. Late Paleocene fossils from the Cerrejón Formation, Colombia, are the earliest record of neotropical rainforest. *Proceedings of the National Academy of Sciences of the United States of America* 106: 18627–18632; Head, J. J., et al. 2009. Giant boid snake from the Palaeocene neotropics reveals hotter past equatorial temperatures. *Nature* 457: 715–717.

11. Johnson, K. R., Ellis, B. 2002. A tropical rainforest in Colorado 1.4 million years after the Cretaceous-Tertiary boundary. *Science* 296: 2379–2383; Morley, R. J. 2000. *Origin and Evolution of Tropical Rain Forests.* Chichester: John Wiley and Sons; Morley, R. J. 2003. Interplate dispersal routes for megathermal angiosperms. *Perspectives in Plant Ecology, Evolution and Systematics* 6: 5–20; Morley, R. J. 2011. Cretaceous and Tertiary climate change and the past distribution of megathermal rainforests. In M. B. Bush, J. R. Flenley, W. D. Gosling (eds.). *Tropical Rainforest Responses to Climatic Change.* Berlin: Springer-Verlag. Pp. 1–34.

12. Goldner, A., Herold, N., Huber, M. 2014. Antarctic glaciation caused ocean circulation changes at the Eocene-Oligocene transition. *Nature* 511: 574–577; Dupont-Nivet, G., Hoorn, C., Konert, M. 2008. Tibetan uplift prior to the Eocene-Oligocene climate transition. Evidence from pollen analysis of the Xining Basin. *Geology* 36: 987–990; Prasad, V., Strömberg, C. A. E., Alimohammadian, H., Sahni, A. 2005. Dinosaur coprolites and the early evolution of grasses and grazers. *Science* 310: 1177–1180; Osborne, C. O. 2008. Atmosphere, ecology and evolution: What drove the Miocene expansion of C_4 grasslands? *Journal of Ecology* 96: 35–45.

13. Metcalfe, S. E., Nash, D. J. 2012. Introduction. In S. E. Metcalfe, D. J. Nash (eds.). *Quaternary Environmental Change in the Tropics.* London: John Wiley & Sons Ltd. Pp. 1–33; Deplazes, G., et al. 2013. Links between tropical rainfall and North Atlantic climate during the last glacial period. *Nature Geoscience* 3: 213–217.

14. Hamon, N., et al. 2012. Growth of subtropical forests in Miocene Europe: The roles of carbon dioxide and Antarctic ice volume. *Geology* 40: 567–570; Cerling, T. E., Wang, Y., Quade, J. 1993. Expansion of C_4 ecosystems as an indicator of global ecological change in the late Miocene. *Nature* 361: 344–345; Feakins, S. J., et al. 2013. Northeast African vegetation change over 12 m.y. *Geology* 41: 295–298; Salzmann, U., Haywood, A. M., Lunt, D. J., Valdes, P. J., Hill, D. J. 2008. A new global biome reconstruction and data-model comparison for the Middle Pliocene. *Global Ecology and Biogeography* 17: 432–447.

15. Martínez-Botí, M. A., et al. 2015. Plio-Pleistocene climate sensitivity evaluated using high-resolution CO_2 records. *Nature* 518: 49–54; Bobe, R., Behrensmeyer, A. K. 2004. The expansion of grassland ecosystems in Africa in relation to mammalian evolution and the origin of the

genus *Homo. Palaeogeography, Palaeoclimatology, Palaeoecology* 207: 399–420; Heaney, L. R. 1991. A synopsis of climatic and vegetational change in Southeast Asia. *Climatic Change* 19: 53–61; Dennell R. W., Roebroeks, W. 2005. Out of Africa: An Asian perspective on early human dispersal from Africa. *Nature* 438: 1099–1104.

16. Roberts, P. 2019. *Tropical Forests in Prehistory, History, and Modernity*. Oxford: Oxford University Press; Corlett, R. T. 2011. Climate change in the tropics: The end of the world as we know it. *Biological Conservation* 151: 22–25; Hooghiemstra, H., Van der Hammen, T. 1998. Neogene and Quaternary development of the neotropical rain forest: The forest refugia hypothesis, and a literature overview. *Earth-Science Reviews* 44: 147–183; Rabett, R. J. 2012. *Human Adaptation in the Asian Palaeolithic*. Cambridge: Cambridge University Press.

17. Koutavas, A., Lynch-Stieglitz, J., Marchitto, T. M., Sachs, J. P. 2002. El Niño–like pattern in ice age tropical Pacific sea surface temperature. *Science* 297: 226–230; Pausata, F. S. R., Messori, G., Zhang, Q. 2016. Impacts of dust reduction on the northward expansion of the African monsoon during the Green Sahara period. *Earth and Planetary Science Letters* 434: 298–307.

CHAPTER 3: "GONDWANAN" FORESTS AND THE DINOSAURS

1. Turner, A. H., Makovicky, P. J., Norell, M. A. 2007. Feather quill knobs in the dinosaur *Velociraptor. Science* 317: 1721.

2. Brusatte, S. 2018. *The Rise and Fall of the Dinosaurs*. London: Picador; Barrett, P. M., Rayfield, E. J. 2006 Ecological and evolutionary implications of dinosaur feeding behaviour. *Trends in Ecology and Evolution* 21: 217–224; Hummel, J., et al. 2008. In vitro digestibility of fern and gymnosperm foliage: Implications for sauropod feeding ecology and diet selection. *Proceedings of the Royal Society B: Biological Sciences* 275. doi: 10.1098/rspb.2007.1728; Colbert, E. H. 1993. Feeding strategies and metabolism in elephants and sauropod dinosaurs. *American Journal of Science* 293A: 1–10.

3. The Paleobiology Database (https://paleobiodb.org).

4. Dunne, E. M., et al. 2018. Diversity change during the rise of tetrapods and the impact of the "Carboniferous rainforest collapse." *Proceedings of the Royal Society B: Biological Sciences* 285. doi: 10.1098/rspb.2017.2730.

5. Irmis, R. B., et al. 2007. A Late Triassic dinosauromorph assemblage from New Mexico and the rise of the dinosaurs. *Science* 317: 358–361; Whiteside, J. H., Grogan, D. S., Olsen, P. E., Kent, D. V. 2011. Climatically driven biogeographic provinces of Late Triassic tropical Pangea. *Proceedings of the National Academy of Sciences of the United States of America* 108: 8972–8977.

6. Whiteside, J. H., et al. 2015. Extreme ecosystem instability suppressed tropical dinosaur dominance for 30 million years. *Proceedings of the National Academy of Sciences of the United States of America* 112: 7909–7913.

7. Salgado, L., et al. 2017. A new primitive Neornithischian dinosaur from the Jurassic of Patagonia with gut contents. *Nature Scientific Reports* 7: 42778; Han, F., Forster, C. A., Xu, X., Clark, J. M. 2017. Postcranial anatomy of *Yinlong downsi* (Dinosauria: Ceratopsia) from the Upper Jurassic Shishugou Formation of China and the phylogeny of basal ornithischians. *Journal of Systematic Palaeontology* 16: 1159–1187.

8. van de Schootbrugge, B., et al. 2009. Floral changes across the Triassic/Jurassic boundary linked to flood basalt volcanism. *Nature Geoscience* 2: 589–594; Volkheimer W., Rauhut O. W. M., Quattrocchio M. E., Martínez, M. A. 2008. Jurassic paleoclimates in Argentina, a review. *Revista de la Asociación Geológica Argentina* 63, 549–556; Van Der Meer, D. G., et al. 2014. Plate tectonic controls on atmospheric CO_2 levels since the Triassic. *Proceedings of the National Academy of Sciences of the United States of America* 111: 4380–4385; Yonetani, T., Gordon, H. B. 2001. Simulated changes in the frequency of extremes and regional features of seasonal/annual temperature and precipitation when atmospheric CO_2 is doubled. *Journal of Climate* 14: 1765–1779.

9. Upchurch, P., Barett, P. M. 2000. The evolution of sauropod feeding mechanisms. In H.-D. Sues (ed.). *Evolution of Herbivory in Terrestrial Vertebrates: Perspectives from the Fossil Record*. Cambridge: Cambridge University Press. Pp. 79–122; Hummel, J., Clauss, M. 2011. Feeding and digestive physiology. In N. Klein, K. Remes, C. T. Gee, P. M. Sander (eds.). *Biology of the Sauropod Dinosaurs: Understanding the Life of Giants*. Bloomington: Indiana University Press. Pp. 11–33; Sander, P. M.,

et al. 2011. Biology of the sauropod dinosaurs: The evolution of gigantism. *Biological Reviews* 86: 117–155; Gee, C. T. 2011. Dietary options for the sauropod dinosaurs from an integrated botanical and paleobotanical perspective. In N. Klein, K. Remes, C. T. Gee, P. M. Sander (eds.). *Biology of the Sauropod Dinosaurs: Understanding the Life of Giants.* Bloomington: Indiana University Press. Pp. 34–57; Upchurch, P., Barrett, P. M. 2005. Phylogenetic and taxic perspectives on sauropod diversity. In K. A. Curry Rogers, J. A. Wilson (eds.). *The Sauropods: Evolution and Paleobiology.* Berkeley: University of California Press. Pp. 104–124; Poulsen, J. R., et al. 2018. Ecological consequences of forest elephant declines for Afrotropical forests. *Conservation Biology* 32: 559–567; Mustoe, G. E. 2007. Coevolution of cycads and dinosaurs. *Cycad Newsletter* 30: 6–9; Leslie, A. 2011. Predation and protection in the macroevolutionary history of conifer cones. *Proceedings of the Royal Society B: Biological Sciences* 278. doi: 10.1098/rspb.2010.2648; Butler, R. J., Barrett, P. M., Kenrick, P., Penn, M. G. 2009. Testing co-evolutionary hypotheses over geological timescales: Interactions between Mesozoic non-avian dinosaurs and cycads. *Biological Reviews* 84: 73–89.

10. Bakker, R. T. 1978. Dinosaur feeding behaviour and the origin of flowering plants. *Nature* 274: 661–663; Weishampel, D. B., Norman D. B. 1989. Vertebrate herbivory in the Mesozoic; jaws, plants, and evolutionary metrics. *Geological Society of America Special Paper* 238:87–101.

11. Barrett, P. M., Willis, K. J. 2001. Did dinosaurs invent flowers? Dinosaur-angiosperm coevolution revisited. *Biological Reviews* 76: 411–447.

12. Barrett, P. M., Willis, K. J. 2001. Did dinosaurs invent flowers? Dinosaur-angiosperm coevolution revisited. *Biological Reviews* 76: 411–447; Weishampel, D. B., Jianu, C.-M. 2000. Plant-eaters and ghost lineages: Dinosaurian herbivory revisited. In H.-D. Sues (ed.). *Evolution of Herbivory in Terrestrial Vertebrates: Perspectives from the Fossil Record.* Cambridge: Cambridge University Press. Pp. 123–143; Erickson, G. M., et al. 2012. Complex dental structure and wear biomechanics in Hadrosaurid dinosaurs. *Science* 338: 98–101; Molnar, R. E., Clifford, H. T. 2000. Gut contents of a small ankylosaur. *Journal of Vertebrate Palaeontology* 20: 194–196.

13. Poulsen, J. R., et al. 2018. Ecological consequences of forest elephant declines for Afrotropical forests. *Conservation Biology* 32: 559–567; Godefroit, P., Golovneva, L., Shcheptov, S., Garcia, G., Alekseev, P. 2009. The last polar dinosaurs: High diversity of latest Cretaceous arctic dinosaurs in Russia. *Naturwussenschaften* 96: 495–501; Paik, I. S., Kim, H. J., Huh, M. 2012. Dinosaur egg deposits in the Cretaceous Gyeongsang Supergroup, Korea: Diversity and paleobiological implications. *Journal of Asian Earth Sciences* 56: 135–146.

14. Voeten, D. F. A. E., et al. 2018. Wing bone geometry reveals active flight in Archaeopteryx. *Nature Communications* 9: 923. doi: 10.1038/s41467-018-03296-8; Barrowclough, G. F., Cracraft, J., Klicka, J., Zink, R. M. 2016. How many kinds of birds are there and why does it matter? *PLOS ONE* 11: e0166307. doi: 10.1371/journal.pone.0166307.

15. Dehling, D. M., et al. 2020. Similar composition of functional roles in Andean seed-dispersal networks, despite high species and interaction turnover. *Ecology* 101(7): e03028. doi: 10.1002/ecy.3028; Gorchov, D. L., Cornejo, F., Ascorra, C., Jaramillo, M. 1993. The role of seed dispersal in the natural regeneration of rain forest after strip-cutting in the Peruvian Amazon. *Vegetatio* 107: 339–349; David, J. P., Manakadan, R., Ganesh, T. 2015. Frugivory and seed dispersal by birds and mammals in the coastal tropical dry evergreen forests of southern India: A review. *Tropical Ecology* 56: 41–55; McConkey, K. R., Meehan, H. J., Drake, D. R. 2004. Seed dispersal by Pacific pigeons (*Ducula pacifica*) in Tonga, western Polynesia. *Emu—Austral Ornithology* 104: 369–376; Bregman, T., et al. 2016. Using avian functional traits to assess the impact of land-cover change on ecosystem processes linked to resilience in tropical forests. *Proceedings of the Royal Society B: Biological Sciences* 283: 20161289. doi: 10.1098/rspb.2016.1289.

CHAPTER 4: "TREE HOUSES" FOR THE FIRST MAMMALS

1. Janis, C. M. 1993. Tertiary mammal evolution in the context of changing climates, vegetation, and tectonic events. *Annual Review of Ecology, Evolution, and Systematics* 14: 467–500; Li, J., Wang, Y., Wang, Y., Li, C. 2001. A new family of primitive mammal from the Mesozoic of western Liaoning, China. *Chinese Science Bulletin* 46: 782–785; Gore, R. 2020. The rise of mammals. *National Geographic.* www.nationalgeographic.com/science/prehistoric-world/rise-mammals; Berg, M. 2016. A

miniscule model for research. *Lab Animal* 45: 133. doi: 10.1038/laban.981; Lockyer, C. 1976. Body weights of some species of large whales. *Journal du Conseil Permanent International pour l'Exploration de la Mer* 36: 259–273.

2. Renne, P. R., et al. 2015. State shift in Deccan volcanism at the Cretaceous-Paleogene boundary, possibly induced by impact. *Science* 350: 76–78; Barnosky, A. D., et al. 2011. Has the Earth's sixth mass extinction already arrived? *Nature* 471: 51–57; Smith, F. A., et al. 2010. The evolution of maximum body size of terrestrial mammals. *Science* 330: 1216–1219.

3. Wiejers, J. W. H., Scouten, S., Sluijs, A., Brinkhuis, H., Damsté, J. S. S. 2007. Warm arctic continents during the Palaeocene-Eocene thermal maximum. *Earth and Planetary Science Letters* 261: 230–238; Sarkar, S., et al. 2019. Late Eocene onset of the Proto-Antarctic Circumpolar Current. *Scientific Reports* 9. doi: 10.1038/s41598-019-46253-1; Zachos, J., Pagani, M., Sloan, L., Thomas, E., Billups, K. 2001. Trends, rhythms, and aberrations in global climate 65 Ma to present. *Science* 292: 686–693; Retallack, G. 2001. Cenozoic expansion of grasslands and climatic cooling. *Journal of Geology* 109: 407–426.

4. Rogers, C. S., et al. 2015. The Chinese Pompeii? Death and destruction of dinosaurs in the Early Cretaceous of Lujiatun, NE China. *Palaeogeography, Palaeoclimatology, Palaeoecology* 427: 89–99; Zhang, F., et al. 2010. Fossilized melanosomes and the colour of Cretaceous dinosaurs and birds. *Nature* 463: 1075–1078; Chen, P-J., Dong, Z-M., Zhen, S-N. 1998. An exceptionally well-preserved theropod dinosaur from the Yixian Formation of China. *Nature* 391: 147–152; Luo, Z-X., Yuan, C-X., Meng, Q-J., Ji, Q. 2011. A Jurassic eutherian mammal and divergence of marsupials and placentals. *Nature* 476: 442–445; Rink, W. J., Thompson, J. W. 2015. *Encyclopedia of Scientific Dating Methods*. Dordrecht: Springer.

5. Maor, R., Dayan, T., Ferguson-Gow, H., Jones, K. E. 2017. Temporal niche expansion in mammals from a nocturnal ancestor after dinosaur extinction. *Nature Ecology & Evolution* 1: 1889–1895; Bhullar, B.-A. S., et al. 2019. Rolling of the jaw is essential for mammalian chewing and tribosphenic molar function. *Nature* 566: 528–532; Rowe, T. B., Macrini, T. E., Luo, Z-X. 2011. Fossil evidence on origin of the mammalian brain. *Science* 332: 955–957; Gill, P. G., et al. 2014. Dietary specializations and diversity in feeding ecology of the earliest stem mammals. *Nature* 512: 303–305.

6. dos Reis, M., et al. 2012. Phylogenomic datasets provide both precision and accuracy in estimating the timescale of placental mammal phylogeny. *Proceedings of the Royal Society B: Biological Sciences* 279: 3491–3500; Zheng, X., Bi, S., Wang, X., Meng, J. 2013. A new arboreal haramiyid shows the diversity of crown mammals in the Jurassic period. *Nature* 500: 199–203; Luo, Z-X., Ji, Q., Wible, J. R., Yuan, C-X. 2003. An Early Cretaceous tribosphenic mammal and metatherian evolution. *Science* 302: 1934–1940; Ji, Q., et al. 2002. The earliest known eutherian mammal. *Nature* 416: 816–822.

7. Maor, R., Dayan, T., Ferguson-Gow, H., Jones, K. E. 2017. Temporal niche expansion in mammals from a nocturnal ancestor after dinosaur extinction. *Nature Ecology & Evolution* 1: 1889–1895; Grossnickle, D. M., Smith, S. M., Wilson, G. O. 2019. Untangling the multiple ecological radiations of early mammals. *Trends in Ecology and Evolution* 34: 936–949; Luo, Z-X., Yuan, C-X., Meng, Q-J., Ji, Q. 2011. A Jurassic eutherian mammal and divergence of marsupials and placentals. *Nature* 476: 442–445; Shattuck, M. R., Williams, S. A. 2010. Arboreality has allowed for the evolution of increased longevity in mammals. *Proceedings of the National Academy of Sciences of the United States of America* 107: 4635–4639; Meng, Q. J., et al. 2015. An arboreal docodont from the Jurassic and mammaliaform ecological diversification. *Science* 347: 764–768; Ji, Q., Luo, Z-X., Yuan, C-X., Tabrum, A. R. 2006. A swimming mammaliaform from the Middle Jurassic and ecomorphological diversification of early mammals. *Science* 311: 1123–1127; Meng, Q-J., et al. 2017. New gliding mammaliaforms from the Jurassic. *Nature* 548: 291–296; Luo, Z-X., et al. 2017. New evidence for mammaliaform ear evolution and feeding adaptation in a Jurassic ecosystem. *Nature* 548: 326–329.

8. Hu, Y., Meng, J., Wang, Y., Li, C. 2005. Large Mesozoic mammals fed on young dinosaurs. *Nature* 433: 149–152; Grossnickle, D. M., Smith, S. M., Wilson, G. O. 2019. Untangling the multiple ecological radiations of early mammals. *Trends in Ecology and Evolution* 34: 936–949; Grossnickle, D. M., Newham, E. 2016. Therian mammals experience an ecomorphological radiation during the Late Cretaceous and selective extinction at the K-Pg boundary. *Proceedings of the Royal Society B: Biological Sciences* 283. doi: 10.1098/rspb.2016.0256; Chen, M., Strömberg, C. A. E., Wilson, G. P. 2019. Assembly of modern mammal community structure driven by Late Cretaceous dental

evolution, rise of flowering plants, and dinosaur demise. *Proceedings of the National Academy of Sciences of the United States of America* 116: 9931–9940; Sun, G., et al. 2002. Archaefructaceae, a new basal angiosperm family. *Science* 296: 899–904; Wilson, G. P., et al. 2012. Adaptive radiation of multituberculate mammals before the extinction of dinosaur. *Nature* 483: 457–460.

9. Lyson, T. R., et al. 2019. Exceptional continental record of biotic recovery after the Cretaceous-Paleogene mass extinction. *Science* 366: 977–983.

10. Nichols, D. J., Johnson, K. R. 2008. *Plants and the K-T Boundary*. Cambridge: Cambridge University Press; Cascales-Miñana, B., Cleal, C. J. 2014. The plant fossil record reflects just two great extinction events. *Terra Nova* 26: 195–200; Kowalczyk, J. B., et al. 2018. Multiple proxy estimates of atmospheric CO_2 from an Early Paleocene rainforest. *Paleoceanography and Paleoclimatology* 33: 1427–1438; Lyson, T. R., et al. 2019. Exceptional continental record of biotic recovery after the Cretaceous-Paleogene mass extinction. *Science* 366: 977–983.

11. Huurdeman, E. P., et al. 2020. Rapid expansion of meso-megathermal rain forests into the southern high latitudes at the onset of the Paleocene-Eocene Thermal Maximum. *Geology.* doi: 10.1130/G47343.1; Janis, C. M. 1989. A climatic explanation for patterns of evolutionary diversity in ungulate mammals. *Palaeontology* 32: 463–481; Prothero, D. R., Foss, S. E. (eds.). 2007. *The Evolution of Artiodactyls*. Baltimore: Johns Hopkins University Press; Gingerich, P. D., ul Haq, M., Zalmout, I. S., Hussain Khan, I., Malkani, M. S. 2001. Origin of whales from early Artiodactyls: Hands and feet of Eocene protocetidae from Pakistan. *Science* 293: 2239–2242.

12. Schaal, S., Ziegler, W. 1993. *Messel: An Insight into the History of Life and of the Earth*. Oxford: Oxford University Press; Collinson, M. E., Manchester, S. R., Wilde, V., Hayes, P. 2010. Fruit and seed floras from exceptionally preserved biotas in the European Paleogene. *Bulletin of Geosciences* 85: 155–162.

13. Jordano, P. 2000. Fruits and frugivory. In M. Fenner (ed.). *The Ecology of Regeneration in Plant Communities*. Wallingford: CAB International. Pp. 125–166; Eriksson, O. 2008. Evolution of seed size and biotic seed dispersal in angiosperms: Paleoecological and neoecological evidence. *International Journal of Plant Sciences* 169: 863–870; Tiffney, B. H. 1984. Seed size, dispersal syndromes, and the rise of the angiosperms: Evidence and hypothesis. *Annals of the Missouri Botanical Garden* 71: 551–576.

14. Kargaranbafghi, F., Neubauer, F. 2018. Tectonic forcing to global cooling and aridification at the Eocene-Oligocene transition in the Iranian plateau. *Global and Planetary Change* 171: 248–254.

15. Solounias, N., Semprebon, G. 2002. Advances in the reconstruction of ungulate ecomorphology with applications to early fossil equids. *American Museum Novitates* 3366: 1–49.

16. Semprebon, G. M., Rivals, F., Janis, C. M. 2019. The role of grass vs. exogenous abrasives in the paleodietary patterns of North American ungulates. *Frontiers in Ecology and Evolution.* doi: 10.3389/fevo.2019.00065; Semprebon, G. M., Rivals, F., Solounias, N., Hulbert Jr., R. C. 2016. Paleodietary reconstruction of fossil horses from the Eocene through Pleistocene of North America. *Palaeogeography, Palaeoclimatology, Palaeoecology* 442: 110–127.

17. Badlangana, N. L., Adams, J. W., Manger, P. R. 2009. The giraffe (*Giraffa camelopardalis*) cervical vertebral column: A heuristic example in understanding evolutionary processes? *Zoological Journal of the Linnean Society* 155: 736–757; Mitchell, G., Skinner, J. D. 2003. On the origin, evolution and phylogeny of giraffes *Giraffa camelopardalis. Transactions of the Royal Society of South Africa* 58: 51–73..

18. Dumont, E. R., et al. 2011. Morphological innovation, diversification and invasion of a new adaptive zone. *Proceedings of the Royal Society B: Biological Sciences* 279. doi: 10.1098/rspb.2011.2005; Eriksson, O. 2014. Evolution of angiosperm seed disperser mutualisms: The timing of origins and their consequences for coevolutionary interactions between angiosperms and frugivores. *Biological Reviews* 91: 168–186; Shilton, L. A., et al. 1999. Old World fruit bats can be long-distance seed dispersers through extended retention of viable seeds in the gut. *Proceedings of the Royal Society B: Biological Sciences* 266. doi: 10.1098/rspb.1999.0625.

19. Beard, K. C., Qi, T., Dawson, M. R., Wang, B., Li, C. 1994. A diverse new primate fauna from Middle Eocene fissure-fillings in southeastern China. *Nature* 368: 604–609; Sussman, R. W., Rasmussen, D. T., Raven, P. H. 2013. Rethinking primate origins again. *American Journal of Primatology* 75: 95–106.

CHAPTER 5: THE LEAFY CRADLES OF OUR ANCESTORS

1. Whiten, A., et al. 1999. Cultures in chimpanzees. *Nature* 399: 682–685; The Chimpanzee Sequencing and Analysis Consortium. 2005. Initial sequence of the chimpanzee genome and comparison with the human genome. *Nature* 437: 69–87.

2. Darwin, C. 2004 (1871). *The Descent of Man, and Selection in Relation to Sex.* London: Penguin; Domínguez-Rodrigo, M. 2014. Is the "Savanna Hypothesis" a dead concept for explaining the emergence of the earliest hominins? *Current Anthropology* 55(1): 59–81; Dennell R. W., Roebroeks, W. 2005. Out of Africa: An Asian perspective on early human dispersal from Africa. *Nature* 438: 1099–1104.

3. Hamon, N., et al. 2012. Growth of subtropical forests in Miocene Europe: The roles of carbon dioxide and Antarctic ice volume. *Geology* 40: 567–570; Nelson, S. 2003. *The Extinction of Sivapithecus: Faunal and Environmental Changes in the Siwaliks of Pakistan.* American School of Prehistoric Research Monographs 1. Boston: Brill Academic Publishers.

4. Macchiarelli, R., Bergeret-Medina, A., Marchi, D., Wood, B. 2020. Nature and relationships of *Sahelanthropus tchadensis. Journal of Human Evolution* 149. doi: 10.1016/j.jhevol.2020.102898; White, T., et al. 2009. *Ardipithecus ramidus* and the paleobiology of early hominids. *Science* 326: 75–86; Haile-Selassie, Y., Suwa, G., White, T. D. 2004. Late Miocene teeth from Middle Awash, Ethiopia, and early hominid dental evolution. *Science* 303: 1503–1505; White, T., et al. 2009. *Ardipithecus ramidus* and the paleobiology of early hominids. *Science* 326: 75–86; Prado-Martinez, J., et al. 2013. Great ape genetic diversity and population history. Nature 499: 471–475.

5. White, T. D., Lovejoy, C. O., Asfaw, B., Carlson, J. P., Suwa, G. 2015. Neither chimpanzee nor human, *Ardipithecus* reveals the surprising ancestry of both. *Proceedings of the National Academy of Sciences of the United States of America* 112: 4877–4884; WoldeGabriel, G., et al. 2009. The geological, isotopic, botanical, invertebrate, and lower vertebrate surroundings of *Ardipithecus ramidus. Science* 326(5949): 65–65e5; White, T., et al. 2009. Ardipithecus ramidus and the paleobiology of early hominids. Science 326: 75–86; Levin, N. E., Simpson, S. W., Quade, J., Cerling, T. E., Frost, S. R. 2008. Herbivore enamel carbon isotopic composition and the environmental context of *Ardipithecus* at Gona, Ethiopia. In J. Quade, J. G. Wynn (eds.). *The Geology of Early Humans in the Horn of Africa.* Geological Society of America Special Paper 446. Boulder: Geological Society of America. Pp. 215–234.

6. Brunet, M., et al. 2002. A new hominid from the Upper Miocene of Chad, Central Africa. *Nature* 418: 145–151; Pickford, M., Senut, B., Gommery, D., Treil, J. 2002. Bipedalism in *Orrorin tugenensis* revealed by its femora. *Comptes Rendus Palevol* 1(4): 191–203.

7. Crompton, R. H., Sellers, W. I., Thorpe, S. K. S. 2010. Arboreality, terrestriality and bipedalism. *Philosophical Transactions of the Royal Society of London B: Biological Sciences* 365: 3301–3314; Elton, S. 2008. The environmental context of human evolutionary history in Eurasia and Africa. *Journal of Anatomy* 212: 377–393; Pusey, A. E., Pintea, L., Wilson, M. L., Kamenya, S., Goodall, J. 2007. The contribution of long-term research at Gombe National Park to chimpanzee conservation. *Conservation Biology* 21: 623–634.

8. Johanson, D. C., Maitland, A. E. 1981. *Lucy: The Beginning of Humankind.* St. Albans: Granada; Latimer, B., Lovejoy, C. O. 1989. The calcaneus of *Australopithecus afarensis* and its implications for the evolution of bipedality. *American Journal of Physical Anthropology* 78(3): 369–386.

9. Harcourt-Smith, W. E. H., Aiello, L. C. 2004. Fossils, feet and the evolution of human bipedal locomotion. *Journal of Anatomy* 204: 403–416; Montgomery, S. 2018. Hominin brain evolution: The only way is up. *Current Biology* 28: R784–R802; Harmand, S., et al. 2015. 3.3-million-year-old stone tools from Lomekwi 3, West Turkana, Kenya. *Nature* 521: 310–315.

10. Lee-Thorp, J. A., van der Merwe, N. J., Brain, C. K. (1994). Diet of *Australopithecus robustus* at Swartkrans from stable carbon isotopic analysis. *Journal of Human Evolution* 27: 361–372.; Sponheimer, M., et al. 2013. Isotopic evidence of early hominin diets. *Proceedings of the National Academy of Sciences of the United States of America* 110(26): 10513–10518.

11. Green, D. J., Alemseged, Z. 2012. *Australopithecus afarensis* scapular ontogeny, function, and the role of climbing in human evolution. *Science* 338: 514–517; Ruff, C. 2009. Relative limb strength and locomotion in *Homo habilis. American Journal of Physical Anthropology* 138(1): 90–100; Sponheimer, M., et al. 2006. Isotopic evidence for dietary variability in the early hominin *Paranthropus robustus. Science* 314: 980–982.

12. Feakins, S. J., et al. 2013. Northeast African vegetation change over 12 m.y. *Geology* 41(3): 295–298; Bonnefille, R. 2010. Cenozoic vegetation, climate changes and hominid evolution in tropical Africa. *Global and Planetary Change* 72: 390–411.

13. Levin, N. E. 2015. Environment and climate of early human evolution. *Annual Review of Earth and Planetary Sciences* 43: 405–429; Levin, N. E., Brown, F. H., Behrensmeyer, A. K., Bobe, R., Cerling, T. E. 2011. Paleosol carbonates form the Omo Group: Isotopic records of local and regional environmental change in East Africa. *Palaeogeography, Palaeoclimatology, Palaeoecology* 307: 75–89; Robinson, J. R., Rowan, J., Campisano, J., Wynn, J. G., Reed, K. E. 2017. Late Pliocene environmental change during the transition from *Australopithecus* to *Homo*. *Nature Ecology & Evolution* 1: 0159; White, T. D., et al. 2006. Asa Issie, Aramis and the origin of *Australopithecus*. *Nature* 440: 883–889; Saylor, B. Z., et al. 2019. Age and context of mid-Pliocene hominin cranium from Woranso-Mille, Ethiopia. *Nature* 573: 220–224; Kingston, J. D., Harrison, T. 2007. Isotopic dietary reconstructions of Pliocene herbivores at Laetoli: Implications for early hominin paleoecology. *Palaeogeography, Palaeoclimatology, Palaeoecology* 243: 272–306; Cerling, T. E., Harris, J. M., Leakey, M. G., Passey, B. H., Levin, N. E. 2010. Stable carbon and oxygen isotopes in East African mammals: Modern and fossil. In L. Werdelin, W. J. Sanders (eds.). *Cenozoic Mammals of Africa*. London: University of California Press. Pp. 941–952.

14. Carbonell, E., et al. 2008. The first hominin of Europe. *Nature* 452: 465–469; Lordkipanidze, D., et al. 2013. A complete skull from Dmanisi, Georgia, and the evolutionary biology of early *Homo*. *Science* 342: 326–331; Ashton, N., et al. 2014. Hominin footprints from Early Pleistocene deposits at Happisburgh, UK. *PLOS ONE* 9(2): e88329; Scott, G. R., Gilbert, L. 2009. The oldest hand-axes in Europe. *Nature* 461: 82–85; Dennell R. W., Roebroeks, W. 2005. Out of Africa: An Asian perspective on early human dispersal from Africa. *Nature* 438: 1099–1104; Zaim, Y., et al. 2011. New 1.5 million-year-old *Homo erectus* maxilla from Sangiran (Central Java, Indonesia). *Journal of Human Evolution* 61(4): 363–376.

15. Smith, R. J., Jungers, W. L. 1997. Body mass in comparative primatology. *Journal of Human Evolution* 32: 523–559; Zhang, Y., Harrison, T. 2017. *Gigantopithecus blacki*: A giant ape from the Pleistocene of Asia revisited. *American Journal of Physical Anthropology, Supplement Yearbook of Physical Anthropology* 162: 153–177; Louys, J., Curnoe, D., Tong, H. 2007. Characteristics of Pleistocene megafauna extinctions in Southeast Asia. *Palaeogeography, Palaeoclimatology, Palaeoecology* 243: 152–173.

16. Marwick B. 2009. Biogeography of Middle Pleistocene hominins in mainland Southeast Asia: A review of current evidence. *Quaternary International* 2002: 51–58; Sutikna, T., et al. 2016. Revised stratigraphy and chronology for *Homo floresiensis* at Liang Bua in Indonesia. *Nature* 532: 366–369; Brumm, A., et al. 2016. Age and context of the oldest known hominin fossils from Flores. *Nature* 534: 249–253; Westaway, K. E., et al. 2009. *Homo floresiensis* and the late Pleistocene environments of eastern Indonesia: Defining the nature of the relationship. *Quaternary Science Reviews* 28: 2897–2912; Sutikna, T., et al. 2018. The spatio-temporal distribution of archaeological and faunal finds at Liang Bua (Flores, Indonesia) in light of the revised chronology for *Homo floresiensis*. *Journal of Human Evolution* 124: 52–74; Weston, E. M., Lister, A. M. 2009. Insular dwarfism in hippos and a model for brain size reduction in *Homo floresiensis*. *Nature* 459: 85–88.

17. Rizal, Y., et al. 2020. Last appearance of *Homo erectus* at Ngandong, Java, 117,000–108,000 years ago. *Nature* 577: 381–385; Sutikna, T., et al. 2016. Revised stratigraphy and chronology for *Homo floresiensis* at Liang Bua in Indonesia. *Nature* 532: 366–369; Louys, J., Roberts, P. 2020. Environmental drivers of megafauna and hominin extinction in Southeast Asia. *Nature* 586: 402–406.

18. Potts, R. 2013. Hominin evolution in settings of strong environmental variability. *Quaternary Science Reviews* 73: 1–13.

CHAPTER 6: ON THE TROPICAL ORIGINS OF OUR SPECIES

1. Hublin, J.-J., et al. 2017. New fossils from Jebel Irhoud, Morocco and the pan-African origin of *Homo sapiens*. *Nature* 546: 289–292; Stringer, C. 2016. The origin and evolution of *Homo sapiens*. *Proceedings of the Royal Society B: Biological Sciences* 371. doi: 10.1098/rstb.2015.0237.

2. Bailey, R., et al. 1989. Hunting and gathering in tropical rain forest: Is it possible? *American Anthropologist* 91: 59–82; Curry, A. 2016. "Green hell" has long been home for humans. *Science* 354: 268–269; Bradfield, J., Lombard, M., Reynard, J., Wurz, S. 2020. Further evidence for bow hunting and its implications more than 60,000 years ago: Results of a use-trace analysis of the bone point from Klasies River main site, South Africa. *Quaternary Science Reviews* 236. doi: 10.1016/j.qua scirev.2020.106295; Tylén, K., et al. 2020. The evolution of early symbolic behaviour in *Homo sapiens. Proceedings of the National Academy of Sciences of the United States of America* 117: 4578–4584; Henshilwood, C. S., et al. 2018. An abstract drawing from the 73,000-year-old levels at Blombos Cave, South Africa. *Nature* 562: 115–118; d'Errico, F., et al. 2017. Identifying early modern human ecological niche expansions and associated cultural dynamics in the South African Middle Stone Age. *Proceedings of the National Academy of Sciences of the United States of America* 114: 7869–7876; Marean, C. W. 2016. The transition to foraging for dense and predictable resources and its impact on the evolution of modern humans. *Philosophical Transactions of the Royal Society of London B: Biological Sciences* 371. doi: 10.1098/rstb.2015.0239; Bird, M., Taylor, D., Hunt, C. 2005. Palaeoenvironments of insular Southeast Asia during the last glacial period: A savanna corridor in Sundaland? *Quaternary Science Reviews* 24: 2228–2242; Mellars, P. 2006. Why did modern human populations disperse from Africa ca. 60,000 years ago? A new model. *Proceedings of the National Academy of Sciences of the United States of America* 103: 9381–9386; Clarkson, C., et al. 2017. Human occupation of northern Australia by 65,000 years ago. *Nature* 547: 306–310; Ardelean, C. F., et al. 2020. Evidence of human occupation in Mexico around the Last Glacial Maximum. *Nature* 584: 87–92. doi: 10.1038/s41586-020-2509-0; Erlandson, J. M., Braje, T. J. 2015. Coasting out of Africa: The potential of mangrove forests and marine habitats to facilitate human coastal expansion via the Southern Dispersal Route. *Quaternary International* 382: 31–41.

3. Cann, R. L., Stoneking, M., Wilson, A. C. 1987. Mitochondrial DNA and human evolution. *Nature* 325: 31–36; Schelebusch, C. M., et al. 2012. Genomic variation in seven Khoe-San groups reveals adaptation and complex African history. *Science* 338: 374–379; Sankararaman, S., et al. 2014. The genomic landscape of Neanderthal ancestry in present-day humans. *Nature* 507: 354–357; Meyer, M., et al. 2012. A high-coverage genome sequence from an archaic Denisovan individual. *Science* 338: 222–226; van de Loosdrecht, M., et al. 2018. Pleistocene North African genomes link Near Eastern and sub-Saharan African human populations. *Science* 360: 548–552; Prendergast, M. E., Sawchuk, E. 2018. Boots on the ground in Africa's ancient DNA "revolution": Archaeological perspectives on ethics and best practices. *Antiquity* 92: 803–815.

4. McDougall, I., Brown, F. H., Fleagle, J. G. 2005. Stratigraphic placement and age of modern humans from Kibish, Ethiopia. *Nature* 433: 733–736; Clark, J. D., et al. 2003. Stratigraphic, chronological and behavioural contexts of Pleistocene *Homo sapiens* from Middle Awash, Ethiopia. *Nature* 423: 747–752; Henshilwood, C. S., et al. 2018. An abstract drawing from the 73,000-year-old levels at Blombos Cave, South Africa. *Nature* 562: 115–118; Potts, R., et al. 2018. Environmental dynamics during the onset of the Middle Stone Age in eastern Africa. *Science* 360(6384): 86–90. doi: 10.1126 /science.aao2200; Henshilwood, C. S., Dubreuil, B. 2011. The Still Bay and Howiesons Poort, 77–59 ka: Symbolic material culture and the evolution of the mind during the African Middle Stone Age. *Current Anthropology* 52: 361–400.

5. Barham, L. 2001. Central Africa and the emergence of regional identity in the Middle Pleistocene. In L. S. Barham, K. Robson-Brown (eds.). *Human Roots: Africa and Asia in the Middle Pleistocene*. Bristol: Western Academic and Specialist Press. Pp. 65–80; Scerri, E., et al. 2018. Did our species evolve in subdivided populations across Africa, and why does it matter? *Trends in Ecology and Evolution* 33: P582–P594; Neubauer, S., Hublin, J.-J., Gunz, P. 2018. The evolution of modern human brain shape. *Science Advances* 4. doi: 10.1126/sciadv.aao5961; Harvati, K., Stringer, C., Grün, R., Aubert, M., Allsworth-Jones, P., Folorunso, C. A. 2011. The Later Stone Age calvaria from Iwo Eleru, Nigeria: Morphology and chronology. *PLOS ONE* 6e24024. doi: 10.1371/journal .pone.0024024; Mazet, O., Rodríguez, W., Grusea, S., Boitard, S., Chikhi, L. 2016. On the importance of being structured: Instantaneous coalescence rates and human evolution—lessons for ancestral population size inference? *Heredity* 116: 362–371; Bergström, A., Stringer, C., Hajdinjak, M., Scerri, E. M. L., Skoglund, P. 2021. Origins of modern human ancestry. *Nature* 590: 229–237.

6. Scerri, E., et al. 2018. Did our species evolve in subdivided populations across Africa, and why does it matter? *Trends in Ecology and Evolution* 33: P582–P594; Tryon, C. 2019. The Middle/Later Stone Age transition and cultural dynamics of Late Pleistocene East Africa. *Evolutionary Anthropology* 28: 267–282.

7. Barham, L. 2001. Central Africa and the emergence of regional identity in the Middle Pleistocene. In L. S. Barham, K. Robson-Brown (eds.). *Human Roots: Africa and Asia in the Middle Pleistocene*. Bristol: Western Academic and Specialist Press. Pp. 65–80; Mercader, J. 2002. Forest people: The role of African rainforests in human evolution and dispersal. *Evolutionary Anthropology* 11: 117–124; Taylor, N. 2016. Across rainforests and woodlands: A systematic reappraisal of the Lupemban Middle Stone Age in Central Africa. In S. C. Jones, B. A. Stewart (eds.). *Africa from MIS 6-2: Population Dynamics and Paleoenvironments*. Dordrecht: Springer. Pp. 273–299.

8. Perry, G. H., Verdu, P. 2017. Genomic perspectives on the history and evolutionary ecology of tropical rainforest occupation by humans. *Quaternary International* 448: 150–157; Shipton, C., et al. 2018. A 78,000-year-old record of tropical adaptation and complex human behaviour in coastal East Africa. *Nature Communications* 9. doi: 10.1038/s41467-018-04057-3; Blome, M. W., Cohen, A. S., Tryon, C. A., Brooks, A. S., Russell, J. 2012. The environmental context for the origins of modern human diversity: A synthesis of regional variability in African climate 150,000–30,000 years ago. *Journal of Human Evolution* 62: 563–592.

9. Gamble, C. 1993. *Timewalkers: The Prehistory of Global Colonization*. Stroud: Alan Sutton.

10. Wedage, O., et al. 2019. Specialized rainforest hunting by *Homo sapiens* 45,000 years ago. *Nature Communications* 10: 739. doi: 10.1038/s41467-019-08623-1.

11. Roberts, P., et al. 2015. Direct evidence for human reliance on rainforest resources in Late Pleistocene Sri Lanka. *Science* 347: 1246–1249; Roberts, P., et al. 2017. Fruits of the forest: Human stable isotope ecology and rainforest adaptations in Late Pleistocene and Holocene (~36 to 3 ka) Sri Lanka. *Journal of Human Evolution* 106: 102–118.

12. Wedage, O., et al. 2019. Specialized rainforest hunting by *Homo sapiens* 45,000 years ago. *Nature Communications* 10: 739. doi: 10.1038/s41467-019-08623-1; Roberts, P. 2017. Forests of plenty: Ethnographic and archaeological rainforests as hotspots of human activity. *Quaternary International* 448. doi:10.1016/j.quaint.2017.03.041.

13. Langley, M. C., et al. 2020. Bows-and-arrows and complex symbolic displays at 48,000 years bp in the South Asian tropics. *Science Advances* 6. doi: 10.1126/sciadv.aba3831.

14. Liu, W., et al. 2015. The earliest unequivocally modern humans in southern China. *Nature* 526: 696–700; Westaway, K. E., et al. 2017. An early modern human presence in Sumatra 73,000–63,000 years ago. *Nature* 548: 322–325; Barker, G., Farr, L. (eds.). 2016. *Archaeological Investigations in the Niah Caves, Sarawak. The Archaeology of the Niah Caves, Sarawak*. Volume 2. Cambridge: McDonald Institute for Archaeological Research.

15. O'Connor, S., Ono, R., Clarkson, C. 2011. Pelagic fishing at 42,000 years before the present and the maritime skills of modern humans. *Science* 334: 1117–1121.

16. Roberts, P., et al. 2020. Direct evidence for initial coastal colonization and subsequent diversification in the human occupation of Wallacea. *Nature Communications* 11. doi: 10.1038/s41467-020-15969-4.

17. Gosden, C. 2010. When humans arrived in the New Guinea Highlands. *Science* 330: 41–42.

18. Summerhayes, G. R., et al. 2010. Human adaptation and plant use in highland New Guinea 49,000 to 44,000 years ago. *Science* 330: 78–81.

19. Kershaw, A. P. 1986. The last two glacial-interglacial cycles from northeastern Australia: Implication for climate change and Aboriginal burning. *Nature* 322: 47–49; Cosgrove, R., Field, J., Ferrier, Å. 2007. The archaeology of Australia's tropical rainforests. *Palaeogeography, Palaeoclimatology, Palaeoecology* 251: 150–173; Ardelean, C. F., et al. 2020. Evidence of human occupation in Mexico around the Last Glacial Maximum. *Nature* 584: 87–92. doi: 10.1038/s41586-020-2509-0; Dillehay, T. D., et al. 2015. New archaeological evidence for an early human presence at Monte Verde, Chile. *PLOS ONE* 10: e0141923. doi: 10.1371/journal.pone.0141923; Roosevelt, A. C., et al. 1996. Paleoindian cave dwellers in the Amazon: The peopling of the Americas. *Science* 272: 373–384; Rademaker, K., et al. 2014. Paleoindian settlement of the high-altitude Peruvian Andes. *Science* 346: 466–469.

20. Nash, D. J., et al. 2010. Going the distance: Mapping mobility in the Kalahari Desert during the Middle Stone Age through multi-site geochemical provenancing of silcrete artefacts. *Journal of Human Evolution* 96: 113–133; Stewart, B. A., Parker, A. G., Dewar, G. I., Morley, M., Allott, L. 2016. Follow the Senqu: Maloti-Drakensberg paleoenvironments and implications for early human dispersals into mountain systems. In Jones, S., Stewart, B. A. (eds.). *Africa from MIS 6-2: Population Dynamics and Paleoenvironments*. Dordrecht: Springer. Pp. 247–271; Ossendorf, G., et al. 2019. Middle Stone Age foragers resided in high elevations of the glaciated Bale Mountains, Ethiopia. *Science* 365: 583–587; Zhang, X. L., et al. 2018. The earliest human occupation of the high-altitude Tibetan Plateau 40 thousand to 30 thousand years ago. *Science* 362: 1049–1051; Pitulko, V. V., et al. 2016. Early human presence in the Arctic: Evidence from 45,000-year-old mammoth remains. *Science* 351: 260–263.

21. Roberts, P., Stewart, B. 2018. Defining the "generalist-specialist" niche for Pleistocene *Homo sapiens*. *Nature Human Behaviour* 2: 542–550. doi: 10.1038/s41562-018-0394-4; Stringer, C., Galway-Witham, J. 2018. When did modern humans leave Africa? *Science* 359: 389–390.

CHAPTER 7: FARMED FORESTS

1. Boivin, N. L., et al. 2016. Ecological consequences of human niche construction: Examining long-term anthropogenic shaping of global species distributions. *Proceedings of the National Academy of Sciences of the United States of America* 113: 6388–6396; Price, T. D., Bar-Yosef, O. 2011. The origins of agriculture: New data, new ideas. *Current Anthropology* 52: S163–S174; Rowley-Conwy, P. 2011. Westward ho! The spread of agriculture from Central Europe to the Atlantic. *Current Anthropology* 52: S431–S451.

2. Denham, T. P. 2018. *Tracing Early Agriculture in the Highlands of New Guinea: Plot, Mound and Ditch*. Oxford: Routledge; Fuller, D., Hildebrand, E. 2013. Domesticating plants in Africa. In P. Mitchell, P. Lane (eds.). *The Oxford Handbook of African Archaeology*. Oxford: Oxford University Press. Pp. 507–525.

3. Iriarte, J., et al. 2020. The origins of Amazonian landscapes: Plant cultivation, domestication and the spread of food production in tropical South America. *Quaternary Science Reviews* 248. doi: 10.1016/j.quascirev.2020.106582.

4. Roberts, P., et al. 2017. The deep human prehistory of global tropical forests and its relevance for modern conservation. *Nature Plants* 3: 17093; Barton, H., Denham, T. P. 2018. Vegecultures and the social-biological transformations of plants and people. *Quaternary International* 489: 17–25; Allen, J., Gosden, C., White, J. P. 1989. Human Pleistocene adaptations in the Tropical Island Pacific: Recent evidence from New Ireland, a Greater Australian outlier. *Antiquity* 63: 548–561.

5. Golson, J., Denham, T., Hughes, P., Swadling, P., Muke, J. 2017. *Ten Thousand Years of Cultivation at Kuk Swamp in the Highlands of Papua New Guinea*. Terra Australis 46. Canberra: Australian National University Press.

6. Golson, J. 1989. The origins and development of New Guinea agriculture. In D. R. Harris, G. C. Hillman (eds.). *Foraging and Farming: The Evolution of Plant Exploitation*. London: Unwin Hyman. Pp. 109–136; Haberle, S. G., Lentfer, C., Denham, T. 2017. Palaeoecology. In J. Golson, T. Denham, P. Hughes, P. Swadling, J. Muke (eds.). *Ten Thousand Years of Cultivation at Kuk Swamp in the Highlands of Papua New Guinea*. Terra Australis 46. Canberra: Australian National University Press. Pp. 145–162; Bulmer, S. 1991. Variation and change in stone tools in the highlands of Papua New Guinea: The witness of Wanelek. In A. Pawley (ed.). *Man and a Half: Essays in Pacific Anthropology and Ethnobiology in Honour of Ralph Bulmer*. Auckland: The Polynesian Society. Pp. 470–478; Denham, T. P. 2016. Revisiting the past: Sue Bulmer's contribution to the archaeology of Papua New Guinea. *Archaeology in Oceania* 51: 5–10.

7. Denham, T. P., et al. 2003. Origins of agriculture at Kuk Swamp in the highlands of New Guinea. *Science* 301: 189–193

8. Roberts, P., Gaffney, D., Lee-Thorp, J., Summerhayes, G. 2017. Persistent tropical foraging in the highlands of Terminal Pleistocene–Holocene New Guinea. *Nature Ecology & Evolution* 1. doi: 10.1038/s41559-016-0044; Gaffney, D., Ford, A., Summerhayes, S. 2015. Crossing the

Pleistocene-Holocene transition in the New Guinea Highlands: Evidence from the lithic assemblage of Kiowa rockshelter. *Journal of Anthropological Archaeology* 39: 223–246.

9. Franklin, L. C. 2013. Corn. In A. F. Smith (ed.). *The Oxford Encyclopedia of Food and Drink in America* (2nd edition). Oxford: Oxford University Press. Pp. 551–558; Piperno, D. R., Ranere, A. J., Holst, I., Iriarte, J., Dickau, R. 2009. Starch grain and phytolith evidence for early ninth millennium BP maize from the Central Balsas River Valley, Mexico. *Proceedings of the National Academy of Sciences of the United States of America* 106: 5019–5024; Kistler, L., et al. 2019. Multiproxy evidence highlights a complex evolutionary legacy of maize in South America. *Science* 362: 1309–1313.

10. Meggers, B. J. 1971. *Amazonia: Man and Culture in a Counterfeit Paradise*. Wheeling: Harlan Davidson; Olsen, K. M. 2002. Population history of *Manihot esculenta* (Euphorbiaceae) inferred from nuclear DNA sequences. *Molecular Ecology* 11: 901–911; Spooner, D. M., McLean, K., Ramsay, G., Waugh, R., Bryan, G. J. 2005. A single domestication for potato based on multilocus amplified fragment length polymorphism genotyping. *Proceedings of the National Academy of Sciences of the United States of America* 102: 14694–14699; Clement, C. R., de Cristo-Araújo, M., d'Eeckenbrugge, G. C., Pereira, A. A., Picanço-Rodrigues, D. 2010. Origin and domestication of native Amazonian crops. *Diversity* 2: 72–106; Scaldaferro, M. A., Barboza, G. E., Acosta, M. C. 2018. Evolutionary history of the chili pepper *Capsicum baccatum* L. (Solanaceae): Domestication in South America and natural diversification in the seasonally dry tropical forests. *Biological Journal of the Linnean Society* 124: 466–478; Zarrillo, S., et al. 2018. The use and domestication of *Theobroma cacao* during the mid-Holocene in the upper Amazon. *Nature Ecology & Evolution* 2: 1879–1888.

11. Iriarte, J., et al. 2020. The origins of Amazonian landscapes: Plant cultivation, domestication and the spread of food production in tropical South America. *Quaternary Science Review* 248. doi: 10.1016/j.quascirev.2020.106582; Denham, T., et al. 2020. The domestication syndrome in vegetatively propagated field crops. *Annals of Botany* 125: 581–597. doi: 10.1093/aob/mcz212; Thomas, E., Alcazar Caicedo, C., McMichael, C. H., Corvera, R., Loo, J. 2015. Uncovering spatial patterns in the natural and human history of Brazil nut (*Bertholletia excelsa*) across the Amazon Basin. *Journal of Biogeography* 42: 1367–1382; Clement, C. R., et al. 2015. The domestication of Amazonia before European conquest. *Proceedings of the Royal Society B: Biological Sciences* 282. doi: 10.1098/rspb.2015.0813; Clement, C. R., et al. 2016. Crop domestication in the upper Madeira River basin. *Boletim do Museu Paraense Emílio Goeldi. Ciências Humanas* 11. doi: 10.1590/1981.81222016000100010.

12. Lombardo, U., et al. 2020. Early Holocene crop cultivation and landscape modification in Amazonia. *Nature* 581: 190–193.

13. Fuller, D. Q. 2006. Agricultural origins and frontiers in South Asia: A working synthesis. *Journal of World Prehistory* 20: 1–86; Asouti, E., Fuller, D. Q. 2008. *Trees and woodlands of South India: Archaeological Perspectives*. Walnut Creek: Left Coast Press; Fuller, D. Q., Boivin, N., Hoogervorst, T., Allaby, R. 2011. Across the Indian Ocean: The prehistoric movement of plants and animals. *Antiquity* 85: 544–558; Morrison, K. 2002. Historicizing adaptation, adapting to history: Forager-traders in South and Southeast Asia. In K. Morrison, L. Junker (eds.). *Forager-Traders in South and Southeast Asia*. Cambridge: Cambridge University Press. Pp. 1–20.

14. Yang, Y. D., Liu, L., Chen, X., Speller, C. F. 2008. Wild or domesticated: DNA analysis of ancient water buffalo remains from North China. *Journal of Archaeological Science* 35: 2778–2785; Wang, M-S., et al. 2020. 863 genomes reveal the origin and domestication of chicken. *Cell Research* 30: 693–701. doi: 10.1038/s41422-020-0349-y..

15. Heckenberger, M., Neves, E. G. 2009. Amazonian archaeology. *Annual Review of Anthropology* 38: 251–266; Barton, H., Denham, T. 2011. Vegeculture and social life in Island Southeast Asia. In G. Barker, M. Janowski (eds.). *Why Cultivate? Anthropological and Archaeological Approaches to Foraging-Farming Transitions in Southeast Asia*. Cambridge: McDonald Institute Monographs. Pp. 17–25.

16. Molina, J., et al. 2011. Molecular evidence for a single evolutionary origin of domesticated rice. *Proceedings of the National Academy of Sciences of the United States of America* 108: 8351–8356; Gutaker, R. M., et al. 2020. Genomic history and ecology of the geographic spread of rice. *Nature Plants* 6: 492–502; Fuller, D. Q., Qin, L. 2009. Water management and labour in the origins and dispersal of Asian rice. *World Archaeology* 41: 88–111.

17. Weber, S., Lehman, H., Barela, T., Hawks, S., Harriman, D. 2010. Rice or millets: Early farming strategies in prehistoric central Thailand. *Archaeological and Anthropological Sciences* 2: 79–88; Deng, Z., et al. 2020. Validating earliest rice farming in the Indonesian archipelago. *Scientific Reports* 10. doi: 10.1038/s41598-020-67747-3; Barron, A., et al. 2020. Sherds as archaeological assemblages: Gua Sireh reconsidered. *Antiquity* 94: 1325–1336; Bellwood, P. 1993 Cultural and biological differentiation in peninsular Malaysia: The last 10,000 years. *Asian Perspectives* 32: 37–60; Krigbaum, J. 2003. Neolithic subsistence patterns in northern Borneo reconstructed with stable carbon isotopes of enamel. *Journal of Anthropological Archaeology* 22: 292–304.

18. Neumann, K., et al. 2011. First farmers in the Central African rainforest: A view from southern Cameroon. *Quaternary International* 249: 53–62; Garcin, Y., et al. 2018. Early anthropogenic impact on western Central African rainforests 2,600 y ago. *Proceedings of the National Academy of Sciences of the United States of America* 115: 3261–3266.

19. Wotzka, H. P. 2019. Ecology and culture of millets in African rainforests: Ancient, historical, and present-day evidence. In B. Eichhorn, A. Höhn (eds.). *Trees, Grasses and Crops: People and Plants in Sub-Saharan Africa and Beyond*. Bonn: Dr. Rudolf Habelt GmbH. Pp. 407–429; Hamilton, A. C., Karamura, D., Kakudidi, E. 2016. History and conservation of wild and cultivated plant diversity in Uganda: Forest species and banana varieties as case studies. *Plant Diversity* 38(1): 23–44; Bleasdale, M., et al. 2020. Isotopic and microbotanical insights into Iron Age agricultural reliance in the Central African rainforest. *Communications Biology* 3. doi: 10.1038/s42003-020-01324-2.

CHAPTER 8: ISLAND PARADISES LOST?

1. Russell, J. C., Kueffer, C. 2019. Island biodiversity in the Anthropocene. *Annual Review of Environment and Resources* 44: 31–60; Blackburn, T. M., Cassey, P., Duncan, R. P., Evans, K. L., Gaston, K. J. 2004. Alien extinction and mammalian introductions on oceanic islands. *Science* 305: 1955–1958.

2. Turvey, S. T., Cheke, A. S. 2008. Dead as a dodo: The fortuitous rise to fame of an extinction icon. *Historical Biology* 20: 149–163; Martin P. S., Steadman D. W. 1999. Prehistoric extinctions on islands and continents. In R. D. E. MacPhee (ed.). *Extinctions in Near Time*. Advances in Vertebrate Paleobiology 2. Boston: Springer; Grayson, D. K. 2001. The archaeological record of human impacts on animal populations. *Journal of World Prehistory* 15: 1–68; Diamond, J. 2005. *Collapse: How Societies Choose to Fail or Survive*. London: Penguin Books; Fordham, D. A., Brook, B. W. 2010. Why tropical island endemics are acutely susceptible to global change. *Biodiversity and Conservation* 19: 329–342.

3. Huebert, J. M., Allen, M. 2020. Anthropogenic forests, arboriculture, and niche construction in the Marquesas Islands (Polynesia). *Journal of Anthropological Archaeology* 57. doi: 10.1016/j.jaa.2019.101122; Kirch, P. V. 1982. Transported landscapes. *Natural History* 91: 32–35.

4. Forster, J., Lake, I. R., Watkinson, A. R., Gill, J. A. 2011. Marine biodiversity in the Caribbean UK overseas territories: Perceived threats and constraints to environmental management. *Marine Policy* 35: 647–657.

5. Napolitano, M. F., et al. 2019. Reevaluating human colonization of the Caribbean using chronometric hygiene and Bayesian modeling. *Science Advances* 5, no. 12. doi: 10.1126/sciadv.aar7806; Siegel, P. E., et al. 2015. Paleoenvironmental evidence for first human colonization of the eastern Caribbean. *Quaternary Science Reviews* 129: 275–295; Pagán-Jiménez, J. R., Rodríguez-Ramos, R., Reid, B. A., van den Bel, M., Hofman, C. L. 2015. Early dispersals of maize and other food plants into the southern Caribbean and northeastern South America. *Quaternary Science Reviews* 123: 231–246.

6. Steadman, D. W., et al. 2005. Asynchronous extinction of Late Quaternary sloths on continents and islands. *Proceedings of the National Academy of Sciences of the United States of America* 102: 11763–11768; Cooke, S. B., Dávalos, L. M., Mychajliw, A. M., Turvey, S. T., Upham, N. S. 2016. Anthropogenic extinction dominates Holocene declines of West Indian mammals. *Annual Review of Ecology, Evolution, and Systematics* 48: 301–327; Rivera-Collazo, I. C. 2015. Por el camino verde: Long-term tropical socioecosystem dynamics and the Anthropocene as seen from Puerto Rico. *Holocene* 25:1604–1611.

7. Fitzpatrick, S. M. 2015. The pre-Columbian Caribbean: Colonization, population dispersal, and island adaptations. *PaleoAmerica* 1(4): 305–331; Keegan, W. F. 2006. Archaic influences in the origins and development of Taíno societies. *Caribbean Journal of Science* 42: 1–10; Giovas, C. M., Lefebvre, M. J., Fitzpatrick, S. M. 2012. New records for prehistoric introduction of Neotropical mammals to the West Indies: Evidence from Carriacou, Lesser Antilles. *Journal of Biogeography* 39: 476–487; Fitzpatrick, S. M., Keegan, W. F. 2007. Human impacts and adaptations in the Caribbean islands: An historical ecology approach. *Earth and Environmental Science Transactions of the Royal Society of Edinburgh* 98: 29–45; Cooke, S. B., Dávalos, L. M., Mychajliw, A. M., Turvey, S. T., Upham, N. S. 2016. Anthropogenic extinction dominates Holocene declines of West Indian mammals. *Annual Review of Ecology, Evolution, and Systematics* 48: 301–327.

8. Newsom, L. A., Wing, E. S. 2004. *On Land and Sea: Native American Uses of Biological Resources in the West Indies*. Tuscaloosa: University of Alabama Press; Turvey, S. T., Weksler, M., Morris, E. L., Nokkert, M. 2010. Taxonomy, phylogeny, and diversity of the extinct Lesser Antillean rice rats (Sigmodontinae: *Oryzomyini*), with description of a new genus and species. *Zoological Journal of the Linnaean Society* 160: 748–772; Giovas, C. M., Clark, M., Fitzpatrick, S. M., Stone, J. 2013. Intensifying collection and size increase of the tessellated nerite snail (*Nerita tessellata*) at the Coconut Walk site, Nevis, northern Lesser Antilles, AD 890–1440. *Journal of Archaeological Science* 40: 4024–4038.

9. Bellwood, P. 2005. *First Farmers*. Oxford: Blackwell; Posth, C., et al. 2018. Language continuity despite population replacement in Remote Oceania. *Nature Ecology & Evolution* 2: 731–740; Kirch, P. V. 2017. *On the Road of the Winds: An Archaeological History of the Pacific Islands Before European Contact*. Berkeley: University of California Press; Montenegro, Á., Callaghan, R. T., Fitzpatrick, S. M. 2016. Using seafaring simulations and shortest-hop trajectories to model the prehistoric colonization of Remote Oceania. *Proceedings of the National Academy of Sciences of the United States of America* 113: 12685–12690.

10. Larson, G., et al. 2007. Phylogeny and ancient DNA of Sus provides insights into Neolithic expansion in Island Southeast Asia and Oceania. *Proceedings of the National Academy of Sciences of the United States of America* 104: 4834–4839; Nogueira-Filho, S. L. G., Nogueira, S. S. C., Fragoso, J. M. V. 2009. Ecological impacts of feral pigs in the Hawaiian Islands. *Biodiversity and Conservation* 18. doi: 10.1007/s10531-009-9680-9; Kirch, P. V. 2001. Pigs, humans, and tropic competition on small Oceania islands. In A. Anderson, T. Murray (eds.). *Australian Archaeologist: Collected Papers in Honour of Jim Allen*. Canberra: Australian National University Press. Pp. 427–439; Kirch, P. V. 1982. Transported landscapes. *Natural History* 91: 32–35; Kinaston, R. L., Bedford, S. B., Spriggs, M., Anson, D., Buckley, H. 2016. Is there a "Lapita diet"? A comparison of Lapita and post-Lapita skeletal samples from four Pacific Island archaeological sites. In M. Oxenham, H. Buckley (eds.). *The Routledge Handbook of Bioarchaeology in Southeast Asia and the Pacific Islands*. London: Routledge. Pp. 427–461; Fall, P. L. 2010. Pollen evidence for plant introductions in a Polynesian tropical island ecosystem, Kingdom of Tonga. In S. G. Haberle, J. Stevenson, M. Prebble (eds.). *Altered Ecologies: Fire, Climate and Human Influence on Terrestrial Landscapes*. Canberra: Australian National University Press. Pp. 253–271; Morrison, A. E., Hunt, T. L. 2007. Human impacts on the nearshore environment: An archaeological case study from Kaua'i, Hawaiian Islands. *Pacific Science* 61: 325–345.

11. Steadman, D. W. 2006. *Extinction and Biogeography of Tropical Pacific Birds*. Chicago: University of Chicago Press; Hunt, T. L. 2007. Rethinking Easter Island's ecological catastrophe. *Journal of Archaeological Science* 34: 485–502.

12. Tromp, M., et al. 2020. Exploitation and utilization of tropical rainforests indicated in dental calculus of ancient Oceanic Lapita culture colonists. *Nature Human Behaviour* 4: 489–495. doi: 10.1038/s41562-019-0808-y; Maxwell, J. J., Howarth, J. D., Vandergoes, M. J., Jacobsen, G. E., Barber, I. G. 2016. The timing and importance of arboriculture and agroforestry in a temperate East Polynesia Society, the Moriori, Rekohu (Chatham Island). *Quaternary Science Reviews* 149: 306–325; Lambrides, A. B. J., Weisler, M. I. 2016. Pacific Islands ichthyoarchaeology: Implications for the development of prehistoric fishing studies and global sustainability. *Journal of Archaeological Research* 24: 275–324.

13. Giovas, C. 2006. No pig atoll: Island biogeography and the extirpation of a Polynesian domesticate. *Asian Perspectives* 45: 69–95; Matisoo-Smith, E. 2007. Animal translocations, genetic

variation, and the human settlement of the Pacific. In J. S. Friedlaender (ed.). *Genes, Language, and Culture History in the Southwest Pacific*. Oxford: Oxford University Press. Pp. 157–170; Ladefoged, T. N., et al. 2011. Agricultural potential and actualized development in Hawai'i: An airborne LiDAR survey of the leeward Kohala field system (Hawai'i Island). *Journal of Archaeological Science* 38: 3605–3619; Kirch, P. V. 2010. *How Chiefs Became Kings: Divine Kingship and the Rise of Archaic States in Ancient Hawai'i*. Berkeley: University of California Press.

14. Kirch, P., Yen, D. 1982. *Tikopia: Prehistory and Ecology of a Polynesian Outlier*. Bulletin 238. Honolulu: Bernice P. Bishop Museum; Ladefoged, T. N., et al. 2010. Soil nutrient analysis of Rapa Nui gardening. *Archaeology in Oceania* 45: 80–85; Rainbird, P. 2002. A message for our future? The Rapa Nui (Easter Island) ecodisaster and Pacific Island environments. *World Archaeology* 33: 436–451.

15. Douglass, K., Zinke, J. 2015. Forging ahead by land and by sea: Archaeology and paleoclimate reconstruction in Madagascar. *African Archaeological Review* 32: 267–299; Hansford, J., et al. 2018. Early Holocene human presence in Madagascar evidenced by exploitation of avian megafauna. *Science Advances* 4: eaat6925. doi: 10.1126/sciadv.aat6925; Anderson, A., et al. 2018. New evidence of megafaunal bone damage indicates late colonization of Madagascar. *PLOS ONE* 13: e0204368. doi: 10.1371/journal.pone.0204368; Douglass, K., et al. 2019. A critical review of radiocarbon dates clarifies the human settlement of Madagascar. *Quaternary Science Reviews* 221: 105878; Burney, D. A., et al. 2004. A chronology for late prehistoric Madagascar. *Journal of Human Evolution* 47: 25–63; Crowley, B. E. 2010. A refined chronology of prehistoric Madagascar and the demise of the megafauna. *Quaternary Science Reviews* 29: 2591–2603; Martin, P. S. 1984. Prehistoric overkill: The global model. In P. S. Martin, R. G. Klein (eds.). *Quaternary Extinctions: A Prehistoric Revolution*. Tucson: University of Arizona Press. Pp. 354–403.

16. Godfrey, L. R., et al. 2019. A new interpretation of Madagascar's megafaunal decline: The "Subsistence Shift Hypothesis." *Journal of Human Evolution* 130: 126–140; Douglass, K., et al. 2018. Multi-analytical approach to zooarchaeological assemblages elucidates Late Holocene coastal lifeways in southwest Madagascar. *Quaternary International* 471: 111–131; Burney, D. A., et al. 2004. A chronology for late prehistoric Madagascar. *Journal of Human Evolution* 47: 25–63; Crowley, B. E. 2010. A refined chronology of prehistoric Madagascar and the demise of the megafauna. *Quaternary Science Reviews* 29: 2591–2603; Li, H., et al. 2020. A multimillennial climatic context for the megafaunal extinctions in Madagascar and Mascarene Islands. *Science Advances* 6: eabb2459. doi: 10.1126/sciadv.abb2459.

17. Godfrey, L. R., et al. 2019. A new interpretation of Madagascar's megafaunal decline: The "Subsistence Shift Hypothesis." *Journal of Human Evolution* 130: 126–140; Crowther, A., et al. 2016. Ancient crops provide first archaeological signature of the westward Austronesian expansion. *Proceedings of the National Academy of Sciences of the United States of America* 113: 6635–6640; Radimilahy, C. M., Crossland, Z. 2015. Situating Madagascar: Indian Ocean dynamics and archaeological histories. *Azania: Archaeological Research in Africa* 50: 495–518; Dewar, R. E., Richard, A. F. 2012. Madagascar: A history of arrivals, what happened, and will happen next. *Annual Review of Anthropology* 41: 495–517.

18. Crowther, A., et al. 2016. Ancient crops provide first archaeological signature of the westward Austronesian expansion. *Proceedings of the National Academy of Sciences of the United States of America* 113: 6635–6640; Schwitzer, C., et al. 2014. Averting lemur extinctions amid Madagascar's political crisis. *Science* 343: 842–843; Kaufmann, J. C. 2004. Prickly pear cactus and pastoralism in southwest Madagascar. *Ethnology* 43: 345–361.

19. Galván, B., et al. 1999. Poblamiento prehistórico en la costa de Buena Vista del Norte (Tenerife). El conjunto arqueológico Fuente-Arena. *Investigaciones Arqueológicas en Canarias* 6: 9–257; Morales, J., Rodríguez, A., Alberto, V., Machado, C., Criado, C. 2009. The impact of human activities on the natural environment of the Canary Islands (Spain) during the pre-Hispanic stage (3rd-2nd century BC to 15th century AD): An overview. *Environmental Archaeology* 14: 27–36; Rando, J. C., Perera, M. A. 1994. Primeros datos de ornitofagia entre los aborígenes de Fuerteventura (Islas Canarias). *Archeofauna* 3: 13–19.

20. Nogué, S., et al. 2013. The ancient forests of La Gomera, Canary Islands, and their sensitivity to environmental change. *Journal of Ecology* 101: 368–377; Machado, C. 1995. Approche paléoécologique et ethnobotanique du site archéologique "El Tendal" (N-E de l'Ile de La Palma,

Archipel des Canaries). In CTHS (eds.). *L'Homme préhistorique et la mer.* 120 Congrés. Aix-en-Provence: CTHS. Pp. 179–186; Gangoso, L., Donázar, J. A., Scholz, S., Palacios, C. J., Hiraldo, F. 2006. Contradiction in conservation of island ecosystems: Plants, introduced herbivores and avian scavengers in the Canary Islands. *Biodiversity and Conservation* 15. doi: 10.1007/s10531-004-7181-4; Crosby, A. W. 1984. An ecohistory of the Canary Islands: A precursor of European colonization in the New World and Australasia 8: 214–235.

21. Boivin, N. L., et al. 2016. Ecological consequences of human niche construction: Examining long-term anthropogenic shaping of global species distributions. *Proceedings of the National Academy of Sciences of the United States of America* 113: 6388–6396.

CHAPTER 9: CITIES IN THE "JUNGLE"

1. Diamond, J. 2005. *Collapse: How Societies Choose to Fail or Survive.* London: Penguin Books.

2. Meggers, B. J. 1971. *Amazonia: Man and Culture in a Counterfeit Paradise.* Wheeling: Harlan Davidson; Webster, D. 2002. *The Fall of the Ancient Maya.* London: Thames & Hudson; Bacus, E. A., Lucero, L. J. (eds.). 1999. *Complex Polities in the Ancient Tropical World.* Archaeological Papers of the American Anthropological Association (AAA) 9. Arlington: AAA; Algaze, G. 2018. Entropic cities: The paradox of urbanism in ancient Mesopotamia. *Current Anthropology* 59: 23–54; Wengrow, D. 2006. *The Archaeology of Early Egypt: Social Transformations in North-East Africa, c. 10,000 to 2,650 BC.* Cambridge: Cambridge University Press; Flad, R. 2018. Urbanism as technology in early China. *Archaeological Research in Asia* 14: 121–134.

3. Childe, V. G. 1950. The Urban Revolution. *Town Planning Review* 21: 3–17; Postgate, J. N. 1992. *Early Mesopotamia: Society and Economy at the Dawn of History.* London: Routledge; Webster, D. 2002. *The Fall of the Ancient Maya.* London: Thames & Hudson.

4. Martin, S., Grube, N. 2008. *Chronicle of the Maya Kings and Queens: Deciphering the Dynasties of the Ancient Maya.* London: Thames & Hudson; Greenwich Park climate station. Met Office. 2020. www.metoffice.gov.uk/research/climate/maps-and-data/uk-climate-averages/u10hb54gm; Scarborough, V. L. 1993. Water management in the southern Maya lowlands: An accretive model for the engineered landscape. *Research in Economic Anthropology* 7: 17–69; Beach, T., Dunning, N., Luzzadder-Beach, S., Cook, D. E., Lohse, J. 2006. Impacts of the ancient Maya on soils and soil erosion in the central Maya lowlands. *Caterna* 65: 166–178; Webster, D. 2002. *The Fall of the Ancient Maya.* London: Thames & Hudson.

5. Fletcher, R. 2012. Low-density, agrarian-based urbanism: Scale, power and ecology. In M. E. Smith (ed.). *The Comparative Archaeology of Complex Societies.* Cambridge: Cambridge University Press. Pp. 285–320.

6. Webster, D., et al. 2007. The Great Tikal Earthwork revisited. *Journal of Field Archaeology* 32: 41–64; Chase, A. F., et al. 2011. Airborne LiDAR, archaeology, and the ancient Maya landscape at Caracol, Belize. *Journal of Archaeological Science* 38: 387–398.

7. Ross, N. J. 2011. Modern tree species composition reflects ancient Maya "forest gardens" in northwest Belize. *Ecological Applications* 21: 75–84; Ford, A., Nigh, R. 2015. *The Maya Forest Garden: Eight Millennia of Sustainable Cultivation of the Tropical Woodlands.* London: Routledge; Lucero, L. J. 1999. Water control and Maya politics in the southern Maya lowlands. In E. A. Bacus, L. J. Lucero (eds.). *Complex Polities in the Ancient Tropical World.* Archeological Papers of the American Anthropological Association (AAA) 9. Arlington: AAA. Pp. 34–49.

8. Medina-Elizalde, M., Rohling, E. J. 2012. Collapse of classic Maya civilization related to modest reduction in precipitation. *Science* 335: 956–959; McNeil, C. L., Burney, D. A., Burney, L. P. 2010. Evidence disputing deforestation as the cause for the collapse of the ancient Maya polity of Copan, Honduras. *Proceedings of the National Academy of Sciences of the United States of America* 107: 1017–1022; Thompson, K. M., Hood, A., Caallaro, D., Lentz, D. L. 2015. Connecting contemporary ecology and ethnobotany to ancient plant use practices of the Maya at Tikal. In D. L. Lentz, N. Dunning, V. Scarborough (eds.). *Tikal: Paleoecology of an Ancient Maya City.* Cambridge: Cambridge University Press. Pp. 124–151; Cook, B. I., et al. 2012. Pre-Columbian deforestation as an amplifier of drought in Mesoamerica. *Geophysical Research Letters* 39: L16706; Douglas, P. M. J., Demarest,

A. A., Brenner, M., Canuto, M. A. 2016. Impacts of climate change on the collapse of lowland Maya civilization. *Annual Review of Earth and Planetary Sciences* 44: 613–645.

9. Hoggarth, J. A., et al. 2016. The political collapse of Chichén Itzá in climatic and cultural context. *Global and Planetary Change* 138: 25–42; Masson, M. A., Peraza Lope, C. 2014. *Kukulkan's Realm: Urban Life at Ancient Mayapán*. Boulder: University Press of Colorado; Chase, D. Z., Chase, A. F. 2006. Framing the Maya collapse: Continuity, discontinuity, method, and practice in the Classic to Postclassic southern Maya lowlands. In G. M. Schwartz, J. J. Nichols (eds.). *After Collapse: The Regeneration of Complex Societies*. Tucson: University of Arizona Press. Pp. 168–187; Coe, M. D. 1999. *The Maya*. London: Thames & Hudson; Ford, A., Nigh, R. 2015. *The Maya Forest Garden: Eight Millennia of Sustainable Cultivation of the Tropical Woodlands*. London: Routledge; Aimers, J. J. 2007. What Maya collapse? Terminal Classic variation in the Maya lowlands. *Journal of Archaeological Research* 15: 329–377.

10. O'Reilly, D. J. W. 2007. *Early Civilizations of Southeast Asia*. New York: Altamira Press; Pottier, C. 2012, Beyond the temples: Angkor and its territory, in A. Haendel (ed.), *Old Myths and New Approaches: Interpreting Ancient Religious Sites in Southeast Asia*, Clayton, Melbourne, Australia: Monash University Publishing, 12–27.

11. Evans, D., et al. 2007. A comprehensive archaeological map of the world's largest preindustrial settlement complex at Angkor, Cambodia. *Proceedings of the National Academy of Sciences of the United States of America* 104: 14277–14282.

12. Evans, D. H., et al. 2013. Uncovering archaeological landscapes at Angkor using lidar. *Proceedings of the National Academy of Sciences of the United States of America* 110: 12595–12600; Higham, C. 2001. *The Civilization of Angkor*. London: Weidenfeld and Nicholson; Fletcher, R., et al. 2008. The water management network of Angkor, Cambodia. *Antiquity* 82: 658–670.

13. Buckley, B. M., et al. 2010. Climate as a contributing factor in the demise of Angkor, Cambodia. *Proceedings of the National Academy of Sciences of the United States of America* 107: 6748–6752; Lucero, L. J., Fletcher, R., Coningham, R. 2015. From "collapse" to urban diaspora: The transformation of low-density, dispersed agrarian urbanism. *Antiquity* 89: 1139–1154.

14. Bandaranayake, S. 2003. The pre-modern city in Sri Lanka: The "first" and "second" urbanisation. In P. J. J. Sinclair (ed.). *The Development of Urbanism from a Global Perspective*. Uppsala: Uppsala University; Conginham, R. A. E. 1999. *Anuradhapura: The British-Sri Lankan Excavations at Anuradhapura Salgaha Watta 2*. British Archaeological Reports, International Series 824. Oxford: Archaeopress Press; BBC News. 2020. Hagia Sophia: Former Istanbul museum welcomes Muslim worshippers. *BBC News*. July 24. www.bbc.com/news/world-europe-53506445; Coningham, R. A. E., et al. 2007. The state of theocracy: Defining an early medieval hinterland in Sri Lanka. *Antiquity* 81: 699–719; Mail, D. 2017. The official boundaries of the city of Liverpool are far too small—and it matters. *CityMonitor*. September 13. https://citymonitor.ai/politics/official-boundaries-city-liverpool-are-far-too-small-and-it-matters-3319.

15. Lucero, L. J., Fletcher, R., Coningham, R. 2015. From "collapse" to urban diaspora: The transformation of low-density, dispersed agrarian urbanism. *Antiquity* 89: 1139–1154; Gilliland, K., et al. 2013. The dry tank development and disuse of water management infrastructure in the Anuradhapura hinterland, Sri Lanka. *Journal of Archaeological Science* 40: 1012–1028.

16. Meggers, B. 1954. Environmental limitations of the development of culture. *American Anthropologist* 56: 801–824; Heckenberger, M. J., et al. 2003. Amazonia 1492: Pristine forest or cultural parkland? *Science* 301: 1710–1714; Heckenberger, M., Neves, E. G. 2009. Amazonian archaeology. *Annual Review of Anthropology* 38: 251–266.

17. Roosevelt, A. 1999. The development of prehistoric complex societies: Amazonia, a tropical forest. In E. A. Bacus, L. J. Lucero, J. Allen (eds.). *Complex Polities in the Ancient Tropical World*. Arlington: American Anthropological Association. Pp. 13–34; Hermenegildo, T., O'Connell, T. C., Guapindaia, V. L. C., Neves, E. G. 2017. New evidence for subsistence strategies of late pre-colonial societies of the mouth of the Amazon based on carbon and nitrogen isotopic data. *Quaternary International* 448: 139–149; Koch, A., Brierley, C., Maslin, M. M., Lewis, S. L. 2019. Earth system impacts of the European arrival and Great Dying in the Americas after 1492. *Quaternary Science Reviews* 207: 13–36.

18. Rostain, S. 2014. *Islands in the Rainforest: Landscape Management in Pre-Columbian Amazonia*. London: Routledge; de Souza, J. G., et al. 2018. Pre-Columbian earth-builders settled along the entire southern rim of the Amazon. *Nature Communications* 9: 1125. https://doi.org/10.1038 /s41467-018-03510-7; Schmidt, M. J., et al. 2014. Dark earths and the human built landscape in Amazonia: a widespread pattern of anthrosol formation. *Journal of Archaeological Science* 42: 152–165.

19. Roosevelt, A. 1999. The development of prehistoric complex societies: Amazonia, a tropical forest. In E. A. Bacus, L. J. Lucero, J. Allen (eds.). *Complex Polities in the Ancient Tropical World*. Arlington: American Anthropological Association. Pp. 13–34; Scarborough, V. L., Lucero, L. 2010. The non-hierarchical development of complexity in the semi-tropics: Water and cooperation. *Water History* 2: 185–205; McIntosh, S. K. 1999. Pathways to complexity: An African perspective. In S. K. McIntosh (ed.). *Beyond Chiefdoms: Pathways to Complexity in Africa*. Cambridge: Cambridge University Press. Pp. 1–30.

20. Carson, J. F., et al. 2014. Environmental impact of geometric earthwork construction in pre-Columbian Amazonia. *Proceedings of the National Academy of Sciences of the United States of America* 111: 10497–10502; Piperno, D. R., McMichael, C., Bush, M. B. 2017. Further evidence for localized, short-term anthropogenic forest alterations across pre-Columbian Amazonia. *Proceedings of the National Academy of Sciences of the United States of America* 114: E4118–E4119.

21. Lucero, L. J., Gonzalez Cruz, J. 2020. Reconceptualizing urbanism: Insights from Maya cosmology. *Frontiers in Sustainable Cities* 2: doi: 10.3389/frsc.2020.00001; Simon, D., Adam-Bradford, A. 2016. Archaeology and contemporary dynamics for more sustainable, resilient cities in the peri-urban interface. *Water Science and Technology* 72: 57–83.

22. Miller, S. W. 2007. *An Environmental History of Latin America*. Cambridge: Cambridge University Press.

CHAPTER 10: EUROPE AND THE TROPICS IN THE "AGE OF EXPLORATION"

1. Associated Press. 2020. Mexico's marks Columbus Day without statue of Columbus, *News18*, October 13. https://apnews.com/article/mexico-mexico-city-columbus-day-60bdc08a76 06b41a4825d467d97c5f6c.

2. Associated Press. 2020. Columbus statue removed in Mexico City, defaced elsewhere. *Northwest Arkansas Democrat Gazette*. October 13. www.nwaonline.com/news/2020/oct/13/columbus -statue-removed-in-mexico-city-defaced; Mineo, L. 2020. A day of reckoning. *Harvard Gazette*. https://news.harvard.edu/gazette/story/2020/10/pondering-putting-an-end-to-columbus-day -and-a-look-at-what-could-follow; Elliott, J. H. 2006. *Empires of the Atlantic World: Britain and Spain in America, 1492–1830*. New Haven: Yale University Press; Steinberg, P. E. 1999. Lines of division, lines of connection: Stewardship in the World Ocean. *Geographical Review* 89: 254–264.

3. Jackson, P. 2000. The Mongol Empire, 1986–1999. *Journal of Medieval History* 26: 189–210; O'Rourke, K. H., Williamson, J. G. 2009. Did Vasco da Gama matter for European markets? *Economic History Review* 62: 655–684; Nichols, D. L., Rodríguez-Alegría, E. 2017. *The Oxford Handbook of the Aztecs*. Oxford: Oxford University Press; Bridges, E. J. 2016. Vijayanagara Empire. Wiley Online Library. January 11. doi: 10.1002/9781118455074.wbeoe424; Cissé, M., McIntosh, S. K., Dussubieux, L., Fenn, T., Gallagher, D., Smith, A. C. 2013. Excavations at Gao Saney: New evidence for settlement growth, trade, and interaction on the Niger Bend in the first millennium CE. *Journal of African Archaeology* 11: 9–37; Koch, A., Brierley, C., Maslin, M. M., Lewis, S. L. 2019. Earth system impacts of the European arrival and Great Dying in the Americas after 1492. *Quaternary Science Reviews* 207: 13–36; Hofman, C., Mol, A., Hoogland, M., Rojas, R. V. 2014. Stage of encounters: Migration, mobility and interaction in the pre-colonial and early colonial Caribbean. *World Archaeology* 46: 590–609; Mitchell, P. 2005. *African Connections: Archaeological Perspectives on Africa and the Wider World*. Lanham: AltaMira Press; Junker, L. L. 1999. *Raiding, Trading, and Feasting: The Political Economy of Philippine Chiefdoms*. Honolulu: University of Hawaii Press.

4. Wolf, E. 1982. *Europe and the People Without History*. Berkeley: University of California Press.

5. Crosby, A. W. 1972. *The Columbian Exchange: Biological and Cultural Consequences of 1492.* Westport: Greenwood Press; Cañizares-Esguerra, J. 2006. *Puritan Conquistadors: Iberianizing the Atlantic, 1550–1700.* Stanford: Stanford University Press; Cavanagh, E. 2016. Corporations and business associations from the commercial revolution to the Age of Discovery: Trade, jurisdiction and the state, 1200–1600. *History Compass* 14: 493–510; Curry, A. 2016. "Green hell" has long been home for humans. *Science* 354: 268–269; Roberts, P. 2017. Forests of plenty: Ethnographic and archaeological rainforests as hotspots of human activity. *Quaternary International* 448. doi:10.1016 /j.quaint.2017.03.041.

6. Ramsey, J. F. 1973. *Spain: The Rise of the First World Power.* Tuscaloosa: University of Alabama Press; Schwarz, G. R. 2008. The Iberian caravel: Tracing the development of a ship of discovery. In F. Vieira de Castro, K. Custer (eds.). *Edge of Empire.* Casal de Cambra: Caleidoscópio—Edićao e Artes Gráficas. Pp. 23–42; Hemming, J. 2004. *The Conquest of the Incas* (2nd edition). New York: Harvest; Livi Bacci, M. 2003. Return to Hispaniola: Reassessing a demographic catastrophe. *Hispanic American Historical Review* 83: 3–51; Vågene, Å. J., et al. 2018. Salmonella enterica genomes from victims of a major sixteenth-century epidemic in Mexico. *Nature Ecology & Evolution* 2: 520– 528; McNeill, W. 1977. *Plagues and Peoples.* London: Anchor; Koch, A., Brierley, C., Maslin, M. M., Lewis, S. L. 2019. Earth system impacts of the European arrival and Great Dying in the Americas after 1492. *Quaternary Science Reviews* 207: 13–36.

7. Nichols, D. L. 2018. Agricultural practices and environmental impacts of Aztec and pre-Aztec Central Mexico. Oxford Research Encyclopedias. doi: 10.1093/acrefore/9780199389414.013.175; Weaver, M. P. 2019. *The Aztecs, Maya, and Their Predecessors: Archaeology of Mesoamerica* (3rd edition). London: Routledge; Daniel, D. A. 1992. Tactical factors in the Spanish conquest of the Aztecs. *Anthropological Quarterly* 65: 187–194; Valdeón, R. A. Doña Marina/La Malinche: A historiographical approach to the interpreter/traitor. *Target* 25: 157–179; Mann, C. 2011. *1493: Uncovering the New World Columbus Created.* New York: Vintage Books; Haskett, R. S. 1991. "Our suffering with the Taxco Tribute": Involuntary mine labour of indigenous society in central New Spain. *Hispanic American Historical Review* 71: 447–475; Diel, L. B. 2010. The spectacle of death in early colonial New Spain in the *Manuscrito del aperreamiento.* In J. Beusterien, C. Cortez (ed.). *Death and Afterlife in the Early Modern Hispanic World. Hispanic Issues* 7: 144–163.

8. Borsdorf, A., Stadel, C. 2015. *The Andes: A Geographical Portrait.* (Translated by B. Scott, C. Stadel.) Dordrecht: Springer; Shimada, I. (ed.). 2015. *The Inka Empire: A Multidisciplinary Approach.* Austin: University of Texas Press; Cieza da León, P. D. 1998. *The Discovery and Conquest of Peru.* (Translated by A. P. Cook and N. D. Cook—original 1553). Durham: Duke University Press; Hemming, J. 2004. *The Conquest of the Incas* (2nd edition). New York: Harvest.

9. de Espinosa, A. 1907. *The Guanches of Tenerife.* (Translated by C. Markham—original 1594). London: Hakluyt Society.

10. Cook, N. D. 1998. *Born to Die: Disease and New World Conquest, 1492–1650.* Cambridge: Cambridge University Press; Cook, N. D. 1993. Disease and depopulation of Hispaniola, 1492–1518. *Colonial Latin American Review* 2: 213–245; Hemming, J. 2009. *Tree of Rivers: The Story of the Amazon.* London: Thames & Hudson; Livi-Bacci, M. 2006. The depopulation of Hispanic America after the conquest. *Population and Development Review* 32: 199–232.

11. Newson, L. A. 2009. *Conquest and Pestilence in the Early Spanish Philippines.* Manoa: University of Hawaii Press; Scott, W. H. 1982. *Cracks in the Parchment Curtain and Other Essays in Philippine History.* Quezon City: New Day Publishers; Acabado, S. B. 2010. Landscapes and the archaeology of the Ifugao agricultural terraces: Establishing antiquity and social organisation. *Hukay: Journal for Archaeological Research in Asia and the Pacific* 15: 31–61; Bassani, E., Fagg, W. 1988. *Africa and the Renaissance: Art in Ivory.* London: Centre for African Art, Prestel-Verlag; Mitchell, P. 2005. *African Connections: Archaeological Perspectives on Africa and the Wider World.* Lanham: AltaMira Press; Fage, J. D. 1980. Slaves and society in western Africa, c. 1445–c.1700. *Journal of African History* 21: 289–310; Nunn, N., Qian, N. 2010. The Columbian exchange: A history of disease, food, and ideas. *Journal of Economic Perspectives* 24: 163–188s.

12. Koch, A., Brierley, C., Maslin, M. M., Lewis, S. L. 2019. Earth system impacts of the European arrival and Great Dying in the Americas after 1492. *Quaternary Science Reviews* 207: 13–36; Adorno, R. 2000. *Guaman Poma: Writing and Resistance in Colonial Peru* (2nd edition). Austin: University of Texas Press.

13. Crosby, A. W. 1972. *The Columbian Exchange: Biological and Cultural Consequences of 1492*. Westport: Greenwood Press; Earle, R. 2010. "If you eat their food . . .": Diets and bodies in early colonial Spanish America. *American Historical Review* 115: 688–713; Earle, R. 2012. The Columbian exchange. In J. M. Pilcher (ed.). *The Oxford Handbook of Food History*. Oxford: Oxford University Press. doi: 10.1093/oxfordhb/9780199729937.013.0019; Dumire, W. 2004. *Gardens of New Spain: How Mediterranean Plants and Foods Changed America*. Austin: University of Texas Press.

14. Melville, E. G. K. 1994. *A Plague of Sheep: Environmental Consequences of the Conquest of Mexico*. Cambridge: Cambridge University Press; Bhattacharya, T., Byrne, R. 2016. Late Holocene anthropogenic and climatic influences on the regional vegetation of Mexico's Cuenca Oriental. *Global and Planetary Change* 138: 56–69; Hooghiemstra, H., et al. 2018. Columbus' environmental impact in the New World: Land use change in the Yaque River valley, Dominican Republic. *The Holocene* 28. doi: 10.1177/0959683618788732; Rifkin, J. 1993. *Beyond Beef: The Rise and Fall of the Cattle Culture*. New York: Plume; Crosby, A. W. 1984. An ecohistory of the Canary Islands: A precursor of European colonialization in the New World and Australasia. *Environmental Review* 8: 214–235; Alexander, R. T., Álvarez, H. H. 2017. Agropastoralism and household ecology in Yucatán after the Spanish invasion. *Journal of Human Palaeoecology* 23: 69–79; Norton, M. 2015. The chicken or the *Iegue*: Human-animal relationships and the Columbian exchange. *American Historical Review* 120: 28–60.

15. Nunn, N., Qian, N. 2010. The Columbian exchange: A history of disease, food, and ideas. *Journal of Economic Perspectives* 24: 163–188; Rain, P. 1992. Vanilla: Nectar of the gods. In N. Foster, L. S. Cordell (eds.). *Chilies to Chocolate: Food the Americas Gave the World*. Tucson: University of Arizona Press. Pp. 35–45; Thornton, E. K., et al. 2012. Earliest Mexican turkeys (*Melagris gallopavo*) in the Maya region: Implications for pre-Hispanic animal trade and the timing of turkey domestication. *PLOS ONE* 7(8). doi: 10.1371/journal.pone.0042630.

16. Amano, N., Bankoff, G., Findley, D. M., Barretto-Tesoro, G., Roberts, P. 2020. Archaeological and historical insights into the ecological impacts of pre-colonial and colonial introductions into the Philippine Archipelago. *The Holocene*. doi: 10.1177/0959683620941152.

17. McCann, J. C. 2005. *Maize and Grace: Africa's Encounter with a New World Crop, 1500–2000*. Cambridge: Harvard University Press; Alpern, S. B. 2008. Exotic plants of western Africa: Where they came from and when. *History in Africa* 35: 63–102; Logan, A. 2020. *The Scarcity Slot: Excavating Histories of Food Security in Ghana*. Oakland: University of California Press; Shanahan, T. M., et al. 2009. Atlantic forcing of persistent drought in West Africa. *Science* 324: 377–380.

18. Logan, A. 2017. Will agricultural technofixes feed the world? Short- and long-term tradeoffs of adopting high-yielding crops. In M. Hegmon (ed.). *The Give and Take of Sustainability: Archaeological and Anthropological Perspectives*. Cambridge: Cambridge University Press. Pp. 109–124; Dotterweich, M. 2013. The history of human-induced soil erosion: Geomorphic legacies, early descriptions and research, and the development of soil conservation—a global synopsis. *Geomorphology* 201: 1–34; Mann, C. 2011. *1493: Uncovering the New World Columbus Created*. New York: Vintage Books.

19. Carney, J., Rosomoff, R. 2009. *In the Shadow of Slavery: Africa's Botanical Legacy in the Atlantic World*. Berkeley: University of California Press; Carney, J. 2001. *Black Rice: The African Origins of Rice Cultivation in the Americas*. Cambridge: Harvard University Press.

20. Crosby, A. F. 2004. *Ecological Imperialism: The Biological Expansion of Europe, 900–1900* (2nd edition). Austin: University of Texas Press; Brown, K. W. 2012. *A History of Mining in Latin America: From the Colonial Era to the Present*. Albuquerque: University of New Mexico Press; de Araujo Shellard, A. H. 2016. History of the colonization of Minas Gerais: An environmental approach. In E. Vaz, C. Joanaz de Melo, L. Pinto Costa (eds.). *Environmental History in the Making*. Environmental History 6. Cham: Springer. Pp. 243–257; Hagan, N., et al. 2011. Estimating historical atmospheric mercury concentrations from silver mining and their legacies in present-day surface soil in Potosí, Bolivia. *Atmospheric Environment* 45: 7619–7626.

21. Miller, S. W. 2007. *An Environmental History of Latin America*. Cambridge: Cambridge University Press; Barretto-Tesoro, G., Hernandez, V. 2017. Power and resilience: Flooding and occupation in a late-nineteenth-century Philippine town. In C. Beaule (ed.). *Frontiers of Colonialism*. Gainesville: University Press of Florida. Pp. 149–178; Cushner, N. P. 1971. *Spain in the Philippines: From Conquest to Revolution*. Quezon City: Ateneo de Manila University; Gerona, D. M. 2001. The

colonial accommodation and reconstitution of native elite in early provincial Philippines, 1600–1795. In M. D. E. Pérez-Grueso, J. M. Fradera, L. A. Alvarez (eds.). *Imperios y naciones en el Pacífico.* Madrid: Consejo, Superior de Investigaciones Científicas. Pp 265–276.

22. Moore, J. W. 2009. Madeira, sugar, and the conquest of nature in the "first" sixteenth century. Part I: From "Island of timber" to sugar revolution, 1420–1506. *Review (Fernand Braudel Center)* 32: 345–390; Moore, J. W. 2000. Sugar and the expansion of the early modern world-economy: Commodity frontiers, ecological transformation, and industrialisation. *Review (Fernand Braudel Center)* 23: 409–433; Dannenfeldt, K. H. 1985. Europe discovers civet cats and civet. *Journal of the History of Biology* 18: 403–431.

23. Polónia, A., Pacheco, J. M. 2017. Environmental impacts of colonial dynamics, 1400–1800: The first global age and the Anthropocene. In G. Austin (ed.). *Economic Development and Environmental History in the Anthropocene: Perspectives on Asia and Africa.* London: Bloomsbury. Pp. 23–49; Teixeira, D. M., Papabero, N. 2010. O tráfico de primatas brasileiros nos séculos XVI e XVII. In L. M. Pessôa, W. C. Tavares, S. Salvatore (eds.). *Mamíferos de restingas e manguezais do Brasil.* Rio de Janeiro: Sociedade Brasileira de Mastozoologia & Museu Nacional da UFRJ. Pp. 253–282; Beaule, C. D., Douglas, J. G. (eds.). 2020. *The Global Spanish Empire: Five Hundred Years of Place Making and Pluralism.* Tucson: University of Arizona Press.

24. Reed, R. R. 1978. *Colonial Manila: The Context of Hispanic Urbanism and Process of Morphogenesis.* Berkeley: University of California Press; Acabado, S. 2017. The archaeology of pericolonialism: Responses of the "unconquered" to Spanish conquest and colonialism in Ifugao, Philippines. *International Journal of Historical Archaeology* 21: 1–26; Bankoff, G. 2013. Deep forestry: Shapers of the Philippine forests. *Environmental History* 18: 523–556.

25. Associated Press. 2020. Columbus statue removed in Mexico City, defaced elsewhere. *Northwest Arkanasas Democrat Gazette.* October 13. www.nwaonline.com/news/2020/oct/13/columbus-statue-removed-in-mexico-city-defaced; Lowrey, M. M. 2020. Why more places are abandoning Columbus Day in favour of Indigenous Peoples' Day. *The Conversation.* October 5. https://theconversation.com/why-more-places-are-abandoning-columbus-day-in-favor-of-indigenous-peoples-day-124481; Vink, M. 2003. "The world's oldest trade": Dutch slavery and slave trade in the Indian Ocean in the seventeenth century. *Journal of World History* 14: 131–177.

26. Wing, J. T. 2015. *Roots of Empire: Forests and State Power in Early Modern Spain, c. 1500–1750.* History of the Environment 4. Leiden: Brill; Miller, S. W. 2007. *An Environmental History of Latin America.* Cambridge: Cambridge University Press; Newson, L. A. 2009. *Conquest and Pestilence in the Early Spanish Philippines.* Manoa: University of Hawaii Press; Bjork, K. 1998. The link that kept the Philippines Spanish: Mexican merchant interests and the Manila trade, 1571–1815. *Journal of World History* 9: 25–50.

CHAPTER 11: GLOBALIZATION OF THE TROPICS

1. Church, J. A., White, N. J., Hunter, J. R. 2006. Sea-level rise at tropical Pacific and Indian Ocean islands. *Global and Planetary Change* 53: 155–186; Alroy, J. 2017. Effects of habitat disturbance on tropical forest biodiversity. *Proceedings of the National Academy of Sciences of the United States of America* 114: 6056–6061; Corlett, R. T., Primack, R. 2011. *Tropical Rain Forests: An Ecological and Biogeographical Comparison.* London: Wiley-Blackwell.

2. Diamond, J. 1997. *Guns, Germs, and Steel: The Fates of Human Societies.* New York: W. W. Norton; Soule, E. B. 2018. From Africa to the Ocean Sea: Atlantic slavery in the origins of the Spanish Empire. *Atlantic Studies* 15: 16–39; von Humboldt, A. 1814–1829. *Personal Narrative.* Volume 4. Pp. 143–144. Cited in A. Wulf. 2015. *The Invention of Nature: The Adventures of Alexander von Humboldt, the Lost Hero of Science.* London: John Murray. Pp. 57–58; Morgan, P. D. 2009. Africa and the Atlantic, c. 1450–1820. In J. P. Greene, P. D. Morgan (eds.). *Atlantic History: A Critical Appraisal.* Oxford: Oxford University Press. Pp. 223–248; Martinez, J. S. *The Slave Trade and the Origins of International Human Rights Law.* Oxford: Oxford University Press.

3. Barquera, R., et al. 2020. Origin and health status of first-generation Africans from early colonial Mexico City. *Current Biology* 30(11): 2078–2091. doi: 10.1016/j.cub.2020.04.002; Schroeder, H., et al.

2015. Genome-wide ancestry of 17th-century enslaved Africans from the Caribbean. *Proceedings of the National Academy of Sciences of the United States of America* 112: 3669–3673; Schroeder, H., O'Connell, T. C., Evans, J. A., Shuler, K. A., Hedges, R. E. M. 2009. Trans-Atlantic slavery: Isotopic evidence for forced migration to Barbados. *American Journal of Physical Anthropology* 139: 547–557.

4. Klein, H. S., Luna, F. V. 2009. *Slavery in Brazil.* Cambridge: Cambridge University Press; Mintz, S. 1986. *Sweetness and Power: The Place of Sugar in Modern History.* London: Penguin; Calderia, A. M. 2011. Learning the ropes in the tropics: Slavery and the plantation system on the island of São Tomé. *African Economic History* 39: 35–71; Borucki, A., Eltis, D., Wheat, D. 2015. Atlantic history and the slave trade to Spanish America. *American Historical Review* 120: 433–461; Mann, C. 2011. *1493: Uncovering the New World Columbus Created.* New York: Vintage Books; Vink, M. 2003. "The world's oldest trade": Dutch slavery and slave trade in the Indian Ocean in the seventeenth century. *Journal of World History* 14: 131–177.

5. Shuler, K. A. 2011. Life and death on a Barbadian sugar plantation: Historic and bioarchaeological views of infection and mortality at Newton Plantation. *International Journal of Osteoarchaeology* 21: 66–81; Bethell, L. 1987. *Colonial Brazil.* Cambridge: Cambridge University Press; Emmer, P. C. 2006. *The Dutch Slave Trade, 1500–1850.* (Translated by C. Emery.) New York: Berghahn Books; Meniketti, M. 2006. Sugar mills, technology, and environmental change: A case study of colonial agro-industrial development in the Caribbean. *IA: The Journal of the Society for Industrial Archaeology* 32: 53–80; Fick, C. 2000. Emancipation in Haiti: From plantation labour to peasant proprietorship. *Slavery & Abolition* 21: 11–40; Morgan, K. 2007. *Slavery and the British Empire: From Africa to America.* Oxford: Oxford University Press; Allen, R. B. 2010. Satisfying the "want for labouring people": European slave trading in the Indian Ocean, 1500–1850. *Journal of World History* 21: 45–73; Russell-Wood, A. J. R. 1977. Technology and society: The impact of gold mining on the institution of slavery in Portuguese America. *Journal of Economic History* 37: 59–83; Bradley, K., Cartledge, P. 2011. *The Cambridge World History of Slavery.* Cambridge: Cambridge University Press; Kelley, S. M. 2019. New world slave traders and the problem of trade in goods: Brazil, Barbados, Cuba, and North America in comparative perspective. *English Historical Review* 134: 302–333; Richardson, D., Schwarz, S., Tibbles, A. 2009. *Liverpool and Transatlantic Slavery.* Liverpool: Liverpool University Press; Barquera, R., et al. 2020. Origin and health status of first-generation Africans from early colonial Mexico City. *Current Biology* 30(11): 2078–2091. doi: 10.1016/j.cub.2020.04.002.

6. Edwards, P., Rewt, P. 1994. *The Letters of Ignartius Sancho.* Edinburgh: Edinburgh University Press, Letter 14, 56.

7. Williamson, K. 2019. Most slave shipwrecks have been overlooked—until now. *National Geographic.* August 23. www.nationalgeographic.com/culture/2019/08/most-slave-shipwrecks-overlooked-until-now; Inikori, Joseph. 2002. *Africans and the Industrial Revolution in England: A Study in International Trade and Economic Development.* Cambridge: Cambridge University Press; MacEachern, S. 2018. *Searching for Boko Haram: A History of Violence in Central Africa.* Oxford: Oxford University Press.

8. Bellagamba, A., Greene, S. E., Klein, M. A. (eds.). 2013. *African Voices on Slavery and the Slave Trade.* Cambridge: Cambridge University Press; Hicks, D. 2020. *The Brutish Museums: The Benin Bronzes, Colonial Violence and Cultural Restitution.* London: Pluto Press; Landers, J. 2007. Slavery in the Spanish Caribbean and the failure of abolition. *Review (Fernand Braudel Center)* 31: 343–371; Thornton, J. 1998. *Africa and Africans in the Making of the Atlantic World, 1400–1800* (2nd edition). New York: Cambridge University Press; DeCorse, C. 2001. Introduction. In C. DeCorse (ed.). *West Africa During the Atlantic Slave Trade: Archaeological Perspectives.* Leicester: Leicester University Press. Pp. 1–13; Behrendt, S. D., Latham, A. J. H., Northrup, D. 2010. *The Diary of Antera Duke: An Eighteenth-Century African Slave Trader.* Oxford: Oxford University Press; Mitchell, P. 2005. *African Connections: Archaeological Perspectives on Africa and the Wider World.* Lanham: AltaMira Press; Williams, H. V. 2010. Queen Nzinga (Njinga Mbande). In L. M. Alexander, W. C. Rucker (eds.). *Encyclopaedia of African American History.* Santa Barbara: ABC-CLIO. Pp. 82–84.

9. Odewale, A. 2019. An archaeology of struggle: Material remnants of a double consciousness in the American South and Danish Caribbean communities. *Transforming Anthropology* 27: 114–132

10. Price, R (ed.). 1996. *Maroon societies: Rebel slave communities in the Americas.* Baltimore: John Hopkins University Press; Richardson, D. 2001. Shipboard revolutions, African authority, and

the Atlantic slave trade. In *New Perspectives on the Transatlantic Slave Trade. William and Mary Quarterly* 58: 69–92; Diouf, S. A. 2016. *Slavery's Exiles: The Story of the American Maroons*. New York: New York University Press; Schwartz, S. B. 2017. Rethinking Palmares: Slave resistance in colonial Brazil. In D. A. Pargas, F. Rosu, (eds.). *Critical Readings on Global Slavery*. Leiden: Brill. Pp. 1294–1325; Odewale, A. 2019. An archaeology of struggle: Material remnants of a double consciousness in the American South and Danish Caribbean communities. *Transforming Anthropology* 27: 114–132; Odewale, A., Foster II, T., Toress, J. M. 2017. In service to a Danish king: Comparing material culture of royal enslaved Afro-Caribbeans and Danish soldiers at the Christianised national historic site. *Journal of African Diaspora Archaeology and Heritage* 6: 1–39; Otele, O. 2020. *African Europeans: An Untold History*. London: Hurst Press; Nunn, N., Qian, N. 2010. The Columbian exchange: A history of disease, food, and ideas. *Journal of Economic Perspectives* 24: 163–188.

11. Hall, C. 2020. The slavery business and the making of "race" in Britain and the Caribbean. *Current Anthropology* 16. doi: 10.1086/709845; Meniketti, M. 2006. Sugar mills, technology, and environmental change: A case study of colonial agro-industrial development in the Caribbean. *IA: The Journal of the Society for Industrial Archaeology* 32: 53–80; Hersh, J., Voth, H.-J. 2009. Sweet diversity: Colonial goods and the rise of European living standards after 1492. SSRN. July 4. http://dx.doi.org/10.2139/ssrn.1443730; Brunache, P. 2019. Mainstreaming African diasporic foodways when academia is not enough. *Transforming Anthropology* 27: 149–163; Brown, C. L. 2012. *Moral Capital: Foundations of British Abolitionism*. Chapel Hill: University of North Carolina Press; Roscoe, P. 2020. How the shadow of slavery still hangs over global finance. *The Conversation*. August 21. https://theconversation.com/how-the-shadow-of-slavery-still-hangs-over-global-finance-144826.

12. Moore, J. W. 2000. Sugar and the expansion of the early modern world-economy: Commodity frontiers, ecological transformation, and industrialization. *Review (Fernand Braudel Center)* 23: 409–433; Barlow, V. 1993. *The Nature of the Islands: Plants and Animals of the Eastern Caribbean*. Dunedin: Chris Doyle Publishing; Meniketti, M. 2006. Sugar mills, technology, and environmental change: A case study of colonial agro-industrial development in the Caribbean. *IA: The Journal of the Society for Industrial Archaeology* 32: 53–80; Dunnavant, J. 2019. A historical ecology of slavery in the Danish West Indies. Talk given for UCI Media and posted to YouTube. www.youtube.com/watch?v=Q8oR_CPxkyQ&feature=youtu.be; Smith, F. H. 2009. *Caribbean Rum: A Social and Economic History*. Gainesville: University of Florida Press.

13. Baptist, E. E. 2016. *The Half Has Never Been Told: Slavery and the Making of American Capitalism*. London: Hachette; Riello, G. 2013. *Cotton: The Fabric That Made the Modern World*. Cambridge: Cambridge University Press; Menon, M., Uzramma. *A Frayed History: The Journey of Cotton in India*. Oxford: Oxford University Press; Beckett, S. 2016. *Empire of Cotton: A Global History*. London: Penguin Books; Tarlo, E. 1996. *Clothing Matters: Dress and Identity in India*. Chicago: University of Chicago Press.

14. Behal, R. P. 2006. Power structure, discipline, and labour in Assam tea plantations under colonial rule. *International Review of Social History* 51: 143–172; Bandarage, A. 1983. *Colonialism in Sri Lanka: The Political Economy of the Kandyan Highlands, 1833–1886*. Berlin: Mouton Publishers; Chatterjee, P. 1995. "Secure this excellent class of labour": Gender and race in labor recruitment for British Indian tea plantations. *Bulletin of Concerned Asian Scholars* 27: 43–56.

15. Topik, S. 1998. Coffee. In S. Topik, A. Wells (eds.). *The Second Conquest of Latin America: Coffee, Henequen and Oil During the Export Boom, 1850–1930*. Austin: University of Texas Press; Pomeranz, K., Topik, S. 2018. *The World That Trade Created: Society, Culture, and the World Economy, 1400 to the Present* (4th edition). New York: Routledge; Topik, S. 1999. Where is the coffee? Coffee and Brazilian identity. *Luso-Brazilian Review* 36: 87–92.

16. Hemming, J. 2009. *Tree of Rivers: The Story of the Amazon*. London: Thames and Hudson; Grandin, G. 2010. *Fordlandia: The Rise and Fall of Henry Ford's Forgotten Jungle City*. London: Picador; Hemming, J. 2004. *Amazon Frontier: Defeat of the Brazilian Indians*. London: Pan MacMillan; Caetano Andrade, V. L., et al. 2019. Growth rings of Amazon nut trees (*Bertholletia excelsa*) as living record of historical human disturbance in central Amazonia. *PLOS ONE* 14(4): e0214128. doi: 10.1371/journal.pone.0214128.

17. Tully, J. 2011. *The Devil's Milk: A Social History of Rubber*. New York: Monthly Review Press; Dean, W. 1987. *Brazil and the Struggle for Rubber: A Study in Environmental History*. New York:

Cambridge University Press; Ross, C. 2017. Developing the rain forest: Rubber, environment and economy in Southeast Asia. In G. Austin (ed.). *Economic Development and Environmental History in the Anthropocene: Perspectives on Asia and Africa.* London: Bloomsbury. Pp. 199–218; Louis, W. R. 1964. Roger Casement and the Congo. *Journal of African History* 5: 99–120; Loadman, J. 2005. *Tears of the Tree.* Oxford: Oxford University Press.

18. Robins, J. E. 2018. Smallholders and machines in the West African palm oil industry, 1850–1950. *African Economic History* 46: 69–103; Mann, K. 2009. Owners, slaves, and the struggle for labour in the commercial transition at Lagos. In R. Law (ed.). *From Slave Trade to "Legitimate" Commerce: The Commercial Transition in Nineteenth-Century West Africa.* Cambridge: Cambridge University Press. Pp. 144–171; Watkins, C. 2011. Dendezeiro: African oil palm agroecologies in Bahia, Brazil, and implications for development. *Journal of Latin American Geography* 10: 9–33.

19. McNeill, W. H. 1999. How the potato changed the world's history. *Social Research* 66: 67–83; Fitzgerald, P., Lambkin, B. 2008. *Migration in Irish History, 1607–2007.* London: Palgrave MacMillan.

20. Morrison, K. D., Hauser, M. W. 2015. Risky business: Rice and inter-colonial dependencies in the Indian and Atlantic Oceans. *Atlantic Studies* 12: 371–392.

21. Moberg, M., Striffler, S. (eds.). 2003. *Banana Wars: Power, Production, and History in the Americas.* Durham: Duke University Press; Chapman, P. 2007. *Bananas: How the United Fruit Company Shaped the World.* Edinburgh: Canongate.

22. Anderson, J. L. 2004. Nature's currency: The Atlantic mahogany trade and the commodification of nature in the eighteenth century. *Early American Studies* 2: 47–80; Bankoff, G. 2013. Deep forestry: Shapers of the Philippine forests. *Environmental History* 18: 523–556; Pearson, M., Lennon, J. 2010. *Pastoral Australia: Fortunes, Failures and Hard Yakka: A Historical Overview.* Victoria, Australia: CSIRO; Frost, W. 1997. Farmers, government, and the environment: The settlement of Australia's "Wet Frontier," 1870–1920. *Australian Economic History Review* 37: 19–38; Ferrier, Å. 2015. *Journeys into the Rainforest: Archaeology of Culture Change and Continuity on the Evelyn Tableland, North Queensland.* Canberra: Australian National University.

23. Brown, L. 2010. Monuments to freedom, monuments to nation: The politics of emancipation and remembrance in the eastern Caribbean. *Slavery & Abolition* 23: 93–116; Cushman, G. T. 2013. *Guano and the Opening of the Pacific World: A Global Ecological History.* Cambridge: Cambridge University Press; Allen, R. B. 2014. Slaves, convicts, abolitionism, and the global origins of the post-emancipation indentured labor system. *Slavery & Abolition* 35: 328–348; Bass, D. 2013. *Everyday Ethnicity in Sri Lanka: Up-country Tamil Identity Politics.* London: Routledge; Firth, S. 1976. The transformation of the labour trade in German New Guinea, 1899–1914. *Journal of Pacific History* 11: 51–65; Ramasamy, P. 1992. Labour control and labour resistance in the plantations of colonial Malaya. *Journal of Peasant Studies* 19: 87–105; Holloway, T. H. 2004. *Immigrants on the Land: Coffee and Society in São Paulo, 1886–1934.* Chapel Hill: University of North Carolina Press; Scharlin, C., Villanueva, L. V. 2000. *Philip Vera Cruz: A Personal History of Filipino Immigrants and the Farmworkers Movement* (3rd edition). Seattle: University of Washington Press; Reyes, M. 2008. *Migration and Filipino Children Left Behind: A Literature Review.* Quezon City: Miriam College /UNICEF. Retrieved from www.unicef.org/philippines/Synthesis_StudyJuly12008.pdf.

24. Hobson, J. A. 2005 [1902]. *Imperialism: A Study.* New York: Cosimo.

25. von Humboldt, A. 1814–1829. *Personal Narrative.* Volume 4. Pp. 143–144. Cited in A. Wulf. 2015. *The Invention of Nature: The Adventures of Alexander von Humboldt, the Lost Hero of Science.* London: John Murray. Pp. 57–58.

26. Hobson, J. A. 2005 [1902]. *Imperialism: A Study.* New York: Cosimo; Wallerstein, I. 2007. The ecology and the economy: What is rational? In A. Hornborg, J. R. McNeill, J. Martinez-Alier (eds.). *Rethinking Environmental History: World-System History and Global Environmental Change.* Plymouth: Altamira Press. Pp. 379–390.

27. McKittrick, K. 2013. Plantation futures. *Small Axe: A Caribbean Journal of Criticism* 17: 1–15; Moulton, A. A., Popke, J. 2017. Greenhouse governmentality: Protected agriculture and the changing biopolitical management of agrarian life in Jamaica. *Environment and Planning D: Society and Space* 35: 714–732; Moulton, A. A., Machado, M. R. 2019. Bouncing forward after Irma and Maria: Acknowledging colonialism, problematizing resilience and thinking climate justice. *Journal of Extreme Events* 6: 1940003. doi: 10.1142/S2345737619400037.

28. Byerlee, D. 2014. The fall and rise again of plantations in tropical Asia: History repeated? *Land* 3: 574–597; Banerjee, A. 2005. History, institutions, and economic performance: The legacy of colonial land tenure systems in India. *American Economic Review* 95: 1190–1213; Rudel, T. K. 2005. *Tropical Forests: Regional Paths of Destruction and Regeneration in the Late Twentieth Century*. New York: Columbia University Press; Pao, H-T., Tsai, C-M. 2010. CO_2 emissions, energy consumption and economic growth in BRIC countries. *Energy Policy* 38: 7850–7860; Houghton, J. T., et al. 2001. *Climate Change 2001: The Scientific Basis. Contribution of Working Group I to the Third Assessment Report of the Intergovernmental Panel on Climate Change*. Cambridge: Cambridge University; Myers N. 1988. Tropical forests and their species: Going, going . . . ? In E. O. Wilson (ed.). *Biodiversity*. Washington, DC: National Academy Press. Pp. 28–35.

29. Movuh, M. C. Y. 2012. The colonial heritage and post-colonial influence, entanglements and implications of the concept of community forestry by the example of Cameroon. *Forest Policy and Economics* 15: 70–77; Barber, S. 2020. Death by racism. *The Lancet Infectious Diseases* 20(8): 903. doi: 10.1016/S1473-3099(20)30567-3; Levang, P., Dounias, E., Sitorus, S. 2012. Out of the forest, out of poverty? *Forests, Trees and Livelihoods* 15: 211–235; Morris, D. A. 1984. *The Origins of the Civil Rights Movement: Black Communities Organizing for Change*. New York: The Free Press; Erfani-Ghettani, R. 2018. Racism, the press, and Black deaths in police custody in the United Kingdom. In M. Bhatia, S. Poynting, W. Tufail (eds.). *Media, Crime and Racism*. Palgrave Studies in Crime, Media and Culture. Cham: Palgrave Macmillan. Pp. 255–275.

30. Wilson, E. O. 1988. *Biodiversity*. Washington, DC: National Academy Press.

CHAPTER 12: A TROPICAL "ANTHROPECENE"?

1. Walker, J. D., Geissman, J. W., Bowring, S. A., Babcock, L. E. 2013. The Geological Society of America geologic time scale. *Geological Society of America Bulletin* 125: 259–272; Malhi, Y. 2017. The concept of the Anthropocene. *Annual Review of Environment and Resources* 42: 77–104; Ronda, M. 2013. Mourning and Melancholia in the Anthropocene. *Post45*. June 10. http://post45.org/2013/06/mourning-and-melancholia-in-the-anthropocene; Zalasiewicz, J., et al. 2017. The Working Group on the Anthropocene: Summary of evidence and interim recommendations. *Anthropocene* 19: 55–60.

2. Crutzen, P. J. 2002. Geology of mankind. *Nature* 415: 23; Ruddiman, W. F. 2003. The Anthropogenic Greenhouse Era began thousands of years ago. *Climatic Change* 61, 261–293; Ruddiman, W. F., Ellis, E. C., Kaplan, J. O., Fuller, D. Q. 2015. Defining the epoch we live in. *Science* 348: 38–39; Ellis, E., Maslin, M., Boivin, N., Bauer, A. 2016. Involve social scientists in defining the Anthropocene. *Nature* 540: 192–193; Roberts, P., Boivin, N., Kaplan, J. 2018. Finding the Anthropocene in tropical forests. *Anthropocene* 23: 5–16; Malhi, Y., Gardner, T. A., Goldsmith, G. R., Silman, M. R., Zelazowski, P. 2014. Tropical forests in the Anthropocene. *Annual Review of Environment and Resources* 39: 125–159.

3. Malhi, Y., Gardner, T. A., Goldsmith, G. R., Silman, M. R., Zelazowski, P. 2014. Tropical forests in the Anthropocene. *Annual Review of Environment and Resources* 39: 125–159.

4. Pimm, S. L., Raven, P. 2000. Extinction by numbers. *Nature* 403: 843–845; Malhi, Y. 2012. The productivity, metabolism and carbon cycle of tropical forest vegetation. *Journal of Ecology* 100: 65–75; Brandon, K. 2014. *Ecosystem Services from Tropical Forests: Review of Current Science* Working Paper 380. Washington DC: Center for Global Development; Wilson, J. B., Peet, R. K., Dengler, J., Pärtel, M. 2012. Plant species richness: The world records. *Journal of Vegetation Science* 23: 796–802; Brugmann, H. 2020. Tree diversity reduced to the bare essentials. *Science* 368: 128–129; DryFlor et al. 2015. Plant diversity patterns in neotropical dry forests and their conservation implications. *Science* 353: 1383–1387.

5. Watts, J. 2020. Tolkien was right: Giant trees have towering role in protecting forests. *Guardian Online*. April 9. www.theguardian.com/environment/2020/apr/09/tolkein-was-right-giant-trees-have-towering-role-in-protecting-forests; Larsen, M. C. 2017. Contemporary human uses of tropical forested watersheds and riparian corridors: Ecosystem services and hazard mitigation, with examples from Panama, Puerto Rico, and Venezuela. *Quaternary International* 448: 190–200; Samarakoon, M. B., Tanaka, N., Iimura, K. 2013. Improvement of effectiveness of existing *Casuarina*

equisetifolia forests in mitigating tsunami damage. *Journal of Environmental Management* 114: 105–114; Powers, J. S., et al. 2009. Decomposition in tropical forests: A pan-tropical study of the effects of litter type, litter placement and mesofaunal exclusion across and precipitation gradient. *Journal of Ecology* 97: 801–811; Rüger, N., et al. 2020. Demographic trade-offs predict tropical forest dynamics. *Science* 368: 165–168; Lutz, J. A., et al. 2018. Global importance of large-diameter trees. *Global Ecology and Biogeography* 27: 849–864.

6. Malhi, Y., Gardner, T. A., Goldsmith, G. R., Silman, M. R., Zelazowski, P. 2014. Tropical forests in the Anthropocene. *Annual Review of Environment and Resources* 39: 125–159; Sheil, D. 2014. How plants water our planet: Advances and imperatives. *Trends in Plant Science* 19: 209–211; Met Office. 2020. Tynemouth. UK Climate Averages. Met Office. www.metoffice.gov.uk/research /climate/maps-and-data/uk-climate-averages/gcybzz9xh; Lawrence, D., Vandecar, K. 2015. Effects of tropical deforestation on climate and agriculture. *Nature Climate Change* 5: 27–36; Nogherotto, R., Coppola, E., Giorgi, F., Mariotti, L. 2013. Impact of Congo Basin deforestation on the African monsoon. *Atmospheric Science Letters* 14: 45–51; Medvigy, D., Walko, R. L., Otte, M. J., Avissar, R. 2013. Simulated changes in northwest U.S. climate in response to Amazon deforestation. *Journal of Climatology* 26: 9115–9136; Werth, D., Avissar, R. 2005. The local and global effects of African deforestation. *Geophysical Research Letters* 32: L12704.

7. Prevedello, J. A., Winck, G. R., Weber, M. M., Nichols, E., Sinervo, B. 2019. Impacts of forestation and deforestation on local temperature across the globe. *PLOS ONE*. doi: 10.1371/journal .pone.0213368; Malhi, Y. 2010. The carbon balance of tropical forest regions, 1990–2005. *Current Opinions in Environmental Sustainability* 2: 237–244; Alkama, R., Cescatti, A. 2016. Biophysical climate impacts of recent changes in global forest cover. *Science* 351: 600–604; Lawrence, D., Vandecar, K. 2015. Effects of tropical deforestation on climate and agriculture. *Nature Climate Change* 5: 27–36.

8. Malhi, Y. 2019. Does the Amazon provide 20% of our oxygen? *Yadvinder Malhi* (blog). August 24. www.yadvindermalhi.org/blog/does-the-amazon-provide-20-of-our-oxygen; Betts, R., Sanderson, M., Woodward, S. 2008. Effects of large-scale Amazon forest degradation on climate and air quality through fluxes of carbon dioxide, water, energy, mineral dust and isoprene. *Philosophical Transactions of the Royal Society of London B: Biological Sciences* 363. doi: 10.1098 /rstb.2007.0027; Hoffmann, W. A., Schroeder, W., Jackson, R. B. 2003. Regional feedbacks among fire, climate, and tropical deforestation. *Journal of Geophysical Research Atmosphere* 108. doi: 10.1029/2003JD003494; Lamb, K. 2019. Indonesian forest fires spark blame game as smoke closes hundreds of Malaysia schools. *Guardian Online*. September 12. www.theguardian.com /world/2019/sep/12/indonesia-forest-fires-spark-blame-game-as-smoke-closes-hundreds-of -malaysia-schools; Barrett, K. S. C., Jaward, F. M., Stuart, A. L. 2019. Forest filter effect for polybrominated diphenyl ethers in a tropical watershed. *Journal of Environmental Management* 248: 109279; Lawrence, D., Vandecar, K. 2015. Effects of tropical deforestation on climate and agriculture. *Nature Climate Change* 5: 27–36.

9. Roberts, P., Petraglia, M. 2015. Pleistocene rainforests: Barriers or attractive environments for early human foragers? *World Archaeology* 47: 718–739; Doughty, C. E., et al. 2015. Megafauna extinction, tree species range reduction, and carbon storage in Amazonian forests. *Ecography* 39: 194–203; Davis, M., Faurby, S., Svenning, J.-C. 2018. Mammal diversity will take millions of years to recover from the current biodiversity crisis. *Proceedings of the National Academy of Sciences of the United States of America* 115: 11262–11267; Fairbairn, A. S., Hope, G. S., Summerhayes, G. R. 2006. Pleistocene occupation of New Guinea's highland and subalpine environments. *World Archaeology* 38: 371–386; Kershaw, A. P. 1986. Climatic change and Aboriginal burning in north-east Australia during the last two glacial/interglacial cycles. *Nature* 322: 47–49; Haberle, S. G., et al. 2010. Paleofire in the wet tropics of northeast Queensland, Australia. *PAGES* 18: 78–80; de Fátima Rossetti, D., de Toledo, P. M., Moraes-Santos, H. M., de Araújo Santos Jr., A. E. 2004. Reconstructing habitats in central Amazonia using megafauna, sedimentology, radiocarbon, and isotope analysis. *Quaternary Research* 61: 289–300.

10. Ruddiman, W. F. 2003. The Anthropogenic Greenhouse Era began thousands of years ago. *Climatic Change* 61, 261–293; Fuller, D. Q., et al. 2011. The contribution of rice agriculture and livestock pastoralism to prehistoric methane levels. *The Holocene* 21: 743–759.

11. Athens, J. S., Ward, J. V. 2004. Holocene vegetation, savanna origins and human settlement of Guam. In V. Attenbrow, R. Fullagar (eds.). *A Pacific Odyssey: Archaeology and Anthropology in the*

Western Pacific. Papers in Honour of Jim Specht. Records of the Australian Museum, Supplement 29. Sydney: Australian Museum. Pp. 15–30; Levis, C., et al. 2018. How people domesticated Amazonian forests. *Frontiers in Ecology and Evolution* 5: doi: 10.3389/fevo.2017.00171; Maezumi, S. Y., et al. 2018. The legacy of 4,500 years of polyculture agroforestry in the eastern Amazon. *Nature Plants* 4: 540–547.

12. Koch, A., Brierley, C., Maslin, M. M., Lewis, S. L. 2019. Earth system impacts of the European arrival and Great Dying in the Americas after 1492. *Quaternary Science Reviews* 207: 13–36; Cook, B. I., et al. 2012. Pre-Columbian deforestation as an amplifier of drought in Mesoamerica. *Geophysical Research Letters* 39: L16706; Chepstow-Lusty, A. J., et al. 1998. Tracing 4000 years of environmental history in the Cuzco area, Peru, from the pollen record. *Mountain Research and Development* 18: 159–172; Levis, C., et al. 2018. How people domesticated Amazonian forests. *Frontiers in Ecology and Evolution* 5: doi: 10.3389/fevo.2017.00171; Arroyo-Kalin, M. 2010. The Amazonian Formative: Crop domestication and anthropogenic soils. *Diversity* 2: 473–504.

13. Penny, D., Hall, T., Evans, D., Polkinghorne, M. 2018. Geoarchaeological evidence from Angkor, Cambodia, reveals a gradual decline rather than a catastrophic 15th-century collapse. *Proceedings of the National Academy of Sciences of the United States of America* 116: 4871–4876; Klein Goldewijk, K., Beusen, A., Doelman, J., et al. 2017. Anthropogenic land use estimates for the Holocene—HYDE 3.2. *Earth System Science Data* 9: 927–953.

14. Koch, A., Brierley, C., Maslin, M. M., Lewis, S. L. 2019. Earth system impacts of the European arrival and Great Dying in the Americas after 1492. *Quaternary Science Reviews* 207: 13–36.

15. *Quotation:* Lewis, S. L., Maslin, M. A. 2015. Defining the Anthropocene. *Nature* 519: 171–180.

16. Anonymous. 1603. Descripcion de la Villa y Minas de Potosí—Ano de 1603. In Ministerio de Fomento (ed.). *Relaciones Geograficas de Indias*. Volume 2. Pp. 113–136. Madrid: Ministerio de Fomento. Quoted in Moore, J. W. 2017. The Capitalocene, Part I: On the nature and origins of our ecological crisis. *Journal of Peasant Studies* 44: 594–630. P. 24; Miller, S. W. 2007. *An Environmental History of Latin America*. Cambridge: Cambridge University Press; Moore, J. W. 2017. The Capitalocene, Part I: On the nature and origins of our ecological crisis. *Journal of Peasant Studies* 44: 594–630. Pp. 24; Wing, J. T. 2015. *Roots of Empire*. Leiden: Brill.

17. Moore, J. W. 2017. The Capitalocene, Part I: On the nature and origins of our ecological crisis. *Journal of Peasant Studies* 44: 594–630. Pp. 24; Wolfe, A. P., et al. 2013. Stratigraphic expressions of the Holocene-Anthropocene transitions revealed in sediments from remote lakes. *Earth Science Reviews* 116: 17–34; Steffen, W., Grinevald, J., Crutzen, P., McNeill, J. 2011. The Anthropocene: Conceptual and historical perspectives. *Philosophical Transactions of the Royal Society of London A: Mathematical, Physical and Engineering Sciences* 369. doi: 10.1098/rsta.2010.0327.

18. Davis, J., Moulton, A. A., Van Sant, L., Williams, B. 2019. Anthropocene, Capitalocene, . . . Platationocene? A manifesto for ecological justice in an age of global crises. *Geography Compass* 13(5): e12438. doi: 10.1111/gec3.12438.

19. Davis, J., Moulton, A. A., Van Sant, L., Williams, B. 2019. Anthropocene, Capitalocene, . . . Platationocene? A manifesto for ecological justice in an age of global crises. *Geography Compass* 13(5): e12438. doi: 10.1111/gec3.12438; Haraway, D. 2015. Anthropocene, Capitalocene, Plantationocene, Chthulucene: Making kin. *Environmental Humanities* 6: 159–165; Yusoff, K. 2018. *A Billion Black Anthropocenes or None*. Minneapolis: University of Minnesota Press.

20. Moore, J. W. 2017. The Capitalocene, Part I: On the nature and origins of our ecological crisis. *Journal of Peasant Studies* 44: 594–630. P. 24.

21. Edgeworth, M., et al. 2015. Diachronous beginnings of the Anthropocene: The lower bounding surface of anthropogenic deposits. *Anthropocene Review* 2: 33–58.

22. Edgeworth, M., et al. 2015. Diachronous beginnings of the Anthropocene: The lower bounding surface of anthropogenic deposits. *Anthropocene Review* 2: 33–58.

23. Craig, O. E., et al. 2013. Earliest evidence for the use of pottery. *Nature* 496: 351–354.

24. Deckard, S. 2016. World-ecology and Ireland: The neoliberal ecological regime. *Journal of World-Systems Research* 22: 145–176; Moore, J. W. 2007. Silver, ecology, and the origins of the modern world, 1450–1640. In A. Hornborg, J. R. McNeill, J. Martinez-Alier (eds.). *Rethinking Environmental History: World-System History and Global Environmental Change*. Plymouth: AltaMira Press.

Pp. 123–142; Woolford, A., Benvenuto, J., Hinton, A. L. (eds.). 2014. *Colonial Genocide in Indigenous North America*. Durham: Duke University Press; Hinkson, M., Vincent, E. (eds.). *Shifting Indigenous Australian Realities: Dispersal, Damage, and Resurgence*. Special issue. *Oceania* 88; Dooling, W. 2007. *Slavery, Emancipation and Colonial Rule in South Africa*. Athens: Ohio University Press.

CHAPTER 13: HOUSES ON FIRE

1. Thunberg, G. 2019. "Our house is on fire": Greta Thunberg, 16, urges leaders to act on climate. *Guardian Online*. January 25. www.theguardian.com/environment/2019/jan/25/our-house -is-on-fire-greta-thunberg16-urges-leaders-to-act-on-climate; Barlow, J., Berenguer, E., Carmenta, R., França, F. 2019. Clarifying Amazonia's burning crisis. *Global Change Biology* 26: 319–321; World Meteorological Organization (WMO). 2019. Australia suffers devastating fires after hottest, driest year on record. WMO. January 7. https://public.wmo.int/en/media/news/australia-suffers-devastat-ing-fires-after-hottest-driest-year-record; Reuters in Brasília. 2020. Brazil's Amazon rainforest suffers worst fires in a decade. *Guardian Online*. October 1. www.theguardian.com/environment/2020 /oct/01/brazil-amazon-rainforest-worst-fires-in-decade.

2. Hansen, M. C., et al. 2013. High-resolution global maps of 21st-century forest cover change. *Science* 342: 850–853; Corlett, R. T., Primack, R. 2011. *Tropical Rain Forests: An Ecological and Biogeographical Comparison*. London: Wiley-Blackwell; Fa, J. E., Peres, C. A., Meeuwig, J. 2002. Bushmeat exploitation in tropical forests: An intercontinental comparison. *Conservation Biology* 16: 232–237; Gibbs, D., Harris, N., Seymour, F. 2018. By the numbers: The value of tropical forests in the climate change equation. World Resources Institute. October 4. www.wri.org/blog/2018/10 /numbers-value-tropical-forests-climate-change-equation.

3. State of the Tropics Project. 2020. *State of the Tropics 2020 Report*. Queensland: James Cook University Press.

4. UNESCO. Wet tropics of Queensland. World Heritage Centre. https://whc.unesco.org/en /list/486.

5. Pimm, S. L., et al. 2014. The biodiversity of species and their rates of extinction, distribution, and protection. *Science* 344: 1246752; Pimm, S. L., Jenkins, C. N., Li, B. V. 2018. How to protect half of Earth to ensure it protects sufficient biodiversity. *Science Advances* 4. doi: 10.1126/sciadv .aat2616; Bastin, J.-F., et al. 2019. The global tree restoration potential. *Science* 365: 76–79; Watson, J. E. M., et al. 2018. The exceptional value of intact forest ecosystems. *Nature Ecology & Evolution* 2: 599–610; Corlett, R. T., Primack, R. 2011. *Tropical Rain Forests: An Ecological and Biogeographical Comparison*. London: Wiley-Blackwell; Ghazoul, J., Chazdon, R. 2017. Degradation and recovery in changing forest landscapes: A multiscale conceptual framework. *Annual Review of Environment and Resources* 42: 161–188.

6. Harrison, R. D. 2011. Emptying the forest: Hunting and the extirpation of wildlife from tropical nature reserves. *BioScience* 61: 919–924; Roberts, P. 2018. Late Pleistocene tropical rainforest forager sustainability and resilience. In N. Sanz (ed.). *Exploring Frameworks for Tropical Forest Conservation: Managing Production and Consumption for Sustainability*. Mexico City: UNESCO. Pp. 116–135; Barker, G., Farr, L. (eds.). 2016. *Archaeological Investigations in the Niah Caves, Sarawak. The Archaeology of the Niah Caves, Sarawak*. Volume 2. Cambridge: McDonald Institute for Archaeological Research; Van Vliet, N., Milner-Gulland, E. J., Bousquet, F., Saqall, M., Nasi, R. 2009. Effect of small-scale heterogeneity of prey and hunter distributions on the sustainability of bushmeat hunting. *Conservation Biology* 24: 1327–1337; Roberts, P. 2019. *Tropical Forests in Prehistory, History, and Modernity*. Oxford: Oxford University Press; Dounias, E. 2016. From subsistence to commercial hunting: Technical shift in cynegetic practices among southern Cameroon forest dwellers during the 20th century. *Ecology and Society* 21: doi: 10.5751/ ES-07946-210123; Corlett, R. T., Primack, R. 2011. *Tropical Rain Forests: An Ecological and Biogeographical Comparison*. London: Wiley-Blackwell; Amano, N., Wang, Y.V., Boivin, N., Roberts, P. 2021. "Emptying Forests?" Conservation Implications of Past Human-primate interactions. *Trends in Ecology and Evolution* 36: 345-359. https://www.cell.com/trends/ecology-evolution/fulltext /S0169-5347(20)30350-5?rss=yes.

7. Logan, A. L., Stump, D., Goldstein, S. T., Orijemie, E. A., Schoeman, M. H. 2019. Usable pasts forum: Critically engaging food security. *African Archaeological Review* 36: 419–438.

8. United Nations. 2011. *Population Distribution, Urbanization, Internal Migration and Development: An International Perspective*. Publication no. ESA/P/WP/223. New York: Department of Economic and Social Affairs, Population Division; Connah, G. 2000. African city walls: A neglected source? In D. M. Anderson, R. Rathbone (eds.). *Africa's Urban Past*. Oxford: James Currey. Pp. 36–51; Simon, D., Adam-Bradford, A. 2014. Archaeology and contemporary dynamics for more sustainable, resilient cities in the peri-urban interface. In B. Maheshwair, V. P., Singh, B. Thoradeniya (eds.). *Balanced Urban Development: Options and Strategies for Liveable Cities*. Cham: Springer International Publishing. Pp. 57–84.

9. Roberts, P., Hunt, C., Arroyo-Kalin, M., Evans, D., Boivin, N. 2017. The deep human prehistory of global tropical forests and its relevance for modern conservation. *Nature Plants* 3: 17093; Carlos Quezada, J., Etter, A., Ghazoul, J., Buttler, A., Guillaume, T. 2019. Carbon neutral expansion of oil palm plantations in the Neotropics. *Science Advances* 5. doi: 10.1126/sciadv.aaw4418; Meijaard, E., Abrams, J. F., Juffe-Bignoli, D., Voigt, M., Sheil, D. 2020. Coconut oil, conservation and the conscientious consumer. *Current Biology* 30: 757–758; Pin Koh, L., Miettinen, J., Chin Liew, S., Ghazoul, J. 2011. Remotely sensed evidence of tropical peatland conversion to oil palm. *Proceedings of the National Academy of Sciences of the United States of America* 108: 5127–5132; Burney, D. A., Burney, L. P. 2007. Paleoecology and "inter-situ" restoration on Kaua'i, Hawai'i. *Frontiers in Ecology and the Environment* 5: 483–490; Boesenkool, S., et al. 2013. Use of ancient sedimentary DNA as a novel conservation tool for high-altitude tropical biodiversity. *Conservation Biology* 28: 446–455; Louys, J., Corlett, R. T., Price, G. J., Hawkins, S., Piper, P. J. 2014. Rewilding the tropics, and other conservation translocations strategies in the tropical Asia-Pacific region. *Ecology and Evolution* 4(22): 4380–4398.

10. Ocheje, P. D. 2007. "In the public interest": Forced evictions, land rights and human development in Africa. *Journal of African Law* 51: 173–214; Stanton, P., Stanton, D., Stott, M., and Parsons, M. 2014. Fire exclusion and the changing landscape of Queensland's Wet Tropics Bioregion 1: The extent and pattern of transition. *Australian Forestry* 77: 51–57. *Also see* Stanton, P., Parsons, M., Stanton, D., and Stott, M. 2014. Fire exclusion and the changing landscape of Queensland's Wet Tropics Bioregion 2: The dynamics of transition forests and implications for management. *Australian Forestry* 77: 58–68; Roberts, P., et al. 2021. Reimagining the relationship between Gondwanan forests and Aboriginal land management in Australia's "Wet Tropics." *IScience* 24: 102190. doi: https://doi.org/10.1016/j.isci.2021.102190; Maezumi, S. Y., et al. 2018. New insights from pre-Columbian land use and fire management in Amazonian Dark Earth forests. *Frontiers in Ecology and Evolution*. doi: 10.3389/fevo.2018.00111.

11. Breuste, J., Dissanyake, L. Socioeconomic and environmental change of Sri Lanka's central highlands. In A. Borsdorf (ed.). *Forschen im Gebirge: Christoph Stadel zum 75. Gebursttag*. Vienna: Verlag der Österreichischen Akademie der Wissenschatfen; Association of Small Island States (AOSIS). 2017. *Rising Tides, Rising Capacity: Supporting a Sustainable Future for Small Island Developing States*. New York: AOSIS, United Nations Development Programme.

12. Schwitzer, C., et al. 2014. Averting lemur extinctions amid Madagascar's political crisis. *Science* 343: 842–843; Douglass, K., Cooper, J. 2020. Archaeology, environmental justice, and climate change on islands of the Caribbean and southwestern Indian Ocean. *Proceedings of the National Academy of Sciences of the United States of America* 117: 8254–8262; Gleeson, J. 2019. *Housing in Our World Cities: London, New York, Paris and Tokyo*. London: Greater London Authority; Wet Tropics Management Authority. 2020. World Heritage Area—facts and figures. Wet Tropics Management Authority. www.wettropics.gov.au/world-heritage-area-facts-and-figures.html.

13. Ferrante, L., Fearnside, P. M. 2020. Protect Indigenous peoples from COVID-19. *Science* 368: 251; Viana, V. 2010. *Sustainable Development in Practice: Lessons Learned from Amazonas*. London: International Institute for Environment and Development.

14. Edwards, G. A. S. 2019. Coal and climate change. *WIREs Climate Change* 10. doi: 10.1002/wcc.607; Finer, M., Orta-Martínez, M. 2010. A second hydrocarbon boom threatens the Peruvian Amazon: Trends, projections, and policy implications. *Environmental Research Letters* 5: 014012;

Meijaard, E., Sheil, D. 2019. The moral minefield of ethical oil palm and sustainable development. *Frontiers in Forests and Global Change* 2: doi: 10.3389/ffgc.2019.00022; Rist, L., Feintrenie, L., Levang, P. 2010. The livelihood impacts of oil palm: Smallholders in Indonesia. *Biodiversity and Conservation* 19: 1009–1024; Müller, H., Rufin, P., Griffiths, P., de Barros Viana Hissa, L., Hostert, P. 2016. Beyond deforestation: Differences in long-term regrowth dynamics across land use regimes in southern Amazonia. *Remote Sensing of Environment* 186: 652–662.

15. Corlett, R. T., Primack, R. 2011. *Tropical Rain Forests: An Ecological and Biogeographical Comparison*. London: Wiley-Blackwell; Ashton, P. S. 2010. Conservation of Borneo biodiversity: Do small lowland parks have a role, or are big inland sanctuaries sufficient? Brunei as an example. *Biodiversity and Conservation* 19: 343–356; Voysey, B. C., McDonald, K. E., Rogers, M. E., Tutin, C. E. G., Parnell, R. J. 1999. Gorillas and seed dispersal in the Lope Reserve, Gabon I: Gorilla acquisition by trees. *Journal of Tropical Ecology* 15: 23–38; DryFlor et al. 2015. Plant diversity patterns in neotropical dry forests and their conservation implications. *Science* 353: 1383–1387.

16. Levis, C., et al. 2020. Help restore Brazil's governance of globally important ecosystem services. *Nature Ecology & Evolution* 4: 172–173; Roberts, P., et al. 2021. Reimagining the relationship between Gondwanan forests and Aboriginal land management in Australia's "Wet Tropics." *IScience* 24: 102190. doi: https://doi.org/10.1016/j.isci.2021.102190.

17. Cooper, J. 2012. Fail to prepare then prepare to fail: Re-thinking threat, vulnerability and mitigation in the pre-Columbian Caribbean. In J. Cooper, P. Sheets (eds.). *Surviving Sudden Environmental Change: Answers from Archaeology*. Boulder: University Press of Colorado. Pp. 91–114; Ferrier, Å. 2015. *Journeys into the Rainforest: Archaeology of Culture Change and Continuity on the Evelyn Tableland, North Queensland*. Canberra: Australian National University; Logan, A. L., Stump, D., Goldstein, S. T., Orijemie, E. A., Schoeman, M. H. 2019. Usable pasts forum: Critically engaging food security. *African Archaeological Review* 36: 419–438.

18. Corlett, R. T., Primack, R. 2011. *Tropical Rain Forests: An Ecological and Biogeographical Comparison*. London: Wiley-Blackwell; Roberts, P., et al. 2021. Reimagining the relationship between Gondwanan forests and Aboriginal land management in Australia's "Wet Tropics." *IScience* 24: 102190. doi: https://doi.org/10.1016/j.isci.2021.102190; Nicholas, A., et al. 2010. Successes in community-based monitoring of cross river gorillas (*Gorilla gorilla diehli*) in Cameroon. *African Primates* 7: 55–60; Sheil, D., Boissière, M., Beaudoin, G. 2015. Unseen sentinels: Local monitoring and control in conservation's blind spots. *Ecology and Society* 20. doi: 10.5751/ES-07625-200239.

19. Putz, F. E., et al. 2012. Sustaining conservation values in selectively logged tropical forests: The attained and the attainable. *Conservation Letters* 5: 296–303; Ebeling, J., Yasué, M. 2009. The effectiveness of market-based conservation in the tropics: Forest certification in Ecuador and Bolivia. *Journal of Environmental Management* 90: 1145–1153; Güemes-Ricalde, F. J., Villanueva-G, R., Echazarreta-González, C., Gómez-Alvarez, R., Pat-Fernández, J. M. 2005. Production costs of conventional and organic honey in the Yucatán Peninsula of Mexico. *Journal of Apicultural Research* 45: 106–111; Soto-Pinto, L., Anzueto, M., Mendoza, J., Jimenez Ferrer, G., de Jong, B. 2010. Carbon sequestration through agroforestry in indigenous communities of Chiapas, Mexico. *Agroforestry Systems* 78. doi: 10.1007/s10457-009-9247-5; Ghazoul, J., Sheil, D. 2010. *Tropical Rain Forest Ecology, Diversity, and Conservation*. Oxford: Oxford University Press.

20. Aguilera, J. 2019. Bolsonaro says he won't accept $20 million to fight Amazon fires unless Macron apologises. *Time*. August 27. https://time.com/5662395/bolsonaro-reject-g7-pledge-amazon-fires; UN-REDD Programme Collaborative Online Workspace. 2020. UN-REDD Programme. www.unredd.net/regions-and-countries/regions-and-countries-overview.html; Saeed, A.R., McDermott, C., Boyd, E. 2018. Examining equity in Ghana's national REDD+ process. *Forest Policy and Economics* 90: 48–58; Ghazoul, J., Sheil, D. 2010. *Tropical Rain Forest Ecology, Diversity, and Conservation*. Oxford: Oxford University Press.

21. Butler, S., Sweeney, M. 2018. Iceland's Christmas TV advert rejected for being political. *Guardian Online*. November 9. www.theguardian.com/media/2018/nov/09/iceland-christmas-tv-ad-banned-political-greenpeace-orangutan.

CHAPTER 14: A GLOBAL RESPONSIBILITY

1. Evenstar, L. A., Stuart, F. M., Hartley, A. J., Tattitch, B. 2015. Slow Cenozoic uplift of the western Andean Cordillera indicated by cosmogenic ^{3}He in alluvial boulders from the Pacific Planation Surface. *Geophysical Research Letters* 42: 8448–8455.

2. Dentan, R. K. 1991. Potential food sources for foragers in Malaysian rainforest: Sago, yams, and lots of little things. *Bijdragen tot de Taal, Land-en Volkenkunde* 147: 420–444; Morcote-Ríos, G., Aceituno, F. J., Iriarte, J., Robinson, M., Chaparro-Cárdenas, J. L. 2020. Colonisation and early peopling of the Colombian Amazon during the Late Pleistocene and the Early Holocene: New evidence from La Serranía La Lindosa. *Quaternary International*. April 29. https://doi.org/10.1016/j.quaint.2020.04.026; Perfecto, I., Vandermeer, J. H., Wright, A. L. 2009. *Nature's Matrix: Linking Agriculture, Conservation and Food Sovereignty*. Sterling: Earthscan; Hawthorne, W. 2010. *From Africa to Brazil: Culture, Identity, and an Atlantic Slave Trade, 1600–1830*. Cambridge: Cambridge University Press; Sonter, L. J., et al. 2017. Mining drives extensive deforestation in the Brazilian Amazon. *Nature Communications* 8. doi: 10.1038/s41467-017-00557-w.

3. O'Neill, S., Nicholson-Cole, S. 2009. "Fear won't do it": Promoting positive engagement with climate change through visual and iconic representations. *Science Communication* 30: 355–379; Sayyid, S. 2017. Post-racial paradoxes: Rethinking European racism and anti-racism. *Patterns of Prejudice* 51: 9–25.

4. Fun facts and trivia, RI.gov, www.ri.gov/facts/trivia.php; Hansen, M. C., et al. 2013. High-resolution global maps of 21st-century forest cover change. *Science* 342: 850–853; Fast facts, NJ.gov, www.nj.gov/nj/about/facts/fastfacts; Strassburg, B. B. N., et al. 2012. Impacts of incentives to reduce emissions from deforestation on global species extinctions. *Nature Climate Change* 2: 350–355.

5. Bullock, E. L., Woodcock, C. E., Souza Jr., C., Olofsson, P. 2020. Satellite-based estimates reveal widespread forest degradation in the Amazon. *Global Change Biology* 26: 2956–2969; Area. 2009. *Annual Abstract of Statistics* 146: 2–4. doi: 10.1057/aas.2010.2; Pearson, T. R. H., Brown, S., Casarim, F. M. 2014. Carbon emissions from tropical forest degradation caused by logging. *Environmental Research Letters* 9. doi: 10.1088/1748-9326/9/3/034017; Hosonuma, N., et al. 2012. An assessment of deforestation and forest degradation drivers in developing countries. *Environmental Research Letters* 7. doi: 10.1088/1748-9326/7/4/044009; Ahrends, A., et al. 2010. Predictable waves of sequential forest degradation and biodiversity loss spreading from an African city. *Proceedings of the National Academy of Sciences of the United States of America* 107: 14556–14561; Romero-Sanchez, M. E., Ponce-Hernandez, R. 2017. Assessing and monitoring forest degradation in a deciduous tropical forest in Mexico via remote sensing indicators. *Forests* 8. doi: 10.3390/f8090302.

6. Cordeiro, R. C., et al. 2014. Palaeofires in Amazon: Interplay between land use change and palaeoclimatic events. *Palaeogeography, Palaeoclimatology, Palaeoecology* 415: 137–151; Marlon, J. R. 2020. What the past can say about the present and future of fire. *Quaternary Research* 96: 66–87; Fonseca, M. G., et al. 2019. Effects of climate and land-use change scenarios on fire probability during the 21st century in the Brazilian Amazon. *Global Change Biology* 25: 2931–2946; Page, S. E., Hooijer, A. 2016. In the line of fire: The peatlands of Southeast Asia. *Proceedings of the Royal Society B: Biological Sciences* 371. doi: 10.1098/rstb.2015.0176; Dutta, R., Das, A., Aryal, J. 2016. Big data integration shows Australian bush-fire frequency is increasing significantly. *Royal Society Open Science* 3. doi: 10.1098/rsos.150241.

7. Lenton, T. M., et al. 2019. Climate tipping points—too risky to bet against. *Nature* 575: 592–596; SIMIP Community. 2020. Arctic sea ice in CMIP6. *Geophysical Research Letters* 47. doi: 10.1029/2019GL086749; State of the Tropics 2020. *State of the Tropics 2020 Report*. Townsville: James Cook University Press; European Environment Agency (EEA). 2020. Global and European temperatures. EEA. September 30. www.eea.europa.eu/data-and-maps/indicators/global-and-european-temperature-10/assessment; King, A. D., Harrington, L. J. 2018. The inequality of climate change from 1.5 to 2°C of global warming. *Geophysical Research Letters* 45: 5030–5033.

8. Lovejoy, T. E., Nobre, C. 2018. Amazon tipping point. *Science Advances* 4: doi: 10.1126/sciadv.aat2340; Ferraz, G., et al. 2003. Rates of species loss from Amazonian forest fragments. *Proceedings of the National Academy of Sciences of the United States of America* 100: 14069–14073; Thatte, P.,

Joshi, A., Vaidyanathan, S., Landguth, E., Ramakrishnan, U. 2018. Maintaining tiger connectivity and minimizing extinction into the next century: Insights from landscape genetics and spatially-explicit simulations. *Biological Conservation* 218: 181–191; Benitez-López, A., Santini, L., Schipper, A. M., Busana, M., Hujibregts, M. A. J. 2019. Intact but empty forests? Patterns of hunting-induced mammal defaunation in the tropics. *PLOS Biology*. doi: 10.1371/journal.pbio.3000247; Ghazoul, J., Sheil, D. 2010. *Tropical Rain Forest Ecology, Diversity, and Conservation*. Oxford: Oxford University Press; Rowland, J., Hoskin, C. J., Burnett, S. 2019. Distribution and diet of feral cats (*Felis catus*) in the Wet Tropics of north-eastern Australia, with a focus on the upland rainforest. *Wildlife Research* 47: 649–659.

9. State of the Tropics 2020. *State of the Tropics 2020 Report*. Townsville: James Cook University Press; United Nations Department of Economic and Social Affairs (DESA). 2019. *2019 Revision of World Population Prospects*. New York: UN DESA; United Nations. 2014. *2014 Revision of World Urbanization Prospects*. New York: United Nations; WHO and United Nations Settlements Programme. 2010. *Hidden Cities: Unmasking and Overcoming Health Inequalities in Urban Settings*. Kobe: World Health Organization Centre for Health Development.

10. State of the Tropics 2020. *State of the Tropics 2020 Report*. Townsville: James Cook University Press; Phelan, B., et al. 2013. Crop expansion and conservation priorities in tropical countries. *PLOS ONE* 8. doi: 10.1371/journal.pone.0051759; Dookhun, A. 2018. *Final Country Report LDN Target Setting Programme—Republic of Seychelles*. Bonn: United Nations Convention to Combat Desertification; Borrelli, P., et al. 2017. An assessment of the global impact of 21st century land use change on soil erosion. *Nature Communications* 8. doi: 10.1038/s41467-017-02142-7.

11. World Bank. 2018. *Poverty and Shared Prosperity 2018: Piecing Together the Poverty Puzzle*. Washington, DC: World Bank.

12. State of the Tropics 2020. *State of the Tropics 2020 Report*. Townsville: James Cook University Press.

13. Diffenbaugh, N. S., et al. 2017. Quantifying the influence of global warming on unprecedented extreme climate events. *Proceedings of the National Academy of Sciences of the United States of America* 114: 4881–4886; Eccles, R., Zhang, H., Hamilton, D. 2019. A review of the effects of climate change on riverine flooding in subtropical and tropical regions. *Journal of Water and Climate Change* 10: 687–707; Alfieri, L., et al. 2017. Global projections of river flood risk in a warmer world. *Earth's Future* 5: 171–182; Zhao, Y., Wang, C., Wang, S. 2005. Impacts of present and future climate variability on agriculture and forestry in the humid and sub-humid tropics. *Climatic Change* 70: 73–116; Larsen, M. C. 2017. Contemporary human uses of tropical forested watersheds and riparian corridors: Ecosystem services and hazard mitigation, with examples from Panama, Puerto Rico, and Venezuela. *Quaternary International* 448: 190–200; Cange, C. W., McGaw-Césaire, J. 2020. Long-term public health responses in high-impact weather events: Hurricane Maria and Puerto Rico as a case study. *Disaster Medicine and Public Health Preparedness* 14: 18–22; Burkett, M. 2011. In search of refuge: Pacific islands, climate-induced migration, and the legal frontier. *AsiaPacific Issues* 98: 1–8; State of the Tropics 2020. *State of the Tropics 2020 Report*. Townsville: James Cook University Press.

14. Carroll, D., et al. 2018. The global virome project. *Science* 359: 872–874; Dobson, A. P., et al. 2020. Ecology and economics for pandemic prevention. *Science* 369: 379–381; Van Heuverswyn, F., Peeters, M. 2007. The origins of HIV and implications for the global epidemic. *Current Infectious Disease Reports* 9: 338–346; Olivero, J., et al. 2017. Recent loss of closed forests is associated with Ebola virus disease outbreaks. *Nature Scientific Reports* 7. doi: 10.1038/s41598-017-14727-9; Zhou, P., et al. 2020. A pneumonia outbreak associated with a new coronavirus of probable bat origin. *Nature* 579: 270–273.

15. Eberhard, D. M., Simons, G. F., Fenning, C. D. (eds.). 2020. *Ethnologue: Languages of the World* (23rd edition). Dallas: SIL International; Camacho, L. D., Gevaña, D. T., Carandang, A. P., Camacho, S. C. 2015. Indigenous knowledge and practices for the sustainable management of Ifugao forests in Cordillera, Philippines. *Journal of Biodiversity Science, Ecosystem Services & Management* 12: 5–13; Benz, B. F., Cevallos E., J., Santana M., F., Rosales A., J., Graf M., S. 2000. Losing knowledge about plant use in the sierra de manantlan biosphere reserve, Mexico. *Economic Botany* 54: 183–191; Stephens, C., Porter, J., Nettleton, C., Willis, R. 2006. Disappearing, displaced, and undervalued: A

call to action for Indigenous health worldwide. *The Lancet* 367: 17–23; Frechette, A., Ginsburg, C., Walker, W. 2018. *A Global Baseline of Carbon Storage in Collective Lands*. Washington, DC: Rights and Resources Initiative.

16. Mitchard, E. T. A. 2018. The tropical forest carbon cycle and climate change. *Nature* 559: 527–534; Sitch, S., et al. 2015. Recent trends and drivers of regional sources and sinks of carbon dioxide. *Biogeosciences* 12: 653–679; Schleussner, C.-F., et al. 2016. Science and policy characteristics of the Paris Agreement temperature goal. *Nature Climate Change* 6: 827–835.

17. Giam, X. 2017. Global biodiversity loss from tropical deforestation. *Proceedings of the National Academy of Sciences of the United States of America* 114: 5775–5777.

18. Alroy, J. 2017. Effects of habitat disturbance on tropical forest biodiversity. *Proceedings of the National Academy of Sciences of the United States of America* 114: 6056–6061; Barnosky, A. D., et al. 2011. Has the Earth's sixth mass extinction already arrived. *Nature* 471: 51–57.

19. Coltart, C. E. M., Lindsey, B., Ghinai, I., Johnson, A. M., Heymann, D. L. 2017. The Ebola outbreak, 2013–2016: Old lessons for new epidemics. *Philosophical Transactions of the Royal Society of London B: Biological Sciences* 372. doi: 10.1098/rstb.2016.0297; World Health Organisation. 2021. Coronavirus disease (COVID-19) dashboard. WHO. https://covid19.who.int; Dobson, A. P., et al. 2020. Ecology and economics for pandemic prevention. *Science* 369: 379–381.

20. State of the Tropics. 2020. *State of the Tropics 2020 Report*. Townsville: James Cook University Press; World Bank. 2019. *Migration and Remittances—Recent Developments and Outlook*. Migration and Development Brief. Washington, DC: World Bank; Biermann, F., Boas, I. 2007. *Preparing for a Warmer World: Towards a Global Governance System to Protect Climate Refugees*. Global Governance Working Paper 33. Amsterdam: Global Governance Project.

21. Roberts, P., Stewart, B. 2018. Defining the "generalist-specialist" niche for Pleistocene *Homo sapiens*. *Nature Human Behaviour* 2: 542–550. doi: 10.1038/s41562-018-0394-4; Barbour, W., Schlesinger, C. 2012. Who's the boss? Post-colonialism, ecological research and conservation management on Australian Indigenous lands. *Ecological Management & Restoration* 13: 36–41.

22. Reality Check team. 2019. Deforestation: Did Ethiopia plant 350 million trees in a day? *BBC News*. www.bbc.co.uk/news/world-africa-49266983; State of the Tropics. 2020. *State of the Tropics 2020 Report*. Townsville: James Cook University Press; Goffner, D., Sinare, H., Gordon, L. J. 2019. The Great Green Wall for the Sahara and the Sahel Initiative as an opportunity to enhance resilience in Sahelian landscapes and livelihoods. *Regional Environmental Change* 19: 1417–1428; Reij, C., Tappan, G., Smale, M. 2009. Re-greening the Sahel: Farmer-led innovation in Burkina Faso and Niger. In Spielman, D. J., Pandya-Lorch, R. (eds.). *Millions Fed: Proven Successes in Agricultural Development*. Washington, DC: International Food Policy Research Institute. Pp. 53–58. http://ebrary .ifpri.org/cdm/ref/collection/p15738coll2/id/130817; Reij, C., Tappan, G., Smale, M. 2009. Agroenvironmental transformation in the Sahel: Another kind of "Green Revolution." IFPRI Discussion Paper 00914. Washington DC: International Food Policy Research Institute; Mbile, P. N., Atangana, A., Mbenda, R. 2019. Women and landscape restoration: A preliminary assessment of women-led restoration activities in Cameroon. *Environment, Development and Sustainability* 21: 2891–2911.

23. Osei, R., Zerbe, S., Beckmann, V., Boaitey, A. 2018. Socio-economic determinants of smallholder plantation sizes in Ghana and options to encourage reforestation. *Southern Forests: A Journal of Forest Science* 81: 49–56; Roberts, P., et al. 2021. Reimagining the relationship between Gondwanan forests and Aboriginal land management in Australia's "Wet Tropics." *IScience* 24: 102190. doi: https://doi.org/10.1016/j.isci.2021.102190; Corlett, R. T., Primack, R. 2011. *Tropical Rain Forests: An Ecological and Biogeographical Comparison*. London: Wiley-Blackwell; Fonseca, M. G., et al. 2019. Effects of climate and land-use change scenarios on fire probability during the 21st century in the Brazilian Amazon. *Global Change Biology* 25: 2931–2946; Hansen, M. C., et al. 2013. High-resolution global maps of 21st-century forest cover change. *Science* 342: 850–853; Barbier, E. B., Lozano, R., Rodríguez, C. M., Troëng, S. 2020. Adopt a carbon tax to protect tropical forests. *Nature* 578: 213–216.

24. Strassburg, B. B. N., et al. 2012. Impacts of incentives to reduce emissions from deforestation on global species extinctions. *Nature Climate Change* 2: 350–355; Morgans, C. L., et al. 2018. Evaluating the effectiveness of palm oil certification in delivering multiple sustainability objectives. *Environmental Research Letters* 13: 064032; Vogelgesang, F., Kumar, U., Sundram, K. 2018. Building

a sustainable future together: Malaysian palm oil and European consumption. *Journal of Oil Palm, Environment and Health* 9: 1–49; Hasanah, N., Komarudin, H., Dray, A., Ghazoul, J. 2019. Beyond oil palm: Perceptions of local communities of environmental change. *Frontiers in Forests and Global Change* 2: doi: 10.3389/ffgc.2019.00041.

25. Warren, M. 2019. Thousands of scientists are backing the kids striking for climate change. *Nature* 567: 291–292; Buchanan, L., Quoctrung, B., Patel, J. K. 2020. Black Lives Matter may be the largest movement in U.S. history. *New York Times*. July 3. www.nytimes.com/interactive/2020/07/03/us/george-floyd-protests-crowd-size.html; Pingue, F. 2020. NFL's Washington team to retire Redskins name and logo. *Reuters*. https://uk.reuters.com/article/uk-football-nfl-washington/nfls-washington-team-to-retire-redskins-name-and-logo-idUKKCN24E1OJ; Elmi, O. 2020. Edward Colston: Why the statue had to fall. *BBC News*. June 8. www.bbc.co.uk/news/uk-england-bristol-52965803; Hurst, B. 2020. Brands backing Black Lives Matter: It might be a marketing ploy, but it also shows leadership. *The Conversation*. June 4. https://theconversation.com/brands-backing-black-lives-matter-it-might-be-a-marketing-ploy-but-it-also-shows-leadership-139874; Batty, D. 2020. Universities criticised for "tokenistic" support for Black Lives Matter. *Guardian Online*. www.theguardian.com/education/2020/jul/06/universities-criticised-for-tokenistic-support-for-black-lives-matter.

26. Marshall, M. 2020. Planting trees doesn't always help with climate change. *BBC News*. May 26. www.bbc.com/future/article/20200521-planting-trees-doesnt-always-help-with-climate-change.

INDEX

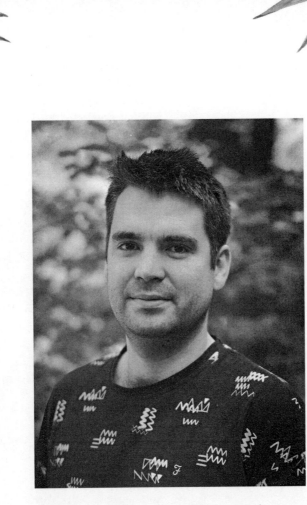

PATRICK ROBERTS is a W2 research group leader at the Max Planck Institute for the Science of Human History. He is a National Geographic Explorer and has received numerous awards for his research on human evolution, including a €1.5 million European Research Council Starting Grant. He lives in Jena, Germany.